AF443330

Chromosomes Today
Volume 11

CHROMOSOMES TODAY

VOLUME 11

A.T. SUMNER AND A.C. CHANDLEY

MRC Human Genetics Unit
Western General Hospital
Edinburgh
UK

CHAPMAN & HALL
London · Glasgow · New York · Tokyo · Melbourne · Madras

Published by Chapman & Hall, 2–6 Boundary Row, London SE1 8HN

Chapman & Hall, 2–6 Boundary Row, London SE1 8HN, UK

Blackie Academic & Professional, Wester Cleddens Road, Bishopbriggs, Glasgow G64 2NZ, UK

Chapman & Hall Inc., 29 West 35th Street, New York NY10001, USA

Chapman & Hall Japan, Thomson Publishing Japan, Hirakawacho Nemoto Building, 6F, 1–7–11 Hirakawa-cho, Chiyoda-ku, Tokyo 102, Japan

Chapman & Hall Australia, Thomas Nelson Australia, 102 Dodds Street, South Melbourne, Victoria 3205, Australia

Chapman & Hall India, R. Seshadri, 32 Second Main Road, CIT East, Madras 600 035, India

First edition 1993

© 1993 The Organizing Committee of the 11th International Chromosome Conference, Edinburgh, UK, 1992

Printed in Great Britain by St Edmundsbury Press Ltd, Bury St Edmunds, Suffolk

ISBN 0 412 47670 3

A catalogue record for this book is available from the British Library

Library of Congress Cataloging-in-Publication data available

Contents

Contents

Contents

Conference members

Abeliovich, D. Department of Human Genetics, Hadassah University Hospital, Ein Kerem, Jerusalem, Israel 91120.

Abulhasan, S.J. 22 Winton Drive, Lister House, Glasgow G12 0QA, UK.

Acar, H. Duncan Guthrie Institute of Medical Genetics, Yorkhill, Glasgow, G3 8SJ, UK.

Adler, I.-D. GSF-Institut für Säugetiergenetik, Ingolstädter Landstr. 1, D-8042, Neuherberg, Germany.

Agoulnik, S. Institut für Biologie, Medizinische Universität zu Lübeck, Ratzeburger Allee 160, D-2400 Lübeck, Germany.

Anamthawat-Jónsson, K. Icelandic Agricultural Research Institute, Keldnaholt, Reykjavik, IS-112, Iceland.

Andersson, A. Department of Genetics, University of Gothenburg, Box 33031, S-40033 Gothenburg, Sweden.

Arana, P. Departamento de Genetica, Facultad de Biologia, Universidad Complutense, 28040 Madrid, Spain.

Archidiacono, N. Istituto di Genetica, Via Amendola 165/A, 70126 Bari, Italy.

Armour, J. Department of Genetics, University of Leicester, Adrian Building, University Road, Leicester, LE1 7RH, UK.

Arnardottir, Ä. Cytogenetics Laboratory, Department of Pathology, University of Iceland, PO Box 1465, IS-121 Reykjavik, Iceland.

Arnason, U. The Wallenberg Laboratory, Box 7031, S-220 07, Lund, Sweden.

Ashley, T. Department of Genetics, Yale University School of Medicine, New Haven, CT 06510, USA.

Avivi, I. Department of Human Genetics, Sackler School of Medicine, Tel Aviv University, Ramat Aviv, 69978 Israel.

Bar-Am, I. Department of Human Genetics, Sackler School of Medicine, Tel Aviv University, Ramat Aviv 69978, Israel.

Belyaev, N. Department of Anatomy, Medical School, University of Birmingham, Edgbaston, Birmingham, B15 2TT, UK.

Bennett, M.D. Jodrell Laboratory, Royal Botanic Gardens, Kew, Richmond, TW9 3DS, UK.

Bennett, S.T. Jodrell Laboratory, Royal Botanic Gardens, Kew, Richmond, TW9 3DS, UK.

Berger, R. INSERM U 301, Institut de Génétique Moléculaire, 27 rue Juliette Dodu, 75010 Paris, France.

Conference members

Bernardi, G.   Institut Jacques Monod, Laboratoire de Génétique Moléculaire, Tour 43, 2 Place Jussieu, 75005 Paris, France.

Beverstock, G.C.   Department of Human Genetics, Sylvius Laboratory, Wassenaarseweg 72, 2333AL Leiden, The Netherlands.

Bianchi, N.O.   IMBICE, 526e/10y11, C.C. 403, 1900 La Plata, Argentina.

Bianchi, U.   Dipartimento di Biologia Animale, Università di Modena, Viale Berengario, u14, 41100 Modena, Italy.

Bickmore, W.A.   MRC Human Genetics Unit, WGH, Crewe Road, Edinburgh, EH4 2XU, UK.

Bizzaro, D.   Dipartimento di Biologia Animale, Viale Berengario 14, 41100 Modena, Italy.

Bocian, E.   Cytogenetics Laboratory, Addenbrooke's Hospital, Cambridge, UK.

Borodin, P.M.   Institute of Cytology & Genetics, Novosibirsk 630090, Russia.

Bougourd, S.M.   Department of Biology, University of York, York YO1 5DD, UK.

Bouza Fernandez, C.   Departamento Biologia Fundamental, Universidad de Santiago, 27002 Lugo, Spain.

Boyd, E.   Department of Medical Genetics, Royal Hospital for Sick Children, Yorkhill, Glasgow G3 8SJ, UK.

Brandham, P.   Jodrell Laboratory, Royal Botanic Gardens, Kew, Richmond, TW9 3DS, UK.

Brandt, C.   Institute of Human Genetics, The Bartholin Building, Aarhus University, DK-8000 Aarhus C. Denmark.

Brice, A.J.   Glaxo Group Research, Park Road, Ware, Herts SG12 0DP, UK.

Britton-Davidian, J.   Laboratoire Génétique et Environnement, Institut des Sciences de l'Evolution, Université Montpellier II, Place E Bataillon, 34095 Montpellier Cedex 05, France.

Brown, T.   Medical Genetics, Medical School, ARI, Foresterhill, Aberdeen AB9 2ZD, UK.

Burgoyne, P.S.   MRC Mammalian Development Unit, Wolfson House, 4 Stephenson Way, London NW1 2HE, UK.

Buys, C.H.C.M.   Department of Medical Genetics, University of Groningen, A Deusinglaan 4, NL 9713 AW Groningen, The Netherlands.

Cabrero Hurtado, J.   Departamento de Genética, Facultad de Ciencias, Universidad de Granada, 18071 Granada, Spain.

Calabrese, G.   Istituto di Biologia e Genetica, Facoltà di Medicina, Via dei Vestini, I-66013 Chieti Scalo, Italy.

Callimassia, M.A.   Jodrell Laboratory, Royal Botanic Gardens, Kew, Richmond, TW9 3DS, UK.

Camacho, J.P.M.   Departamento de Genética, Facultad de Ciencias, Universidad de Granada, 18071 Granada, Spain.

Capanna, E.   Dipartimento di Biologia Animale e dell' Uomo, Università di Roma 'La Sapienza', Via A. Borelli 50, I-00161 Rome, Italy.

Ceoloni, C. Department of Agrobiology and Agrochemistry, University of Tuscia, 01100 Viterbo, Italy.

Chaki, R. Genetic Institute, Cheba Medical Center, Tel Hashomer 52621, Israel.

Chandley, A.C. MRC Human Genetics Unit, WGH, Crewe Road, Edinburgh, EH4 2XU, UK.

Cinti, C. Istituto di Citomorfologia Normale e Patologica del CNR, c/o Istituto di Ricerche 'Codivilla Putti', Via di Barbiano 1/10,40136 Bologna, Italy.

Clarke, C. Cytogenetics Laboratory, Middlesbrough General Hospital, Middlesbrough, Cleveland, TS5 5AZ, UK.

Claussen, U. Institut für Humangenetik und Anthropologie, Schwabachanlage 10, 8520 Erlangen, Germany.

Cohn, M. Department of Molecular Genetics, Sölvegatan 29, S-223 62 Lund, Sweden.

Coulaud, J. E.S.V. Bât 362, Université Paris Sud, F 91405 Orsay Cedex, France.

Couzin, D. Medical Genetics, Polwarth Building, Foresterhill, Aberdeen, AB9 2ZD, UK.

Dahoun-Hadorn, S. Division de Genetique Medicale, CMU, 1 rue Michel Servet, 1211 Geneva 4, Switzerland.

Danford, N. Microptic Ltd, University Innovation Centre, Singleton Park, Swansea, SA2 8PP, UK.

Darmency, M. Institut National de Recherche Agronomique, Amelioration des Plantes, BV 1540, 21034 Dijon, France.

De Bragança, K. Institut für Humangenetik, Wilhelmstrasse 31, D-5300 Bonn 1, Germany.

De Cabo, S.F. Departamento Biologia, Unidad de Genetica, Modulo C-XV, Facultad de Ciencias, Universidad Autonoma de Madrid, Cantoblanco, 28049 Madrid, Spain.

De France, H.F. Klinisch Genetisch Centrum Utrecht, Postbus 18009, 3501 CA Utrecht, The Netherlands.

Del Coco, R. Istituto di Citomorfologia Normale e Patologica del CNR, c/o Istituto di Ricerche 'Codivilla Putti', Via di Barbiano 1/10, 40136 Bologna, Italy.

Del Porto, G. Cattedra di Genetica Medica Universita 'La Sapienza', Via Portuense, 292, 00149 Rome, Italy.

Donini, P. Department of Agrobiology and Agrochemistry, University of Tuscia, 01100 Viterbo, Italy.

Dotan, A. Department of Human Genetics, Sackler School of Medicine, Tel Aviv University, Ramat Aviv, Tel Aviv, Israel.

Douar, A.M. INSERM U342, Hospital St Vincent de Paul, 75014, Paris, France.

Duncan, A.M.V. Cytogenetics Laboratory, Doran I., Kingston General Hospital, 76 Stuart Street, Kingston, Ontario, K7L 2V7, Canada.

Conference members

Earnshaw, W. Department of Cell Biology and Anatomy, John Hopkins School of Medicine, 725 North Wolfe Street, Baltimore, MD 21205, USA.

Eichenlaub-Ritter, U. Universität Bielefeld, Fakultät für Biologie, Gentechnologie-Mikrobiologie, Postfach 100131, 4800 Bielefeld 1, Germany.

Eizenga, G.C. USDA-ARS, Agronomy Department, University of Kentucky, Lexington, KY 40546-0091, USA.

Ellis, M. Digital Scientific, 14 Millers Yard, Mill Lane, Cambridge CB2 1RS, UK.

Ellis, P.M. Lothian Area Cytogenetics Laboratory, Royal Hospital for Sick Children, Edinburgh, EH9 1LF, UK.

Evans, E.P. MRC Radiobiology Unit, Chilton, Didcot, Oxon OX11 0RD, UK.

Evans, H.J. MRC Human Genetics Unit, WGH, Crewe Road, Edinburgh, EH4 2XU, UK.

Everett, C.A. MRC Radiobiology Unit, Chilton, Didcot, Oxon OX11 0RD, UK.

Falconi, M. Istituto di Citomorfologia Normale e Patalogica, c/o Istituto di Ricerca 'Codivilla Putti', Via di Barbiano 1/10, 40136 Bologna, Italy.

Falistocco, E. Istituto Miglior Genetico Piante Foraggere CNR, Borgo XX Giugno, Perugia, Italy.

Fantes, J. MRC Human Genetics Unit, WGH, Crewe Road, Edinburgh, EH4 2XU, UK.

Farabegoli, F. c/o Prof. F. Lampert, Kinderklinik, Feulgen Strasse 12, 6300 Giessen, Germany.

Fasciani, R. Strada Colle Scorrano 98, 65100 Pescara (PE), Italy.

Febles, R. Jardin Botanico 'Viera y Clavijo', Apartado 14, de Tafira Alta, 35017 Las Palmas de Gran Canaria (Canary Islands), Spain.

Feldman, M. Department of Plant Genetics, The Weizmann Institute of Science, Rehovot 76 100, Israel.

Fennell, S.J. Department of Cytogenetics, Royal Manchester Childrens Hospital, Pendlebury, Manchester M27 1HA, UK.

Ferguson-Smith, M.A. Department of Pathology, Cambridge University, Tennis Court Road, Cambridge CB2 1QP, UK.

Ferguson-Smith, M.E. Department of Cytogenetics, Addenbrooke's Hospital, Hills Road, Cambridge CB2 2QQ, UK.

Fernandez Garcia, R.M. Departamento de Psicologia, Facultad de Humanidades, La Zapateira s/n, La Coruña, Spain.

Fernandez-Peralta, A.M. Departamento de Biologia, Unidad de Genetica, Facultad de Ciencias, C-XV, Universidad Autonoma de Madrid, Cantoblanco, 28049 Madrid, Spain.

Ferraro, M. Dipartimento di Genetica e Biologia Molecolare, Università di Roma 'La Sapienza', P.le Aldo Moro 5, 00185 Rome, Italy.

Fitchett, M. Department of Medical Genetics, Churchill Hospital, Headington, Oxford OX3 7LJ, UK.

Flavell, R.B. John Innes Institute, John Innes Centre, Norwich Research Park, Colney, Norwich NR4 7UH, Norfolk, UK.

Fletcher, H. School of Biology and Biochemistry, The Queens University of Belfast, Medical Biology Centre, 97 Lisburn Road, Belfast BT9 7BL, UK.

Forsyth, L. Cytogenetics Department, Royal Northern Infirmary, Ness Walk, Inverness IV2 3UJ, UK.

Fossey, A. Department of Genetics, University of Pretoria, Pretoria, 0002, South Africa.

Fredga, K. Department of Genetics, Box 7003, S-750 07, Uppsala, Sweden.

Gabriel-Robez, O. Faculté de Médecine, Institut d'Embryologie, 11 rue Humann, F-67085 Strasbourg Cedex, France.

Gahlmann, R. Institute of Industrial Toxicology, Bayer AG, PO Box 101709, D-5600 Wuppertal 1, Germany.

Gahmberg, N. Department of Obstetrics and Gynaecology, Laboratory of Prenatal Genetics, Helsinki University Central Hospital, Haartmaninkatu 2A, 00290 Helsinki 29, Finland.

Gelman-Kohan, Z. Tchernichowsky 49/1, Rehovot 76529, Israel

Gerresheim, F. Institute of Pathology and Cytogenetics, Weyertal 76, 5000 Köln 41, Germany.

Ghidoni, A. Dipartimento di Genetica, Via Celoria 26, 20133 Milano, Italy.

Gillies, C.B. Macleay Building A.12, University of Sydney, Sydney, NSW 2006, Australia.

Godelle, B. E.S.V. Bât 362, Université de Paris-Sud, F-91405 Orsay, Cedex, France.

Goldman, A. DNA Laboratory, Yardley Green Unit, East Birmingham Hospital, Birmingham B9 5PX, UK.

Gonzalez-Aguilera, J.J. Departamento de Biologia, Unidad de Genetica, Facultad de Ciencias, C-XV, Universidad Autonoma de Madrid, Cantoblanco, 28049 Madrid, Spain.

Guerra, M. Departamento de Genética, CCB, Universidade Federal de Pernambuco, 50 739, Recife, PE, Brazil.

Gustavino, B. Biology Department, 2nd University of Rome 'Tor Vergata', Via E. Carnevale, 00173 Rome, Italy.

Gustavsson, I. Department of Animal Breeding and Genetics, Swedish University of Agricultural Sciences, Box 7023, S-750 07 Uppsala 7, Sweden.

Haapala, A.K. Department of Clinical Genetics, Tampere University Hospital, PL2000, 33521 Tampere, Finland.

Hall, K.J. School of Biological Sciences, Queen Mary and Westfield College, London E1 4NS, UK.

Harbott, J. Universitäts Kinderpoliklinik, Chromosomenlabor, Feulgenstr. 12, 6300 Gießen, Germany.

Harnden, D.G. Paterson Institute for Cancer Research, Christie Hospital, Wilmslow Road, Manchester M20 9BX, UK.

Hartman, T.P.V.   School of Biology, The University, Edgbaston, Birmingham B15 2TT, UK.

Hassold, T.   Department of Paediatrics, Emory University School of Medicine, 2040 Ridgewood Drive N.E., Atlanta, Georgia 30322, USA.

Hauffe, H.C.   Department of Zoology, University of Oxford, South Parks Road, Oxford OX1 3PS, UK.

Heim, S.   Department of Medical Genetics, Odense University, Winsløw-parken 15, DK-5000 Odense C, Denmark.

Henegariu, O.   Department of Molecular Human Genetics, Institute for Human Genetics and Anthropology, Im Neuenheimer Feld 328, 6900 Heildelberg 1, Germany.

Hernandez-Verdun, D.   Institut Jacques Monod, 2 Place Jussieu, F-75 251 Paris, France.

Heslop-Harrison, J.S.   Karyobiology Group, John Innes Centre for Plant Science Research, Colney Lane, Norwich NR4 7UJ, UK.

Hewitt, G.M.   Biological Sciences, University of East Anglia, Norwich NR4 7TJ, UK.

Hickey, I.   Division of Genetic Engineering, The Queens University of Belfast, Medical Biology Centre, Lisburn Road, Belfast BT9 7BL, UK.

Hidas, A.   Research Institute for Small Amimal Breeding, P.O. Box 65, H-2101 Gödöllö, Hungary.

Higgs, D.   MRC Molecular Haematology Unit, Institute of Molecular Medicine, John Radcliffe Hospital, Headington, Oxford OX3 9DU, UK.

Hindkjaer, J.   Institute of Human Genetics, The Bartholin Building, Aarhus University, 8000 Aarhus C., Denmark.

Holmberg, K.   Environmental Medicine Unit, CNT/Noviem, Karolinska Institutet, 14157 Huddinge, Sweden.

Holmes, D.   Department of Biology, University of York, Heslington, York YO1 5DD, UK.

Holmquist, G.   City of Hope-Biology, Duarte, CA 91010, USA.

Hook, G.J.   NIEHS, MD E403, PO Box 12233, Research Triangle Park, NC 27709, USA.

Hultén, M.   Regional Genetic Services, East Birmingham Hospital, Bordesley Green East, Birmingham B9 5ST, UK.

Hunt, C.   4511 Forest Park Boulevard, St Louis, MO 63108, USA.

Hunt, P.A.   Department of Genetics, Case Western Reserve University School of Medicine, Cleveland, Ohio 44106, USA.

Iwasaki, S.   NODAI Research Institute, Tokyo University of Agriculture, Sakuragaoka 1–1–1, Setagaya-ku, Tokyo 156, Japan.

Jamilena Quesada, M.   Departamento de Genetica, Facultad de Ciencias, Universidad de Granada, 18071 Granada, Spain.

Jauhar, P.   USDA-ARS, Northern Crop Science Lab., 1307 18th Street North, PO Box 5677, State University Station, Fargo, ND 58105, USA.

Jenkins, G.   Department of Agricultural Sciences, University College of Wales, Penglais, Aberyswyth, Dyfed SY23 3DD,UK.

Jeppesen, P.   MRC Human Genetics Unit, Western General Hospital, Crewe Road, Edinburgh EH4 2XU, UK.

Jimenez, M.M.   Departamento de Genética, Facultad de Biologia, Universidad Complutense, 28040 Madrid, Spain.

Johannisson, R.   Institut für Pathologie, Ratzeburger Allee 160, 2400 Lübeck, Germany.

Johnson, M.   Jodrell Laboratory, Royal Botanic Gardens, Kew, Richmond, Surrey TW9 3DS, UK.

Jones, R.N.   Department of Agricultural Sciences, University College of Wales, Penglais, Aberystwyth SY23 3DD, UK.

Joshua, S.K.   School of Biological Sciences, Queen Mary and Westfield College, London E1 4NS, UK.

Just, W.   Department of Clinical Genetics, University of Ulm, PO Box 4066, D-7900 Ulm, Germany.

Kadhim, M.A.   MRC Radiobiology Unit, Harwell, Didcot, Oxon OX11 0RD, UK.

Kähkönen, M.   Department of Clinical Genetics, Oulu University Central Hospital, SF-90220 Oulu, Finland.

Kenton, A.   Jodrell Laboratory, Royal Botanic Gardens, Kew, Richmond, Surrey, TW9 3DS, UK.

Kingston, S.T.   Department of Biological Sciences, Room D606, Polytechnic Southwest, Plymouth PL4 8AA, UK.

Kolin-Gerresheim, I.   Institute of Pathology and Cytogenetics, Weyertal 76, 5000 Köln 41, Germany.

Kølvraa, S.   Institute for Human Genetics, University of Aarhus, 8000 Aarhus C, Denmark.

Koopman, P.   Centre for Molecular Biology and Biotechnology, The University of Queensland, Brisbane 4072, Australia.

Koppers, B.   Institut für Humangenetik der Universität, Vesaliusweg 12/14, 4400 Münster, Germany.

Koulischer, L.   Service de Génétique Humaine, CHU-Tour de Pathologie, 4000 Sart-Tilman, Liège, Belgium.

Krishna Murthy, D.S.   Medical Genetics Centre, Maternity Hospital, PO Box 4080, Safat, 13041, Kuwait.

Kubota, H.   Department of Microbial Genetics, Research Institute for Microbial Diseases, Osaka University 3-1, Yamadaoka, Suita, Osaka, Japan.

Kynast, R.G.   Institute of Plant Genetics and Crop Research, Corrensstr. 3, D-O-4325 Gatersleben, Germany.

Lacàdena, J-R.   Departamento de Genética, Facultad de Ciencias Biológicas, Universidad Complutense, 28040 Madrid, Spain.

Lafi, A.   Building 113 QACCP, Wellcome Foundation, Langley Court, Beckenham, Kent BR3 3BS, UK.

Conference members

Laing, N.G. Australian Neuromuscular Research Institute, Queen Elizabeth II Medical Centre, Nedlands 6009, Western Australia.

Lavender, J.S. Anatomy Department, Medical School, Vincent Drive, Edgbaston, Birmingham B15 2TT, UK.

Lawrie, M. DNA Laboratory, Yardley Green Unit, East Birmingham Hospital, Yardley Green Road, Birmingham B9 5PX, UK.

Le Beau, M. University of Chicago, 5841 South Maryland Avenue, MC 2115, Chicago, IL 60637, USA.

Lee, C.L.Y. Izaak Walton Children Hospital, 5850 University Avenue, PO Box 3070, Halifax, Nova Scotia B3J 3G9, Canada.

Leipoldt, I. Institut für Humangenetik, Wilhelmstrasse 31, D-5300 Bonn 1, Germany.

Leitch, A. Karyobiology Group, John Innes Centre for Plant Science Research, Colney Lane, Norwich NR4 7UH, UK.

Leitch, I. Karyobiology Group, John Innes Centre for Plant Science Research, Colney Lane, Norwich NR4 7UH, UK.

Lemeunier, F. Laboratoire de Biologie et Génétique Evolutives, Centre National de la Recherche Scientifique, 91198 Gif sur Yvette, France.

Levan, G. Department of Genetics, Guldhedsg. 23, S.41390 Gothenburg, Sweden.

Lichter, P. Deutsches Krebsforschungszentrum, Abt. Organisation Komplexer Genome, ATV/Hochhaus 6. OG INF 280, D-6900 Heildelberg, Germany.

Litmanovich, T. 5, Geva St., Givathim 53315, Israel.

Loidl, J. Institute of Botany, University of Vienna, Rennweg 14, A-1030 Vienna, Austria.

Lovett-Doust, J. Department of Biological Sciences, University of Windsor, Windsor, Ontario N9B 3P4, Canada.

Lowther, G.W. Duncan Guthrie Institute for Medical Genetics, Royal Hospital for Sick Children, Yorkhill, Glasgow G3 8SJ, UK.

MacDonald, D.A. MRC Radiobiology Unit, Harwell, Didcot, Oxon OX11 0RD, UK.

McGregor, D. International Agency for Research on Cancer, 150 Cours Albert-Thomas, 69372 Lyon, France.

Macgregor, H. Zoology Department, University of Leicester, Leicester LE1 7RH, UK.

MacKay, J.M. ICI Central Toxicology Laboratory, Alderley Park, Macclesfield, Cheshire, SK10 4TJ, UK.

MacLeod, R.A.F. DSM-Deutsche Sammlung von Mikroorganismen und Zellkulturen GmbH, Mascheroder Weg 1b, D-3300 Braunschweig, Germany.

Madan, K. Cytogenetics Laboratory, Free University Hospital, PO Box 7057, 1007 MB Amsterdam, The Netherlands.

Mahadevaiah, S.K. MRC Mammalian Development Unit, 4 Stephenson Way, London NW1, UK.

Malet, P. Laboratoire Cytogénétique, Faculté de Médecine, B.P.38, 63001

Clermont-Ferrand, Cedex, France.

Malheiro, J.B.   Rua dos Açores, 265–1o Esqo, P-4200 Porto, Portugal

Malheiro, M.I.   Rua dos Açores, 265-1o Esqo, P-4200 Porto, Portugal.

Manicardi, G.C.   Dipartimento di Biologia Animale, Viale Berengario 14, 41100 Modena, Italy.

Martin, R.H.   Genetics Clinic, Alberta Children's Hospital, 1820 Richmond Road, Calgary, Alberta, Canada T2T 5C7.

Martinez, P.   Departamento Biologia Fundamental, Area de Genetica, Universidad de Santiago, 27002 Lugo, Spain.

Matsukuma, S.   Kanagawa Cancer Center, Nakao-cho, 54-2, Asaki-ku, Yokohama 241, Japan.

Mazoti, L.B.   CC4, 1836 Llavallol, Buenos Aires, Argentina.

Mazzotti, G.   Istituto Anatomia Umana Normale, Università, via Irnerio 48, 40127 Bologna, Italy.

Meehan, R.   Institute of Cell Biology and Molecular Biology, Darwin Building, King's Buildings, Mayfield Road, Edinburgh EH9 3JR, UK.

Mendez Felpeto, J.   Facultad de Ciencias, Universidad de la Coruña, La Zapateira, s/n La Coruña, Spain.

Meneely, P.   Fred Hutchinson Cancer Research Center, 1124 Columbia Street, Seattle, WA 98104, USA.

Mercer, S.   Department of Medical Genetics, Churchill Hospital, Old Road, Headington, Oxford OX3 7LJ, UK.

Miller, K.   Abteilung Humangenetik, Medizinische Hochschule Hannover, PO Box 610180, D-3000 Hanover 61, Germany.

Miller, T.E.   Cambridge Laboratory, John Innes Centre, Colney Lane, Norwich, Norfolk NR4 7UJ, UK.

Myllyharju, J.   Laboratory of Genetics, Department of Biology, University of Turku, SF-20500 Turku, Finland.

Nacheva, E.   University of Cambridge, Department of Haematology, Hills Road, MRC Centre, Cambridge, UK.

Narain Navarro, Y.   Wadham College, Oxford OX1 3PS, UK.

Naranjo, C.A.   CC4, 1836 Llavallol, Buenos Aires, Argentina.

Neitzel, H.   Institute of Human Genetics, Free University Berlin, Heubnerweg 6, D-1000 Berlin, Germany.

Nicol, L.   MRC Human Genetics Unit, WGH, Crewe Road, Edinburgh, EH4 2XU, UK.

Nieuwint, A.W.M.   Cytogenetics Laboratory, Free University Hospital, PO Box 7057, 1007 MB Amsterdam, The Netherlands.

Nokkala, S.   Laboratory of Genetics, Department of Biology, University of Turku, SF-20500 Turku, Finland.

O'Neill, S.G.   Division of Genetic Engineering, School of Biology and Biochemistry, Medical Biology Centre, 97 Lisburn Rd, Belfast BT9 7BL, UK.

Palka, G.   Istituto di Biologia e Genetica, Università di Chieti, Via dei Vestini, I–66013 Chieti Scalo, Italy.

Conference members

Papes, D. Department of Molecular Biology, Faculty of Science, University of Zagreb, Rooseveltov trg 6, 41000 Zagreb, Croatia.

Parker, J.S. School of Biological Sciences, Queen Mary and Westfield College, London E1 4NS, UK.

Parokonny, A. Jodrell Laboratory, Royal Botanic Gardens, Kew, Richmond, Surrey TW9 3DS, UK.

Pelliccia, F. Dipartimento di Genetica e Biologia Molecolare, Università 'La Sapienza' di Roma, Ple. Aldo Moro 5, 00185 Roma, Italy.

Percy, M. Genetic Engineering Division, School of Biology and Biochemistry, The Queen's University of Belfast, Belfast BT9 7BL, UK.

Perry, P.E. MRC Human Genetics Unit, WGH, Crewe Road, Edinburgh, EH4 2XU, UK

Perticone, P. Istituto di Genetica, Università 'La Sapienza' di Roma, 00185 Roma, Italy.

Pienkowska, B. Department of Tumor Biology, M. Sklodowska Curie Institute of Oncology, Wawelska 15, 00-973 Warsaw, Poland.

Pierluigi, M. Centro Regionale di Genetica Umana, E.O. Ospedali Galliera, Via A. Volta 10r, 16128 Genova, Italy.

Plas, G. Vrije Universiteit Brussel, Lab. voor Antropogenetica, Pleinlaan 2, B-1050, Brussels, Belgium.

Plowman, A. Department of Biology, University of York, York YO1 5DD, UK.

Poggio, L. CC4, 1836 Llavallol, Buenos Aires, Argentina.

Polani, S. Dipartimento di Genetica e Biologia Molecolare, Università 'La Sapienza', Rome, Italy.

Puertas, M. Departamento de Genética, Facultad de Biología, Universidad Complutense, 28040 Madrid, Spain.

Quack, B. Laboratoire de Cytogénétique, Centre Hospitalier, BP 1125, 73011 Chambery Cedex, France.

Radford, I.R. Molecular Science Group, Peter MacCallum Cancer Institute, 481 Lt Lonsdale Street, Melbourne 3000, Australia.

Raimondi, E. Dipartimento di Genetica e Microbiologia, Via Abbiategrasso 207, 27100 Pavia, Italy.

Rajcan-Separovic, E. Agriculture Canada, Centre for Food and Animal Research, K.W. Neatby Building, Ottawa K1A 0CG, Canada.

Ramsay, G. Scottish Crop Research Institute, Invergowrie, UK.

Raska, I. Czechoslovak Academy of Sciences, Institute of Experimental Medicine, Albertov 4, 12000 Prague, Czechoslovakia.

Ravia, Y. Genetic Institute, Sheba Medical Center, Tel Hashomer, Ramat-Gan 52621, Israel.

Rebollo, E. Departamento de Genetica, Facultad de Biologia, Universidad Complutense, 28040, Madrid, Spain.

Recco-Pimentel, S.M. Department of Cell Biology, Institute of Biology, C.P. 6109, University of Campinas, 13081 Campinas (SP), Brazil.

Reik, W.   Institute of Animal Physiology, Babraham, Cambridge CB2 4AT, UK.

Renzetti, D.   Via Montanara No. 35, 65123 Pescara (PE), Italy.

Renzi, L.   Department of Biology, Via Trieste, 75, 35121 Padova, Italy.

Reynaert, J.   Abdydreef 13, 3070 Kortenberg, Belgium.

Ribas, M.   Istituto di Biologia Generale, Università di Cagliari, Via Ospedale 119, 09100 Cagliari, Italy.

Richards, R.I.   Department of Cytogenetics and Molecular Genetics, Adelaide Children's Hospital, North Adelaide, South Australia 5006, Australia.

Rickards, G.K.   School of Biological Sciences, Victoria University of Wellington, Wellington, New Zealand.

Ricroch, A.   Université Paris-Sud, Laboratoire d'Evolution et Systematique des Végétaux, Bâtiment 362, 91405 Orsay Cedex, France.

Rocchi, A.   Dipartimento di Genetica e Biologia Molecolare, Università 'La Sapienza', 00185 Roma, Italy.

Rocchi, M.   Istituto di Genetica, Via Amendola 165/A, 70126 Bari, Italy.

Roitzheim, B.   Institut für Humangenetik, Wilhelmstrasse 31, D-5300 Bonn 1, Germany.

Rosenmann, A.   19, Bialik Str., Jerusalem 96221, Israel.

Ross, A.R.   MRC Human Genetics Unit, WGH, Crewe Road, Edinburgh, EH4 2XU, UK.

Rossi, A.R.   Dipartimento BAU, Via Borelli 50, 00161 Roma, Italy.

Rovira, C.   Department of Molecular Genetics, Solvegatan 29, S-223 62, Lund, Sweden.

Roy, K.L.   Department of Microbiology, University of Alberta, Edmonton, Alberta, Canada T6G 2E9.

Ruiz Rejon, C.   Departamento de Genetica, Facultad de Ciencias, Universidad de Granada, Granada 18071, Spain.

Ruiz Rejon, M.   Departamento de Genetica, Facultad de Ciencias, Universidad de Granada, Granada 18071, Spain.

Rumpler, Y.   Faculté de Médecine, Institut d'Embryologie, 11 rue Humann, F-67085 Strasbourg Cedex, France.

Russo, A.   Department of Biology, Via Trieste 75, 35121 Padova, Italy.

Safronova, L.D.   Institute of Evolution, Animal Morphology and Ecology, Russian Academy of Sciences, 33 Leninsky Prospekt, Moscow, Russia.

Sanchez, L.   Departamento de Biologia Fundamental, Area de Genetica, Universidad de Santiago, 27002 Lugo, Spain.

Sarri, C.   Institute of Child Health, Agia Sophia Children's Hospital, GR 115 27, Athens, Greece.

Schempp, W.   Institut für Humangenetik und Anthropologie, Albert-Ludwigs Universität, Breisacherstr. 33, D-7800 Freiburg, Germany.

Schoefer, C.   Histologisch-Embryologisches Institut, Schwarzspanierstr. 17, A-1090 Vienna, Austria.

Schriever-Schwemmer, G.   GSF-Forschungszentrum für Umwelt und Gesundheit, Institut für Säugetiergenetik, Ingolstädter Landstr. 1, D-8042 Neuherberg, Germany.

Schubert, I. Institute of Plant Genetics and Crop Plant Research, 0-4325 Gatersleben, Germany.

Schubert, R. Institut für Humangenetik, Wilhelmstrasse 31, D-5300 Bonn 1, Germany.

Schüler, H.M. Institut für Humangenetik, Wilhelmstrasse 31, D-5300 Bonn 1, Germany.

Schwanitz, G. Institut für Humangenetik, Wilhelmstrasse 31, D-5300 Bonn 1, Germany.

Schwarzacher, H.G. Histologisch-Embryologisches Institut, Schwarzspanierstr. 17 A-1090 Vienna, Austria.

Schwarzacher, T. Karyobiology Group, John Innes Centre for Plant Science Research, Colney Lane, Norwich NR4 7UH, UK.

Scriven, P.N. Paediatric Research Unit, 8th Floor Guy's Tower, Guy's Hospital, St. Thomas' Street, London SE 1 9RT, UK.

Searle, J. Department of Zoology, University of Oxford, South Parks Road, Oxford OX1 3PS, UK.

Serville, F. Centre de Génétique, Hopital d'Enfants-Pellegrin, 33076 Bordeaux, France.

Shi Liming, Kunming Institute of Zoology, Academia Sinica, Kunming, China.

Siljak-Yakovlev, S. Université Paris-Sud, Laboratoire d'Evolution et Systematique des Végétaux, Bâtiment 362, 91405 Orsay Cedex, France.

Singh, A.P. Institute of Human Genetics, Free University of Berlin, Heubnerweg 6, 1000 Berlin 19, Germany.

Sites, J.W. c/o Dr. Scott K. Davis, Faculty of Genetics – Animal Science, Texas A & M University, College Station, TX 77843, USA.

Slater, H. Cytogenetics Laboratory, Victorian Clinical Genetics Services, The Murdoch Institute, Royal Childrens Hospital, Parkville 3052, Australia.

Smeets, D.F.C.M. Department of Human Genetics, University Hospital, Nijmegen, PO Box 9101, 6500 HB Nijmegen, The Netherlands.

Sola, L. Department of Human and Animal Biology, University of Rome 'La Sapienza', Via A. Borelli 50, 00161 Rome, Italy.

Sperling, K. Institute of Human Genetics, Free University of Berlin, Heubnerweg 6, D-1000 Berlin 19, Germany.

Spielvogel, H. Institut für Humangenetik, Universität Erlangen, D-8520 Erlangen, Germany.

Spina, R. Via Campo Sportivo No.27, 65010 Spoltore (PE), Italy.

Spyropoulos, B. Department of Biology, York University, 4700 Keele St., Downsview, Ontario M9N 1C4, Canada.

Stacey, G. ECACC, PHLS-CAMR, Porton Down, Salisbury, Wiltshire SP4 0JG, UK.

Stack, S. Department of Biology, Colorado State University, Fort Collins, Colorado 80523, USA.

Steinarsdottir, M. Cytogenetics Laboratory, Department of Pathology, University of Iceland, PO Box 1465, 1S-121 Reykjavik, Iceland.

Stemp, G. Genetic and Reproductive Toxicology, Glaxo Group Research Ltd., Ware, Herts, SG12 0DP, UK.

Stoll, C. Institut de Puericulture, 23 Rue de la Porte de l'Hôpital, 67091 Strasbourg Cedex, France.

Stuppia, L. Istituto di Biologia e Genetica, Università 'G. d'Annunzio', Via dei Vestini, Chieti, Italy.

Suja, J.A. Unidad de Biologia Celular, Departamento de Biologia, Facultad de Ciencias, Universidad Autonoma de Madrid, 28049 Madrid, Spain.

Sumner, A.T. MRC Human Genetics Unit, WGH, Crewe Road, Edinburgh, EH4 2XU, UK.

Swoboda, P. Friedrich Miescher Institut, PO Box 2543, CH-4054 Basel, Switzerland.

Tagarro, I. Departamento de Biologia, Unidad de Genética, Facultad de Ciencias C-XV, Universidad Autónoma de Madrid, 28049 Madrid, Spain.

Tease, C. MRC Radiobiology Unit, Chilton, Didcot, Oxon, OX11 0RD, UK.

Telenius, H. Department of Pathology, University of Cambridge, Cambridge, UK.

Temperani, P. Rome, Italy.

Thornhill, A.R. MRC Mammalian Development Unit, Wolfson House, 4 Stephenson Way, London NW1 2HE, UK.

Toder, R. Institut für Humangenetik und Anthropologie, Albert-Ludwigs Universität, Breisacherstr. 33, D-7800 Freiburg, Germany.

Thomson, E. MRC Human Genetics Unit, WGH, Crewe Road, Edinburgh, EH4 2XU, UK.

Tosi, S, c/o Professor Masera, Clinica Pediatrica dell' Università di Milano, Ospedale S. Gerardo, Via Donizetti 106, 20052 Monza MI, Italy.

Traut, W. Institut für Biologie, Medizinische Universität, Ratzeburger Allee 160, D-2400 Lübeck, Germany.

Tupler, R. Biologia Generale e Genetica Medica, Via Forlanini 14, 27100 Pavia, Italy.

Turner, B.M. Anatomy Department, Birmingham University, Medical School, Edgbaston, Birmingham B15 2TT, UK.

Valentino, F. Centro di Genetica Evoluzionistica, c/o Dipartimento di Genetica e Biologia Molecolare, Università la Sapienza, P. le Aldo Moro 5, 00185 Roma, Italy.

Valerio, D. Laboratorio di Citogenetica e Diagnosi Prenatale, Viale Gramsci 17B, 80121 Napoli, Italy.

Van der Veen, A.Y. Department of Medical Genetics, University of Groningen, A. Deusinglaan 4, NL 9713 AW Groningen, The Netherlands.

Van Hemel, J.O. Department of Clinical Genetics, University Hospital, Dijkzigt, Erasmus University, PO Box 1738, 3000 DR Rotterdam, The Netherlands.

Vekemans, M. Service d'Histologie, Embryologie, Cytogénétique, Hôpital Necker, Enfants Malades, 149 Rue de Sèvres, 75015 Paris, France

Conference members

Veuskens, J. Plant Genetics, Free University of Brussels, Paardenstraat 65, B-1640 St. Genesius Rode, Belgium.

Viegas, W.S. Departamento Botanica, Instituto Superior de Agronomia, Tapada de Ajuda, 1399 Lisboa Codex, Portugal.

Viersbach, R. Institut für Humangenetik, Wilhelmstrasse 31, D-5300 Bonn 1, Germany.

Viinikka, Y. Department of Biology, University of Turku, 20500 Turku, Finland.

Villagomez, D.A.F. Department of Animal Breeding and Genetics, Swedish University of Agricultural Sciences, PO Box 7023, S-750 07 Uppsala 7, Sweden.

Vinas Dias, A.M. Departamento de Biologia Fundamental (Area de Genetica), Facultad de Veterinaria, Universidad de Santiago, Campus de Lugo 27002, Lugo, Spain.

Vincent, J. School of Biological Sciences, Queen Mary & Westfield College, London E1 4NS, UK.

Vogt, P. Section of Molecular Human Genetics, Institute of Human Genetics and Anthropology, University of Heidelberg, Im Neuenheimer Feld 328, D-6900 Heidelberg, Germany.

Volobouev, V. CNRS-URA 620, Structure et Mutagenèse Chromosomiques, Institut Curie, Section de Biologie, 26 rue d'Ulm, 75231 Paris Cedex 05, France.

Volpi, E.V. Imperial Cancer Research Fund, 44 Lincoln's Inn Fields, London WC2A 3PX, UK.

Von Koskull, H. Helsinki University Central Hospital, Departments of Obstetrics and Gynaecology, Laboratory of Prenatal Genetics, Haartmaninkatu 2, SF-00290 Helsinki, Finland.

Wachtler, F.J. Histologisch-Embryologisches Institut der Universität Wien, Schwarzspanierstr. 17, A-1090 Wien, Austria.

Wahls, W.P. Fred Hutchinson Cancer Research Center, 1124 Columbia Street, Seattle, WA 98104, USA.

Wallace, B. School of Biological Sciences, University of Birmingham, Edgbaston, Birmingham B15 2TT, UK.

Wandall, A. Institute of Medical Genetics, Panum, Blegdamsvej 3, DK-2000, Copenhagen, Denmark.

Wang, M. Cambridge Laboratory, John Innes Centre for Plant Science Research, Colney Lane, Norwich NR4 7UJ, UK.

Wlodarska, I. Centre for Human Genetics, University of Leuven, Herestraat 49, B-3000 Leuven, Belgium.

Wolf, K. Institute of Biology, Medical University Lübeck, Ratzeburger Allee 160, W-2400 Lübeck 1, Germany.

Wootton, A. Genetic and Reproductive Toxicology, Glaxo Group Research Ltd., Park Road, Ware, Herts SG12 0DP, UK.

Zuccotti, M. MRC Mammalian Development Unit, Wolfson House, 4 Stephenson Way, London NW1 2HE, UK.

Preface

When the late Professor C.D. Darlington founded what developed into the International Chromosome Conferences in Oxford in 1964, he was concerned that scientists who worked on different aspects of chromosomes, or who studied them in different ways, should have the opportunity of "discussing the fundamental problems of chromosomes with one another". The fact that well over 300 scientists with a wide variety of interests came to Edinburgh in August 1992 for the 11th International Chromosome Conference shows that there is still the same need, and also the desire among chromosomologists to have such discussions. The present volume contains almost all the invited contributions, and attests to the diversity of approaches and applications in chromosomal studies. A few years ago it may have seemed to some that chromosome studies were being superseded by molecular biology, but the molecular biologists have now realized that they need to know about chromosomes, and indeed an important, if ill-defined discipline of 'molecular cytogenetics' has grown up in recent years. We are pleased that in planning the Conference and this book, so much of the work presented is at the interface between cytogenetics and molecular biology. This will surely continue in the future, as boundaries between disciplines are largely artificial, and each has much to learn from the others.

When the venue for the 11th International Chromosome Conference was being discussed, it was felt that it was time to return to Britain, its country of origin, and we in Edinburgh had the honour of hosting this Conference. Although it was hard work, it was also enjoyable. It could not, however, have taken place successfully without the support of a large number of people and organizations: the International Scientific Advisory Committee, our Local Organizing Committee, a large band of volunteer helpers, the University of Edinburgh and the various other organizations that helped us financially or in kind. We should like to take this opportunity of putting on record our thanks to them all.

The next Chromosome Conference will be held in Madrid, Spain, in 1995. At the First (Oxford) Chromosome Conference, Darlington indicated that the long-term purpose of such meetings should be the establishment of a permanent body to arrange such meetings. In fact, there is no standing committee, and each Chromosome Conference has been organized individually. There was a feeling at the Edinburgh Conference that this was perhaps not the ideal way to ensure the future of the Conferences, and as well as wishing Maria Puertas and her colleagues success in organizing the 12th International Chromosome Conference, we hope to be able to pass on some of our experience.

Adrian T. Sumner
Ann C. Chandley

Opening lecture

1 From chromosome number to chromosome map: the contribution of human cytogenetics to genome mapping

M.A. FERGUSON-SMITH

Cambridge University, UK

1 Introduction

Human chromosomes made virtually no contribution to the discipline of cytogenetics until the late 1950's. This can be attributed, not to a lack of interest, but to inadequacy of the available material and technology. Mitoses could only be studied in histological sections of actively dividing tissues such as cancer material and samples of testis. Probably the first recorded description of human chromosomes was by Arnold in 1879 using mitotic cancer cells (Table 1). Hansemann in 1891 and Flemming in 1898 attempted to count the number of chromosomes in serial sections, obtaining crude estimates of approximately 24.

Quite different results were first reported by de Winiwarter in 1912. He was probably the first to study gonadal material and he found a count of 47 in testis and 48 in ovary. He concluded that man was similar to some invertebrate species that had an X:XX sex determining mechanism. Nine years later Painter (1921) contradicted this by demonstrating the presence of a Y chromosome in primary spermatocytes. Painter followed earlier work and assumed that both sexes had a chromosome number of 48 and an XX:XY sex determining mechanism. Interestingly in his 1921 paper he states that he could count only 46 chromosomes in some of the clearest mitotic figures. He also predicted, in a paper published in 1923, the existence of individuals with unusual combinations of sex chromosomes, in particular intersexes with an XXY sex chromosome complement. Apparently no one attempted to test this prediction until 1942 when Severinghaus described an XY sex chromosome complements in an XY female.

During the period that de Winiwarter and Painter were studying human chromosomes there was intense activity in plant and invertebrate cytogenetics. The results of both cytology and genetic linkage were used by Morgan and Bridges and others (e.g. Bridges, 1916) to establish the chromosomal theory of heredity. The polytene chromosomes of Drosophila, discovered by Balbiani in 1881, were later used to provide some of the first detailed genetic maps (Bridges, 1938). Cytological techniques improved and in the 1930's more attention was given to mammalian species by a number of cytologists, notably Koller and Darlington (1934). They described the detailed behaviour of the mammalian sex bivalent in the prophase of first meiosis, distinguishing the presence of "pairing" and "non-pairing" or differential segments of the X and Y chromosomes. Koller (1937) confirmed the presence of an X-Y pairing segment in human meiosis and speculated about the possibility of partial sex linkage - proof of which had to wait for another 40 years.

In the early days of human cytogenetics there seemed to be little interest in the possibility that pathological conditions might be associated with chromosome abnormality. The important exception was the prophetic work of Theodor Boveri on cancer cytogenetics. He became convinced that chromosomal rearrangements were important in the aetiology of malignancy and described his ideas in a classic monograph first published in 1914.

Chromosomes Today   Volume 11. Edited by A.T. Sumner and A.C. Chandley. Published in 1993 by Chapman & Hall, London. ISBN 0 412 47670 3

Table 1 Some Milestones in Human Cytogenetics

A   Early Investigations 1879-1954

1879	Chromosomes noted in tumour cells	Arnold
1891	Chromosome number about 24	Hansemann
1912	47 in testis, 48 in ovary	De Winiwarter
1914	Chromosomal basis of malignancy	Boveri
1921	Y chromosome seen; 2n = 48 in both sexes	Painter
1937	X-Y pairing segment	Koller
	Linkage: colour blindness and haemophilia	Bell & Haldane
1942	XY Female	Severinghaus
1949	Sex chromatin body	Barr
1951	Linkage: lutheran and secretor	Mohr
1952	Hypotonic pre-treatment	Hsu, Makino
1954	Male nuclear sex in Turner syndrome	Polani

B   Awakening interest 1956-1961

1956	2n = 46 in somatic cells	Tjio & Levan
	2n = 46 in testis	Ford & Hamerton
1959	Trisomy 21 in Down's syndrome	Lejeune
	45,X in Turner syndrome	Ford
	47, XXY in Klinefelter syndrome	Jacobs
	Barr body formed by single X	Ohno
1960	Trisomies 13 and 18	Patau, Edwards
	Triploidy	Hook
	Philadelphia chromosome in CML	Nowell
	Phytohaemagglutinin for lymphocytes	Moorhead
1961	X-inactivation	Lyon

C   Diagnosis and development 1962-69

1962	Secondary constrictions and chromosome polymorphism	
	Satellite association	
	Spontaneous abortions	
	Chromosome specific replication patterns	
	More structural aberrations	
1964	Pachytene analysis	
1965	Monosomy for X-Y loci in Turners	
	Incomplete X-inactivation	
1966	X-Y Interchange in XX males	
1967	Interspecific somatic cell hybrids	
	TK1 maps to chromosome 17	
1968	Linkage chromosome 1 and Duffy	
1969	Prenatal diagnosis of chromosome aberrations	
	In situ hybridisation of DNA repeats	
	C-banding	

Table 1 (continued)

D   Molecular era 1970-85

1970	Quinacrine banding (Zech)
1971	Giemsa banding (Sumner)
	Trypsin banding (Seabright)
	Reverse banding (Dutrillaux)
1972	Ribosomal DNA maps to satellite stalks
1973	Deletion mapping: ACP1 to 2pter
	First human gene mapping workshop
1975	Chromosome analysis by flow cytometry
1977	Alpha globin to Chr 16
	DNA sequencing
1978	Beta globin to Chr 11
	Restriction fragment length polymorphisms
1981	In situ mapping of single genes: insulin, IGK
	Sequence of the mitochondrial genome
1983	Huntington's disease to Chr 4
1984	X-AUT translocation and muscular dystrophy
1985	Polymerase chain reaction
	DNA fingerprinting
	VNTR markers

E   Gene cloning 1986-92

1986	Cloning Duchenne muscular dystrophy locus
1987	Fluorescence in situ hybridisation
1988	Chromosome painting
	Interphase mapping
	Meiotic recombination in sperm
1989	Cloning cystic fibrosis
	Microdissection of chromosomes
1990	Multicolour FISH
1991	Cloning fragile X syndrome
1992	Universal DNA priming
	Reverse painting

An early suggestion that Down's syndrome (Mongolism) might be due to a numerical chromosomal aberration resulting from non-disjunction was made by Waardenburg in 1932 . This prompted Mittwoch (1952) to study testis material from a Down's syndrome patient. She counted an average of 24 bivalents at diakinesis (presumably 23 bivalents + 1 univalent) and about 48 chromosomes in spermatogonial mitoses and concluded that the chromosome number was normal. Had hypotonic pre-treatment and control material from a normal individual been available to her, it is likely that she would have discovered the trisomic condition seven years earlier than Lejeune and his colleagues.

2 The modern era of human cytogenetics

The modern era of human cytogenetics can be regarded as starting from the discovery of the correct chromosome number by Tjio and Levan in 1956. Both were experienced cytologists who exploited the availability in their laboratory of fetal tissue cultures (from abortus material). Levan had earlier introduced into plant cytogenetics the use of colchicine to arrest and accumulate mitoses and he appreciated the value of pre-fixation treatment with hypotonic salt solutions. The latter had been discovered independently by T.C. Hsu and S. Makino in 1952. The accidental use of water instead of isotonic saline before fixation produced improved separation of the chromosomes, which could be dispersed from one another more readily by the squash technique then in current use. Ford and Hamerton (1956) were quick to confirm the human diploid number in testis material in the same year and Ford et al (1958) two years later published a method for exploiting rapidly dividing bone marrow cells for chromosome preparations. This was used extensively by British cytologists over the next two years.

2.1 Nuclear sex in the Turner & Klinefelter's syndromes

In the 1950ies the major impetus to study human chromosomes came from the paradoxical findings between nuclear sex and phenotypic sex in the Turner and Klinefelter syndromes. In 1949 Barr and Bertram had observed the presence of a densely staining mass of chromatin, the "nucleolar satellite" (later called the sex chromatin body or Barr body) in the nuclei of female but not male neurones of the cat. This nuclear sex dimorphism was quickly confirmed in other mammalian species including man where it found clinical application in the diagnosis and management of intersex conditions. In 1954 Polani and colleagues discovered that female patients with the Turner syndrome had apparently "male" nuclear sex in that they lacked the sex chromatin body. As the incidence of colour blindness in these patients approximated to a male rather than female incidence, it was postulated that these patients might be sex reversed males. Similarly, when in 1956 male patients with Klinefelter's syndrome (small testes, infertility and gynaecomastia) were found to have "female" nuclear sex it was thought that they might be sex reversed females (Bradbury et al, 1956). The only apparent inconsistency was the observation that some chromatin- positive Klinefelter patients showed complete spermatogenesis in rare seminiferous tubules; a normal XY sex bivalent was noted in one early case (Ferguson-Smith and Munro 1958). The question of sex reversal and whether or not a Y chromosome was present could only be resolved by analysing the chromosomes of these patients and this was finally achieved for both conditions in 1959, Turner patients having a 45,X karyotype (Ford et al, 1959) and Klinefelter patients a 47,XXY karyotype (Jacobs & Strong, 1959). (The first Klinefelter patient to be tested proved to be 46,XX (Ford et al, 1958) and may have been the first XX male). The importance of these observations lay in the conclusions that in humans and all other mammalian species the Y chromosome carried dominant male determinants.

At that time there was much speculation about the nature of the sex chromatin body and whether or not it was formed by heterochromatin from two X chromosomes. Ohno (1959) suggested that the sex chromatin arose from one heteropycknotic X chromosome and this seemed to be supported by the sex chromatin findings from individuals with three or more X chromosomes (Ferguson-Smith et al, 1960); the maximum number of sex chromatin bodies per nucleus was always one less than the number of X chromosomes. Lyon (1961) correctly postulated that X chromosome condensation was associated with random inactivation of one X chromosome and resulted in dosage compensation for X-linked genes; this was demonstrated elegantly in female mice heterozygous for coat colour genes which had a characteristic mosaic phenotype. X-inactivation appeared to involve all X-linked loci in the mouse, and the monosomy X (XO) mouse was only distinguishable from the normal female by mild subfertility and a male pattern of coat colour genes.

It was not long before human X-linked genes such as G6PD were noted to exhibit random X-inactivation. However, studies of patients with Turner's syndrome and its variants resulting from structural abnormalities of the X and Y chromosomes led to the conclusion that X-inactivation was not complete in our own species (Ferguson-Smith, 1965). It was postulated that there were X-linked loci which had active homologues in the Y chromosome and which escaped inactivation. Monosomy for these loci whether on the X or Y chromosome led to the short stature and associated malformation of Turner's syndrome. Only recently has it been confirmed that there are a number of X-Y homologous loci which escape X-inactivation in man but are inactivated in the mouse (Ashworth et al, 1992). These include steroid sulphatase (STS), zinc finger protein on the Y (ZFY), ribosomal protein S4 (RPS4) and the Kallmann gene (Franco et al, 1991). Other determinants, including those for stature and the prevention of Turner stigmata which have yet to be characterised, are presumed to belong to the same class of X-Y homologous genes. One of the loci determining stature may be located within the pairing segment (Ogata et al, 1992), but it is clear that others map to the differential segments (Ferguson-Smith, 1991).

B 2.2 Down's syndrome and other autosomal abnormalities

While the British cytologists were preoccupied with unravelling the paradoxical findings in the Klinefelter and Turner syndromes, Lejeune and his colleagues in France were re-examining the chromosomes in patients with Down's syndrome. Their discovery of trisomy for one of the smallest autosomes was announced in 1958 although not published until January 1959 (Lejeune et al, 1959). On the basis that chromosome 21 was satellited and chromosome 22 was not, the French workers assumed that chromosome 21 was trisomic in Down's syndrome. It was later shown that all 5 pairs of acrocentric chromosomes in the human karyotype were satellited (Ferguson-Smith & Handmaker, 1961), and careful measurements using camera lucida drawings revealed that it was the smallest acrocentric that was trisomic in Down's syndrome (Levan & Ferguson-Smith, unpublished) . Nonetheless, in recognition of the pioneering work of Lejeune, the smallest autosome is today still numbered 21 and not 22.

It came as a surprise to many human geneticists that such gross chromosome aberrations as trisomy 21 could be associated with viability. There followed intense activity among paediatricians and pathologists to determine whether other dysmorphic conditions were due to recognisable chromosomal syndromes. Trisomies 13 and 18 were quickly recognised followed by several chromosomal duplications and deletions. Translocation Down's syndrome and the 5p deletion causing the cri-du-chat syndrome were notable early discoveries. Then followed the discovery of the Philadelphia chromosome in chronic myeloid leukaemia (Nowell & Hungerford 1960), an important confirmation of Boveri's theory on the aetiology of malignancy. Carr (1965) later showed that a large proportion of fetal wastage in the form of spontaneous abortion, was due to a wide range of gross chromosome aberrations, mostly autosomal trisomies. A common aetiological factor appeared to be increased maternal age - a feature long recognised to be associated with the occurrence of Down's syndrome.

The search for new chromosomal syndromes was accompanied by strenuous attempts to improve the identification of individual chromosomes in order that smaller structural changes could be recognised. The most important of these earlier technical developments was the introduction of phytohaemagglutinin to permit chromosome preparations to be made from peripheral blood samples (Moorhead et al 1960). This agent was originally used to clear red cells from preparations of peripheral lymphocytes, but it was noted serendipidously that a proportion of the lymphocytes (T cells) underwent transformation and started dividing. With the use of colchicine to arrest and accumulate cells in mitosis from short term cultures it was possible to make air-dried drop preparations of metaphases far superior to anything that had been achieved previously. The technique was simple, and its introduction was undoubtedly responsible for the rapid development of human cytogenetics in the 1960's. In those days only chromosomes 1, 2, 3, 16, and the Y chromosome could be confidently distinguished from the remainder. Attempts to improve the identification of other chromosomes by careful measurement of length and contromere index, by the recognition of heterochromatic "secondary" constrictions (Ferguson- Smith et al, 1962) which were the forerunners of chromosome banding, and by the establishment of patterns of DNA replication by pulse labelling with tritiated thymidine (German, 1962), undoubtedly increased the resolution of chromosome analysis. The disadvantage of these methods was that they were difficult and time-consuming. However, they revealed that there was considerable normal variation in chromosome size and centromere position, which was heritable and of no clinical significance. These chromosomal heteromorphisms mainly involved the centromeres of chromosomes 1, 9, 16 and the Y, and the short arms and satellites of the five pairs of acrocentric chromosomes; chromosomes 3, 4 and 17 were less commonly heteromorphic.

The study of meiotic chromosomes, which had played such a prominent part in the early days of human cytogenetics, was not neglected during the later period and those fortunate enough to have access to testicular biopsies taken for the investigation of male infertility were able to use some of the newer cytological techniques to study the behaviour of human chromosomes in meiosis. Of particular interest was the earlier observations of chromosome patterns at pachytene and the identity of the nucleolar chromosomes (Yerganian 1957). The chromosome patterns proved to be sufficiently reproducible to allow a tentative identification of many of the autosomal bivalents (Hungerford, 1971), particularly chromosomes 1, 9 and 16 and the acrocentric chromosomes. The latter could be seen in preparations made without hypotonic pre-treatment to be attached to prominent nucleoli (Ferguson-Smith, 1964). Many cells showed evidence of non-homologous pairing between the nucleolar organiser regions of acrocentric bivalents, similar to the common phenomenon of satellite association in mitosis, and providing an explanation for the relatively high frequency of chromosomal rearrangement at these sites on the basis of abnormal recombination events. It was possible using these techniques to recognise quadrivalents and trivalents at diakinesis and pachytene in individuals heterozygous for structural rearrangements (Ferguson-Smith & Page, 1973). However, the most precise characterisation of these rearrangements had to await the elegant electron microscopic preparations of synaptenemal complexes developed by Moses et al (1975). These revealed the clearest evidence of inversion loops (Saadallah & Hulten, 1986) and complex translocations and allowed the most detailed analysis of the pairing and differential segments of the XY bivalent (Chandley et al, 1984).

2.3 Chromosome banding and the development of human chromosome maps

During the first half of the 20th century progress in human genetics was severely limited by the lack of genetic maps with which to study the genetic basis of normal and pathological variation. Efforts thus tended to be directed at population and biochemical genetics. Maps were much easier to develop in experimental species particularly Drosophila, maize and the mouse, in which inbred strains and controlled experimental crosses could be exploited for genetic linkage studies. <u>Drosophila</u> had the special advantage of large polytene chromosomes in which duplications and deletions associated with characteristic phenotypes could be readily mapped. Although the sex linkage of a number of traits (e.g. colour blindness) and disease (e.g. haemophilia) had been recognised since the previous century, the measurement of linkage between two X-linked traits (again between the loci for colour blindness and haemophilia) had to await Bell and Haldane's classic publication (1937) The first human autosomal linkage (between Lutheran and secretor) was described by Mohr in 1951. It took another 17 years before an autosomal locus was assigned to a specific chromosome. Donahue and colleagues (1968) exploited a heritable heteromorphism of chromosome 1 to demonstrate linkage of the Duffy blood group locus to the centromere of chromosome 1.

The previous year saw another major breakthrough in human genetics namely the emergence of cultures of interspecific somatic cell hybrids for gene mapping. Weiss and Green (1967) found that in somatic cell hybrids made by fusing human and rodent cells, the human chromosomes were selectively lost. Man-mouse hybrids made with a thymidine kinase deficient mouse cell line grown over an extended period in selective medium requiring thymidine kinase, preferentially retained one of the small human autosomes (later shown to be chromosome 17), thus mapping the TK1 locus to chromosome 17. Over the next few years the somatic cell hybrid technology led to over 300 assignments of enzyme and other genetic loci to specific chromosomes. Regional localisations were possible by using chromosomal rearrangements or by radiation induced chromosome breakage in the human parental cell line. In fact, any human biochemical product which could be distinguished electrophoretically from its rodent counterpart could be mapped by these methods.

Contributions to the human gene map continued to be made by genetic linkage studies using familial chromosome aberrations and polymorphisms in association with biochemical polymorphism; a notable early example was the assignment of the haptoglobin locus to chromosome 16 (Robson et al, 1969). Unbalanced chromosome aberrations were also exploited for mapping by seeking evidence of hemizygosity from allele loss or dosage effect. The method was particularly effective for the loci for red cell enzyme polymorphisms, the first being red cell acid phosphatase which mapped to the short arm of chromosome 2 (Ferguson-Smith et al, 1973).

These advances occurred at the time two other major technical advances in cytogenetics were being established. The first was the demonstration by Pardue and Gall (1970) that satellite (repetitive) DNA could be annealed to complementary DNA sequences in fixed chromosome preparations. They showed by autoradiography that isotopically labelled mouse major satellite DNA hybridised to the paracentric regions of mouse chromosomes. The denaturation of the chromosomes by alkali and heat required for the hybridisation process was noted by these authors to produce preferential staining of the centromeric regions when Giemsa staining was used. Various modifications of the denaturation and staining process yielded a series of dark-staining or Giemsa bands (G banding) along each chromosome which was found to be specific and useful in chromosome identification. A variety of different denaturing agents were then applied to human chromosome preparations including heat (Sumner et al 1971), alkali, trypsin (Seabright, 1971), and detergents producing a series of staining reactions referred to as G bands, R (reverse) bands, C (centromeric) bands etc. all of which proved to be extremely valuable for identification. In fact, using these methods it was possible to identify unambiguously each human chromosome. As more bands could be identified in prophase and early metaphase than at later stages of mitosis, methods of prometaphase banding soon became established. These yielded up to 1000 distinctive bands with which many previously unrecognised minor duplications and deletions could now be recognised (ISCN, 1981).

In the early days of C banding, Caspersson et al (1970) independently discovered that quinacrine compounds which intercalate in DNA yielded bright fluorescence bands along the chromosome using fluorescence microscopy. These Q-bands were at first more reproducible than those produced by G-banding, but yielded virtually the same banding pattern and were equally valuable for chromosome identification. In addition the quinacrine stained preferentially the distal end of the long arm of the Y chromosome and certain centric and satellite polymorphisms, not only in metaphase but also interphase. The fluorescent Y-body was in vogue for a short time as a complementary technique to Barr body analysis for nuclear sexing, but was found to be less reliable

2.4 The contribution of in situ hybridisation to human gene mapping

In situ hybridisation was first applied to the chromosomal localisation of highly repetitive satellite DNA which mapped mainly to centromeric regions and to the short arms of acrocentric chromosomes. The technique was then used to map the moderately repetitive ribosomal DNA to the nucleolar organiser regions in the satellite stalks of the five human acrocentric chromosomes (Henderson et al, 1972). The first attempts to apply the method for mapping single copy genes used globin messenger RNA from rabbit reticulocytes but this was unsuccessful largely because it proved impossible to incorporate sufficient radioisotope into the RNA probe (Price et al, 1972). The method was largely abandoned for this purpose until recombinant DNA techniques had become established, when it was possible to clone in phage or plasmid vectors large quantities of complemetary DNA from specific genes such as alpha and beta globin. The cloned DNA could also be heavily labelled with radioactive isotopes including tritium or 125 iodine. It was soon confirmed that not only did the alpha and beta globin genes map to chromosomes 16 and 11 respectively, but that they could be assigned to the short arms of both chromosomes (Malcolm et al 1981a). Localisation of genes not previously mapped followed, including the Kappa light chain locus to chromosome 2p (Malcolm et al, 1981b) and the insulin locus to chromosome 11p (Harper et al, 1981) and the method became firmly established as the successor to somatic cell hybrids for human gene mapping. For many human cytogeneticists, this advance provided their first introduction to the powerful techniques of molecular genetics and thus was an important milestone in the development of the discipline.

While procedures using radioisotopes and autoradiography remain very sensitive for in situ mapping, the method lacked precision in that the silver grains were often scattered over an area surrounding the hybridisation site. Also, the long exposure times (sometimes up to three weeks) meant that there was considerable delay between setting up an experiment and finding out whether or not it had been unsuccessful and had to be repeated. These considerations led to the search for alternative and more efficient methods for labelling DNA probes, and to what has become the era of molecular cytogenetics. The basis of the new approach has been to label the nucleotides used in the construction of DNA probes with a non-isotopic molecule. The one most commonly used is biotin (Langer et al, 1981), but others including digoxigenin (Heiles et al, 1988), mercury (Bauman et al., 1983), and acetylaminofluorene (Landegent et al, 1984) are also important. The labels are detected by reporter molecules which are usually antibodies to which various fluorochromes are coupled. In the case of biotin, streptavidin coupled to a fluorochrome could be used. The signal could also be intensified by using additional layers of mouse or goat antibody reporter molecules. Instead of fluorochromes such as FITC, Texas red or rhodamine, enzyme molecules including horseradish peroxidase or alkaline phosphatase can be incorporated and detected using appropriate substrates. However, fluorescence in situ hybridisation (FISH) using U.V. microscopy is undoubtedly the most versatile system as it allows the simultaneous use of different coloured reporter molecules for detecting each signal. FISH is now the most widely used technique for assigning specific DNA sequences to their respective chromosomal locations, and is arguably the most important technique in human gene assignment.

3   Towards constructing a complete physical map of the human genome

At the time of the first international Human Gene Mapping Workshop in 1973 only about 60 gene loci had been assigned to autosomes and most of these were due to the successful exploitation of interspecific somatic cell hybrids. At the time of writing, after 11 HGM Workshops and nearly twenty years later (Table 2) over 2300 human gene loci have been mapped together with an additional 8000 anonymous DNA sequences (D-segments) which act as markers interspersed virtually randomly throughout the human genome. Large sections of the human genome including the Y chromosome, parts of the X chromosome, and substantial regions of chromosomes 21, 16 and 19 have been cloned and physical maps constructed which comprise large sections of contiguous cloned sequences (contigs). These contigs have been assembled using DNA vectors, especially cosmids and yeast artificial chromosomes (YACs), capable of incorporating large inserts of human DNA from 40 to 1000 kilobases in length. The largest YACs can now carry as much as 1.4 million base pairs and, with these, it should be possible to construct a complete physical map of the human genome within the next 3-5 years. The role of molecular cytogenetics will be to map these YAC clones to chromosomal regions, so that they may be assembled more easily into contigs, and to check the order of clones along the chromosome and detect inconsistencies due to cross-hybridisation.

Table 2 Human Gene Mapping Workshops

			Assignments	
			Autosomal	X-linked
1	New Haven	1973	64	152
2	Rotterdam	1974	86	161
3	Baltimore	1975	125	175
4	Winnipeg	1977	176	198
5	Edinburgh	1979	230	213
6	Oslo	1981	345	234
7	Los Angeles	1983	488	254
8	Helsinki	1985	633	280
9	Paris	1987	928	305
10	New Haven	1989	1446*	171*
11	London	1991	2104*	221*

*Excluding D-segments. With D-segments the total is over 10,000.

3.1 Positional cloning and applications of the human gene map

The human gene map is of interest to the medical geneticist first because it allows the study of the molecular pathology and pathogenesis of genetic disease, for example thalassaemia, and secondly because it can be used in the diagnosis and carrier detection of inherited disorders such as cystic fibrosis. The ultimate aim is somatic gene therapy. The human gene map is also vital for the study of the physiology of cells and tissues as it can lead to the identification of normal genes and their products, and to the elucidation of gene regulation essential to an understanding of cellular differentiation. The identification of the sex determining gene on the Y chromosome is a good example (Sinclair et al, 1990). There have been notable successes in identifying and characterising disease genes using the gene mapping approach, now usually referred to as positional cloning. The various steps are straightforward. First, the disease mutation is mapped by testing for genetic linkage with a series of provisionally assigned polymorphic DNA markers strategically interspersed throughout the genome. This is the genome screen, which assigns the mutation successively to its chromosome, its chromosome region and then to its exact location within the region. The closest markers which flank the mutation are identified and, if these are within about 500 kilobases of one another, they can be used to isolate clones containing the marker sequence from genomic DNA libraries constructed in YAC and cosmid vectors. The second step is to sub-clone the YAC DNA containing the disease locus and search among the subclones for possible candidate genes. These may be identified first by their expression in appropriate tissues, their conservation in closely related species, and possibly by their homology with related genes within the human genome database. The final step is to sequence the candidate gene (or its cDNA) and look for differences between the mutant and normal DNA. A DNA mutation which is consistently present in the candidate gene and absent in normal controls provides strong evidence that the disease gene has been cloned. The next step is to express the normal gene, raise antibodies to its product and study its expression in normal and diseased tissues.

Positional cloning has been outstandingly successful in characterising the disease gene in disorders such as cystic fibrosis (Rommens et al, 1989), and myotonic dystrophy (Harley et al, 1992). It is, however, an extremely laborious and expensive procedure. The gene locus for Huntington's disease, has eluded this approach although it was mapped to the distal end of chromosome 4 nearly ten years ago (Gusella et al, 1983). Other disease loci have proved less refractory either because there has been an obvious candidate gene in the vicinity (e.g. rhodopsin in the case of one variety of retinitis pigmentosa (Dryja et al 1990) and fibrillin (Dietz et al, 1991) in the case of Marfan's disease) or because a gross chromosome aberration has been found in association with the disease phenotype. In the latter case, deletion or disruption of the disease locus has occurred at one of the breakpoints in the chromosome aberration. Cloning the breakpoint will then often identify the disease locus. The fortuitous occurrence of such chromosome aberrations has been responsible for the identification of a number of important disease genes including Duchenne muscular dystrophy (Koenig et al, 1987), familial adenomatous polyposis (Groden et al, 1991), neurofibromatosis (Wallace et al, 1990) and the fragile X syndrome of mental retardation (Verkerk et al, 1991). A number of other examples of the successful cloning of disease genes using gene mapping strategies are given in Table 3.

While the location of disease genes are important landmarks for the human gene map, linkage analysis for positional cloning depends largely on polymorphic DNA sequences which are not necessarily located within known genes. Restriction fragment length polymorphisms are detected by Southern blot analysis and are variations in sequence length between specific restriction cleavage sites. Point mutations which interrupt cleavage sites or create new ones, are associated with differences in DNA fragment size detected by gel electrophoresis using a probe from within the fragment (Southern blot). Some cloned DNA sequences contain a variable number of tandem repeats, so that the consensus sequence can be used as a probe to distinguish variations in sequence length, again by electrophoresis. Other DNA sequences contain variable numbers of di- tri- or tetranucleotide repeats and these can be detected using flanking sequences as primers in the polymerase chain reaction (PCR) (Saiki et al, 1988). PCR exploits the enzyme Taq polymerase to amplify DNA between two primer sequences. Differences in size between amplified sequences containing microsatellites can be identified by PCR without the need for Southern blotting, and so are easier to use in genetic linkage studies as well as being usually more highly polymorphic than earlier DNA markers. PCR based markers now contribute essential information to the human genetic map on which the final physical map of overlapping contigs is being assembled.

4   Current developments in molecular cytogenetics

It should be appreciated from the above that human cytogenetics has not only been an important ingredient in the development of human genetics over the past 35 years but has also made substantial technological contributions to animal and plant cytogenetics through the development of banding techniques, fluorescence in situ hybridisation and other mapping technologies. The rapid development of genetic maps in domestic and farm animals owes much to the exploitation of homologies in genetic linkage and DNA sequence with the genetic map of our own species. There are signs that human cytogenetics has not even now reached its zenith and is leading the way in developments which have important applications in genetics in general and genome mapping in particular.

Up to now FISH has been used mainly to map and order DNA sequences in metaphase chromosomes (Lichter et al, 1990). Plasmid, cosmid and YAC vectors are used, and typically each chromatid shows a distinct signal at the hybridisation site (Fig. 1). Distances between DNA sequences greater than one megabase can be resolved; smaller distances at metaphase require the use of translocation breakpoints to separate two closely linked sequences. Recently it has become possible to resolve distances between sequences smaller than one megabase using the extended chromosomes found in interphase (Trask et al, 1989). Chromosomes at interphase are generally believed to be extended approximately ten times their length at metaphase. Using two or more fluorochromes and different

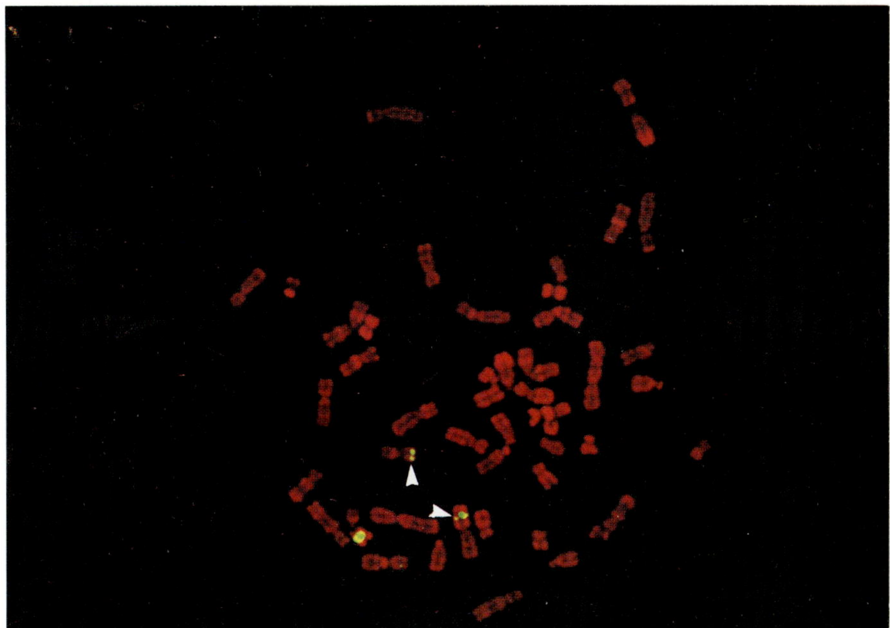

Fig. 1   Fluorescence in situ hybridisation of a YAC clone containing the gene locus for multiple self-healing squamous epithelioma (Goudie et al, 1992) hybridised to a metaphase showing a balanced translocation (46,XY,t(6;9)(p23;q22.3).   The hybridisation spans the breakpoint of the translocation and can be seen in both translocation derivatives (arrows).   The normal chromosome 9 shows the site of hybridisation of the entire YAC insert to 9q22.3.

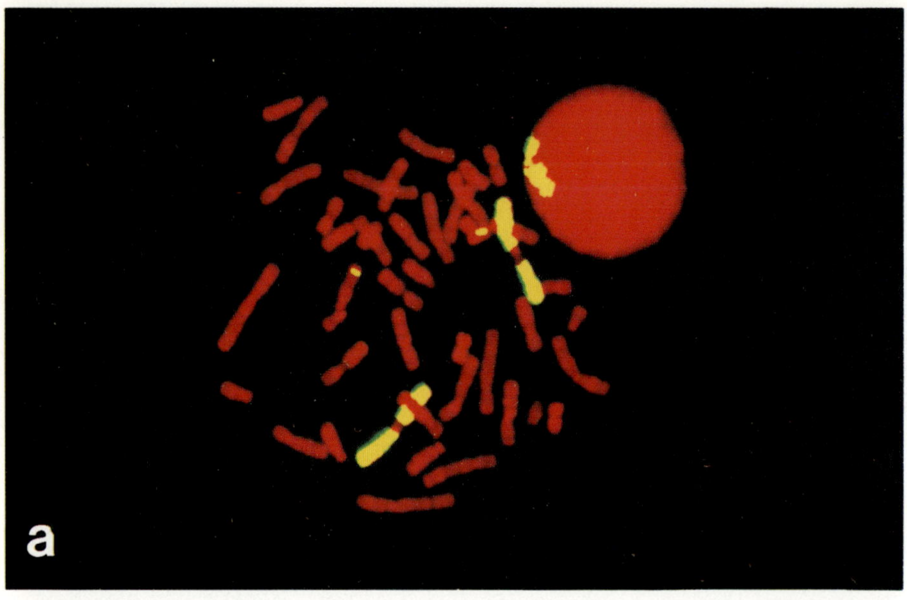

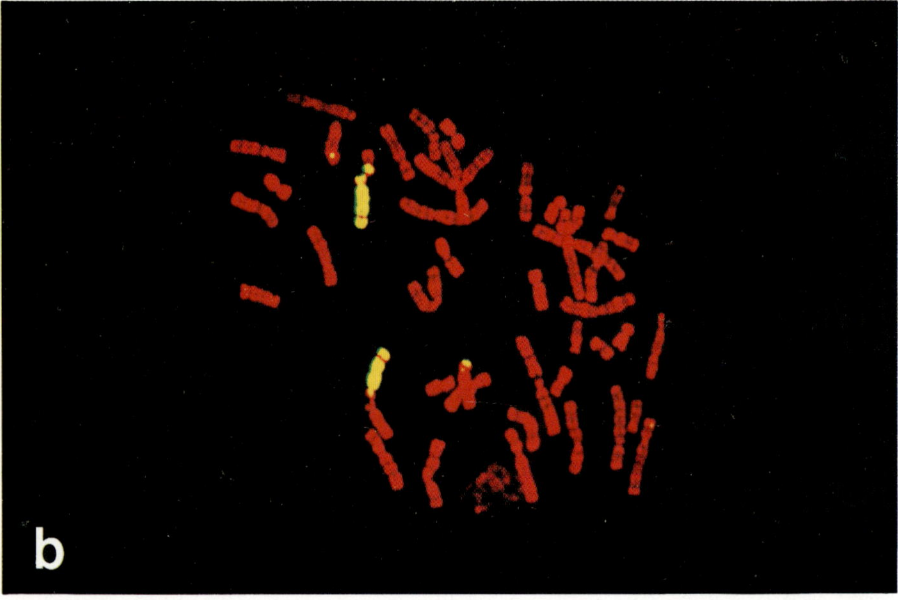

Fig. 3 Chromosome paints derived by flow sorting the derivative chromosomes shown in Fig. 2 have been hybridised to a normal male metaphase. A. The chromosome 1 derivative probe paints both chromosomes 1 and that part of chromosome 13 involved in the insertion. B. The chromosome 13 derivative probe paints all of both normal chromosomes 13 except for that part which is inserted into chromosome 1.

Table 3 Disease genes cloned by gene mapping

Year	Disease	Regional Assignment	Associated Chromosome Aberration	Local Candidate Gene Known
1986	Chronic granulomatous disease	Xp21.1	Deletion	No
"	Duchenne muscular dystrophy	Xp21.2	Deletion, X-AUT	No
"	Retinoblastoma	13q14.2	Deletion	No
1989	Cystic fibrosis	7q31-32	No	No
1990	Wilm's tumour/Drash syndrome	11p13	Deletion	No
"	Retinitis pigmentosa 4	3q21-24	No	Rhodopsin
"	Neurofibromatosis 1	17q11.2	Deletion, Translocation	No
"	Malignant hyperthermia	19q13	No	Ryanodine receptor
"	Testis determining factor	Yp11.3	X-Y Interchange	No
"	Li-Fraumeni syndrome	17p13.1	No	p53 protein
"	Choroideraemia	Xq21.1	Deletion	No
1991	Alzheimer disease	21q21.2	No	Amyloid P.P.
"	Retinitis pigmentosa	6p12	No	Peripherin
"	Fragile X syndrome	Xq27.3	Fragile Site	No
"	Familial adenomatous polyposis	5q21	Translocation	No
"	X-linked spinal musc. atrophy	Xq11.2	No	Androgen receptor
"	Marfan syndrome	15q21.1	No	Fibrillin
"	Aniridia	11p13	Deletion	PAX-6
"	Kallmann syndrome	Xp22.32	Deletion	No
1992	Waardenburg syndrome	2q35-37	Inversion	PAX-3
"	Myotonic dystrophy	19q13.3	No	No
"	Charcot-Marie-Tooth disease	17p11.2	No	Peripheral Myelin P.
"	Norrie disease	Xp11.4	Deletion	No
"	Lowe's Oculocerebrorenal synd.	Xq25-26	X-AUT	No

DNA labels it is possible to order three or more DNA sequences which map to the same chromosome provided the distance between each locus is at least 50 kilobases and no more than one megabase. Closer distances can be resolved by extending the DNA molecule even further using histone depletion techniques which cause the DNA fibre (which forms the Laemmli loop) to spread out from the nucleus. These DNA "haloes" can resolve the DNA within separate cosmids which form part of a cosmid contig provided each cosmid clone is labelled with a different fluorochrome (Wiegant, personal communication). As small fluorescence signals are weak and tend to fade on exposure to U.V. (despite antifade agents), special optical systems have been developed for their detection and analysis. These involve cooled charged-coupled device (CCD) cameras, optical disc storage and image analysis systems. CCD cameras are extremely sensitive and the recorded signals can be analysed at leisure once they are recorded on optical disc. The use of these systems has led to the development of multicolour FISH whereby 5 or more colours can be used simultaneously to identify DNA probes in situ (Ellis et al, pers. comm.). This is achieved by the technique of "ratioing" in which two or three fluorochromes in different ratios are used to label each individual DNA probe (Nederlof et al, 1990). The image analysis system is capable of converting the different ratios into different "false" colours so identifying each DNA probe with its distinct colour. The potential of this exciting technique for ordering gene loci at interphase has yet to be fully realised, but it should be emphasised that the method has a higher resolving capacity than genetic linkage analysis using polymorphic DNA markers in family studies. Even pulsed field gel electrophoresis cannot match multicolour FISH for resolution. As linkage studies in families depend on the total number of meiotic events, it is often difficult to determine accurately recombination frequencies of less than one percent. However, the recent introduction of multiplex analysis in single sperm (Li et al, 1988; Furlong et al, submitted) now allows the analysis of much larger numbers of meiotic events (theoretically many more than a thousand events) and so it may be possible to extend the use of genetic linkage beyond its present limitations.

The above description of techniques of chromosome analysis have involved the use of conventional, fluorescence or electron microscopy, but this is not the only way in which human chromosomes have been studied. Flow cytometry is an important complementary technique which can be used to determine the DNA content and to separate specific chromosomes from one another (Carter et al, 1990). A dual laser fluorescence activated cell sorter is modified to analyse chromosomes rather than cells in fluid suspension. Chromosomes are stained in suspension using Hoechst 33258 and chromomycin A3 and sorted on the basis of size and base pair ratio. All chromosomes, except the four pairs 9-12 (which have similar DNA characteristics) can be separated by this technique to generate a flow karyotype (Fig. 2) and sometimes even individual homologues can be sorted if they are sufficiently heteromorphic. The various applications of flow cytogenetics can be summarised:-

(i) The measurement of DNA content in normal and abnormal chromosomes (Harris et al, 1986). Differences due to deletion, duplication and translocation of as little as 0.5 of a megabase can now be resolved.

(ii) The construction of chromosome specific libraries (Davies et al, 1981). Sufficient numbers of chromosomes of a particular type (e.g. 2 million chromosomes 21) can be sorted into a separate tube, digested and used to create a genomic library in a suitable phage, plasmid or cosmid vector.

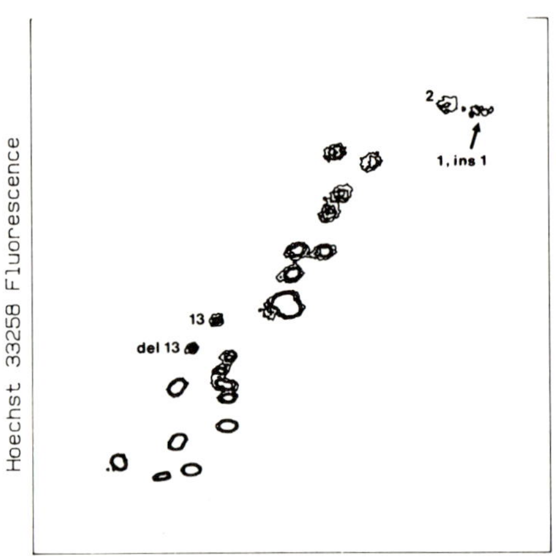

Fig. 2   Flow karyotype from a carrier of an insertional translocation 46,XX,ins(1;13)(p36;q32q34). The peaks of the chromosome 1 with the insertion and the chromosome 13 with the deletion are indicated.

(iii)   The mapping of DNA sequences to their specific chromosome. Two methods can be used. 10,000 chromosomes of each type are sorted directly onto nitrocellulose filters (dot blots) and used in Southern analysis to map a ^{35}P labelled probe to its respective chromosome.   Alternatively, if sequence information on the DNA probe is available, PCR analysis can be used directly in reaction tubes containing as little as 300-500 sorted chromosomes.

(iv)   The production of chromosome specific paints.   300-500 sorted chromosomes of a particular type are sufficient to produce a label which will hybridise by FISH along the entire length of a chromosome in routine air direct metaphases.   The sorted chromosomes are digested in solution, amplified by the PCR reaction using a degenerate oligonucleotide primed method and biotinylated for use with FISH (Telenius et al, 1992). Specific chromosome paints have been made for almost all chromosomes, and are proving extremely valuable in the cytogenetic analysis of chromosome rearrangements (Fig. 3), whether constitutional or in neoplastic cells.

(v)   The reverse painting technique (Carter et al, 1992).   This valuable approach to the analysis of complex chromosome rearrangements depends on the ability to sort 300-500 copies of the **abnormal** chromosome, produce a paint of the sorted chromosomes as in (iv) and then hybridise the chromosome paint onto **normal** metaphase preparations using FISH.   This will reveal the chromosomal origins of the rearranged chromosome and clearly define the breakpoints involved. The method will provide the most detailed cytogenetic analysis of any rearranged chromosome and is particularly important in the analysis of **de novo** chromosome abnormalities.

A further recent innovation contributed by human cytogenetics is the technique of chromosome microdissection. This involves the manipulation of micropipettes to "chisel" individual chromosomes or parts of individual chromosomes from microscope slides (Ludecke et al, 1989), and use the material in PCR reactions to make genomic libraries or chromosome paints (Meltzer et al, 1992). Alternatively, a laser beam can be used to remove parts of the chromosome not required in the reaction.   The method, which requires considerable skill, is very valuable for the production of genomic libraries from small regions of a chromosome containing, for example, a gene which is to be cloned.   It may also be used to produce chromosome band- specific paints which may have application in chromosome identification.

5   Conclusions

This article has reviewed major landmarks in the brief history of human cytogenetics, and their contribution to human genome mapping.   However, it should not be forgotten that much of the driving force for these developments has come from the search for improved diagnosis of chromosomal syndromes and the need to understand the molecular basis of genetic disease.   The FISH techniques which have proved so valuable in the precise mapping of DNA sequences also have other applications, for example in preimplantation diagnosis, in aneuploidy detection in fetal cells isolated from the maternal circulation, in the monitoring of remission and relapse in the treatment of leukaemia, and in the carrier detection of X-linked muscular dystrophy.

These are only a few of the medical applications arising out of new developments in human cytogenetics.   Others of even greater significance are to be expected in the future and one might mention an improved understanding of the pathogenesis of malignancy from the more detailed analysis of cancer chromosomes now possible by reverse painting which should provide clues to the location of cancer genes.   Further into the future one can speculate on the likely role of cytogenetics in the construction of mammalian artificial chromosomes (MACs) as vehicles for successful gene therapy.

Throughout the past 30 years it has been commonplace to hear the uninitiated advise young human cytogeneticists that their field has no future in "modern" genetics. The above account should illustrate that cytogenetics has been at the heart of modern genetics since the chromosomal theory of heredity was established, and that it will continue to be an essential part of genetics in the future. Molecular biologists and others who ignore cytogenetics do so at their peril as, without it, they are likely to have an incomplete understanding of the fundamentals of genetics.

6   References

Arnold, T. (1879) Beobachtungen uber Kentheilungen in den Zellen der Geschwulste. **Virchows Arch**. 78, 279-301

Ashworth, A.et al(1991) X- chromosome inactivation may explain the difference in viability of XO humans and mice. **Nature** 351, 406-408

Barr, M.L. and Bertram, E.G. (1949) A morphological distinction between neurones of the male and female, and the behaviour of the nucleolar satellites during accelerated nucleo-protein synthesis. **Nature**, 163, 676

Bauman, J.G.J., Wiegant, J. and van Duijn, P. (1983) The development, using poly (Hg-U) in a model system, of a new method to visualise hybridisation in fluorescence microscopy. **J. Histochem. Cytochem**. 31, 571-578

Bell, J. and Haldane, J.B.S. (1937) The linkage between the genes for colour-blindness and haemophilia in man. **Proc. Roy. Soc. B.**, 123, 119-150

Boveri, T. (1914) Zur Frage der Entstehung maligner Tumoren. Jena. Verlag von Gustav Fischer (Translation by Boveri, M (1929) The origin of malignant tumours. Baltimore, Williams & Wilkins)

Bradbury, J.T., Bunge, R.G. and Boccabella, R.A. (1956) Chromatin test in Klinefelter's Syndrome. **J. clin. Endocrin. Metab**. 16, 689

Bridges, C.B. (1916) Nondisjunction as a proof of the chromosome theory of heredity **Genetics** 1, 1-52, 107-163

Bridges, C.B. (1938) A revised map of the salivary gland X chromosome of Drosophila melanogaster. **J. Heredity** 29, 11-13

Carr, D.H. (1965) Chromosome studies in spontaneous abortions. **Obstet. Gynec**. 26, 308-326

Carter, N.P. et al (1990) A study of X chromosome abnormality in XX males using bivariate flow karyotype analysis and flow sorted dot blots. **Cytometry** 11, 202-207

Carter, N.P. et al (1992) Reverse chromosome painting: a method for the rapid analysis of aberrant chromosomes in clinical cytogenetics. **J. Med. Genet**. 29, 299-307

Caspersson, T., Zech, L. and Johansson, C. (1970) Differential banding of alkylating fluorochromes in human chromosomes. **Exp. Cell Res**. 60, 315-319

Chandley, A.C. et al. (1984) On the nature and extent of XY pairing at meiotic prophase in man. **Cytogenet. Cell Genet**. 38, 241-247

Davies, K.E. et al (1981) Cloning of a representative genomic library of the human X chromosome after sorting by flow cytometry. **Nature** 293, 374-376

De Winiwarter, H. (1912) Etudes sur la spermatogenese humaine. **Arch. Biol. (Liege)** 27, 91-189

Dietz, H.C. (1991) Marfan syndrome caused by a recurrent de novo missense mutation in the fibrillin gene. **Nature** 352, 337-339

Donahue, R.P. et al (1968) Probable assignment of the Duffy blood group locus to chromosome 1 in man. **Proc. Natl. Acad. Sci. U.S.A**. 61, 949-955

Dryja, T.P. et al. (1990) A point mutation of the rhodopsin gene in one form of retinitis pigmentosa **Nature** 343, 364-366

Ferguson-Smith, M.A. (1964) The sites of nucleolus formation in human pachytene chromosomes. **Cytogenetics**, 3, 124-134

Ferguson-Smith, M.A. (1965) Karyotype-phenotype correlations in gonadal dysgenesis and their bearing on the pathogenesis of malformations. **J. Med. Genet.**, 2, 142-155

Ferguson-Smith, M.A. and Page B.M. (1973) Pachytene analysis in a human reciprocal (10;11) translocation. **J. med. Genet.** 10, 282-287

Ferguson-Smith, M.A. (1991) Genotype-phenotype correlations in individuals with disorders of sex determination and development including Turner's Syndrome. **Seminars in Developmental Biology,** 2, 265-276

Ferguson-Smith, M.A. et al (1962) The sites and relative frequencies of secondary constrictions in human somatic chromosomes. **Cytogenetics,** 1, 325- 343

Ferguson-Smith, M.A. and Handmaker, S.D. (1961) Observations on the satellited human chromosomes. **Lancet,** i, 638-640

Ferguson-Smith, M.A., Johnston, A.W. and Handmaker, S.D. (1960) Primary amentia and micro-orchidism associated with an XXXY sex-chromosome constitution. **Lancet,** ii, 184-187

Ferguson-Smith, M.A. and Munro, I.B. (1958) Spermatogenesis in the presence of female nuclear sex. **Scot. med. J.,** 3, 39-42

Ferguson-Smith, M.A., et al (1973) Assignment by deletion of human red cell acid phosphatase gene locus to the short arm of chromosome 2. **Nature New Biol.,** 243, 271-274

Flemming, W. (1898) Ueber die Chromosomenzahl beim Menshen. **Anatomischer Anzeiger,** 14, 171-174

Ford, C.E. and Hamerton, J.L. (1956) The chromosomes of man. **Nature,** 178, 1020

Ford, C.E. et al (1958) Human somatic chromosomes. **Nature,** 181, 1565-1568.

Ford, C.E. et al (1959) A sex-chromosome anomaly in a case of gonadal dysgenesis (Turner's Syndrome). **Lancet,** i, 711-713

Franco, B. et al (1991) A gene deleted in Kallmann's syndrome shares homology with neural cell adhesion and axonal path-finding molecules. **Nature,** 353, 529-535

German, J.L. (1962) DNA synthesis in human chromosomes. **Trans. N.Y. Acad. Sci.,** 24, 395-407

Goudie, D.R. et al. Localisation of the gene responsible for mutliple self-healing squamous epitheliomata (ESS1) to 9q22.1-q31 by genetic linkage in families of common descent. Nature Genetics (in press)

Groden, J. et al. (1991) Identification and characterisation of the familial adenomatous polyposis coli gene. **Cell,** 66, 589-600

Gusella, J.E. et al (1983) A polymorphic DNA marker genetically linked to Huntington's disease. **Nature,** 306, 234-238

Hansemann, D. (1891) Uber Pathologische Mitosen. **Virchows Arch.,** 123, 356

Harley, H.G. et al (1992) Expansion of an unstable DNA region and phenotypic variation in myotonic dystrophy. **Nature,** 355, 545-546

Harper, M.E., Ullrich, A. and Saunders, G.F. (1981) Localisation of the human insulin gene to the distal end of the short arm of chromosome 11. **Proc. Natl. Acad. Sci. USA.,** 78, 4458-4460

Harris, P. et al (1986) Determination of the DNA content of human chromosomes by flow cytometry. **Cytogenet. Cell Genet.,** 41, 14-21

Heiles, H.B.J. et al (1988) In situ hybridisation with digoxigenin-labelled DNA of papilloma viruses (HPV 16/18) in HeLa and Sitla cells. **Biotechniques,** 6, 978-981

Henderson, A.S., Warburton, D. and Atwood, K.C. (1972) Location of rDNA in the human chromosome complement. **Proc. Natl. Acad. Sci. USA.,** 69, 3394-3398

Hungerford, D.A. (1971) Chromosome structure and function in man. I. Pachytene mapping in the male, improved methods and general discussion of initial results. **Cytogenetics,** 10, 23-32

Hsu, T.C. (1952) Mammalian chromosomes in vitro I. The karyotype of man. **J. Hered.,** 43, 167-172

ISCN (1981) An international system for human cytogenetics nomenclature: high resolution banding. **Cytogenet. Cell Genet.** 31. 1-23

Jacobs, P.A. and Strong, J.A. (1959) A case of human intersexuality having a possible XXY sex determining mechanism. **Nature,** 183, 302

Koenig, M. et al (1987) Complete cloning of the Duchenne muscular dystrophy (DMD) cDNA and preliminary genomic organisation of the DMD gene in normal and affected individuals. **Cell,** 50, 509-517

Koller, P.C. (1937) The genetical and mechanical properties of sex chromosomes. III Man **Proc. Roy. Soc. Edinb. B.,** 57, 194-214

Koller, P.C. and Darlington, C.D. (1934) The genetical and mechanical properties of the sex chromosomes. I. Rattus norvegicus. J. Genet., 29, 159-173

Landegent, J.E. et al (1984) 2-Acetylaminofluorene-modified probes for the indirect hybridocytochemical detection of specific nucleic acid sequences. **Exp. Cell. Res.,** 153, 61-72

Langer, P.R., Waldrop, A.A. and Ward, D.C. (1981) Enzymatic synthesis of biotin labelled polynucleotides: novel nucleic acid affinity probes. **Proc. Natl. Acad. Sci. USA.,** 78, 6633-6637

Lejeune, J., Gautier, M. and Turpin, R. (1959) Les chromosomes somatique des enfants mongoliens. **Comptes Rend. Acad. des Sc. Paris.,** 248, 1721

Li, H., et al (1988) Amplification and analysis of DNA sequences in single human sperm. **Nature,** 335, 414-417

Lichter, P. et al (1990) High resolution mapping of human chromosome 11 by in situ hybridisation with cosmid clones. **Science,** 247, 64-69

Ludecke, H.J., Senger, G., Claussen, U. and Horsthemke, B. (1989) Cloning defined regions of the human genome by microdissection of banded chromosomes and enzymatic amplification. **Nature,** 338, 348-350

Lyon, M. (1961) Gene action in the mammalian X chromosome of the mouse (**Mus musculus L). Nature,** 190, 372-373

Makino, S. and Nishimura, I. (1952) Water pre-treatment squash technique. **Stain Technol.,** 27, 1-7

Malcolm, S. et al (1981a) Chromosomal localisation of a single copy gene by in situ hybridisation: human β-globin genes on the short arm of chromosome 11. **Ann. Hum. Genet.,** 45, 135-141

Malcolm, S. et al (1981b) Assignment of a IGKV locus for immunoglobulin light chains to the short arm of chromosome 2 (2 cen - p13) by in situ hybridisation using a cRNA probe of H101 CH4A. **Cytogenetics & Cell Genetics,** 32, 296

Meltzer, P.S.et al (1992) Rapid generation of region specific probes by chromosome microdissection and their application. **Nature Genetics,** 1, 24-28

Mittwoch, U. (1952) The chromosome complement in a mongolian imbecile. **Ann. Eugen. Camb.** 17, 37

Mohr, J. (1951) Estimation of linkage between the Lutheran and the Lewis blood groups. **Acta Path. Microbiol. Scand.,** 29, 339-344

Moorhead, P.S. et al (1960) Chromosome preparations of leukocytes cultures from human peripheral blood. **Exp. Cell Res.,** 20, 613-616

Moses, M.J., Counce, S.J. and Paulson, D.F. (1975) Synaptonemal complex complement of man in spreads of spermatocytes, with details of the sex chromosome pair. **Science,** 187, 363-365

Nederlof, P.M. et al (1990) Multiple fluorescence in situ hybridization. Cytometry, 11, 126-131

Nowell, P.C. and Hungerford, D.A. (1960) A minute chromosome in human chronic granulocytic leukaemia. **Science,** 132, 1497

Ogata, T. et al (1992)   Short stature in a girl with a terminal Xp deletion distal to DXYS15: localisation of a growth gene(s) iun the pseudoautosomal region. **J. Med. Genet.,** 29, 455-459

Ohno, S., Kaplsan, W.D. and Kinosita,, R. (1959) Formation of the sex chromatin by a single X-chromosome in liver cells of Rattus norvegicus. **Exp. Cell Res.,** 18, 415-418

Painter, T.S. (1921) The Y chromosome in mammals. **Science,** 53, 503

Painter, T.S. (1923) Studies in mammalian spermatogenesis II. The spermatogenesis in man. **J. exp. Zool.,** 37, 291

Pardue, M.L. and Gall, J.G. (1970) Chromosomal localisation of mouse satellite DNA. **Science,** 168, 1356-1358

Polani, P.E., Hunter, W.F.and Lennox, B. (1954) Chromosomal sex in Turner's syndrome with coarctation of the aorta. **Lancet, 2,** 120

Price, P.M., Conner, J.H. and Hirschhorn, K. (1972) Chromosomal localisation of human haemoglobin structural genes. **Nature,** 237, 340-342

Robson, E.B. (1969) Probable assignment of the alpha locus of haptoglobin to chromosome 16 in man. **Nature,** 223, 1163-1165

Rommens, J.M. et al (1989) Identification of the cystic fibrosis gene: chromosome walking and jumping. **Science,** 245, 1059-1065

Saadallah, N. and Hulten, M. (1986) EM investigations of surface spread synaptonemal complexes in a human male carrier of a pericentric inversion inv(13)(p12q14): the role of heterosynapsis for spermatocyte survival. **Ann. Hum. Genet.,** 50, 369-383

Saiki, R.K., et al(1988) Primer-directed enzymatic amplication of DNA with a thermostable DNA polymerase. **Science,** 239, 487-491

Seabright, M. (1971) A rapid banding technique for human chromosomes. **Lancet,** 2, 971-972

Severinghaus, A.E. (1942) Sex chromosomes in a human intersex. **Amer. J.Anat.,** 70, 73-92

Sinclair, A.H. et al (1990) A gene from the human sex determining region encodes a protein with homology to a conserved DNA-binding motif. **Nature,** 346, 240-244

Sumner, A.T., Evans, H.J. and Buckland, R.A. (1971) New technique for distinguishing between human chromosomes. **Nature,** 232, 31-32

Telenius, H. et al (1992) Cytogenetic analysis by chromosome painting using DOP-PCR amplified flow-sorted chromosomes. **Genes, Chromosomes & Cancer,** 4, 257-263

Tjio, J.H. and Levan, A. (1956) The chromosome number of man. **Hereditas,** 42, 1-6

Trask, B.J., Pinkel, D. and van den Eng, G. (1989) The proximity of DNA sequences in interphase cell nuclei in correlated to genomic distance and permits ordering of cosmids spanning 250 kilobase pairs. **Genomics,** 5, 710-717

Verkerk, A.J.M.H. et al (1991) Identification of a gene (FMR-1) containing a CGG repeat coincident with a breakpoint cluster region exhibiting length variation in fragile X syndrome. Cell, 65, 905-914

Waardenburg, P.J. (1932) Das menschliche Auge und seine Erbanlargen. **Haag: Martinas Nijhoff,** pp. 47-48

Wallace, M.R. et al (1990) Type 1 neurofibromatosis gene: identification of a large transcript disrupted in three NF1 patients. **Science,** 249, 181-186

Weiss, M. and Green, H. (1967) Human-mouse hybrid cell lines containing partial complements of human chromosomes and functioning human genes. **Proc. Natl. Acad. Sci. US,** 58, 1104-1111

Yerganian, G. (1957) Cytologic maps of some isolated human pachytene chromosomes. **Amer. J. Hum. Genet.,** 9, 42-54

Chromosome structure

2 Molecular cloning and characterization of human centromeric autoantigen CENP-C: a component of the inner kinetochore plate

W.C. EARNSHAW, H. SAITOH, J.E. TOMKIEL, C.A. COOKE, R.L. BERNAT, H. RATRIE, M. MAURER and N.F. ROTHFIELD
John Hopkins University School of Medicine, USA

1. Introduction

Chromosomes move in mitosis as a result of interactions between the centromere region and microtubules of the mitotic spindle. It is currently believed that a specialized chromosomal substructure located at the surface of the centromere is responsible for these movements. This structure, the kinetochore, is typically described as a three-layered disk composed of two dense plates with a space between them (Rieder, 1982). The inner dense plate is congruent with the surface of the centromeric heterochromatin, and under some circumstances is not clearly resolved from the surrounding chromatin. In human the three layers are all about 35 nm thick and 430 nm in diameter (diameter estimated from Rieder (1982) assuming 20 microtubules per chromosome).

Much recent attention has focused on the dense outer plate of the kinetochore, as this is where the spindle microtubules attach. In addition, the outer surface of this plate is associated with an amorphous structure, termed the fibrous corona, that is seen under conditions when there are no microtubules bound to the plate (in early prometaphase and in the presence of microtubule-depolymerizing drugs, such as colcemid and nocodazole). The fibrous corona appears to contain mechanochemical motors such as cytoplasmic dynein (Pfarr et al., 1990; Steuer et al., 1990; Wordeman et al., 1991) and members of the kinesin superfamily (Sawin et al., 1992). Studies of the earliest prometaphase chromosomal movements in newt lung cells reveal that the chromosomes slide along the surface of the microtubules, making contact solely in the region of the fibrous corona.

Thus, we have a reasonable understanding of possible functions for the fibrous corona (chromosome movement) and outer dense plate (microtubule binding). At the present time, we know much less about the function of the inner kinetochore plate and the interzone between the two plates. However, as will be seen below, emerging studies from our laboratory suggest that one essential function of the inner kinetochore plate may be to assemble a solid substratum for the outer force-producing regions of the kinetochore. Surprisingly, this structural solidity appears to be required for mitotic cells to execute a timely transition from metaphase to anaphase.

The studies from our laboratory have exploited naturally occurring autoantibodies to centromeric proteins that are present in the sera of approximately 20% of patients with scleroderma spectrum disease (Moroi et al., 1980; Earnshaw et al., 1986). We and others have shown that these antibodies recognize four polypeptides, which we termed CENP-A (CENtromere Protein A - M_r 17 kDa), CENP-B (M_r 80 kDa), CENP-C (M_r 140 kDa) (Earnshaw and Rothfield, 1985). CENP-D (M_r 50 kDa) was subsequently recognized (Kingwell and Rattner, 1987). cDNAs encoding three of these proteins (CENPs B-D) are now cloned, and limited peptide sequence is also available for the fourth member (CENP-A).

Chromosomes Today Volume 11. Edited by A.T. Sumner and A.C. Chandley. Published in 1993 by Chapman & Hall, London. ISBN 0 412 47670 3

CENP-A is a centromere-specific histone subtype that is distantly related to histone H3. Certain regions of the protein show great similarity to H3, while others are unique (Palmer et al., 1991). This variability is unheard of for H3, which is extremely highly conserved throughout evolution. CENP-A appears to be present in nucleosomal particles derived from the centromeres of human chromosomes (Palmer and Margolis, 1987). The distribution of this protein within centromeres has not yet been determined by immunoelectron microscopy, and its role in centromere function is unknown.

CENP-B migrates anomalously in SDS-PAGE: its true molecular weight is closer to 65,000 Da (Earnshaw et al., 1987). CENP-B contains two highly acidic domains that, together with a proline-rich hinge region, divide the protein into four domains. Immunoelectron microscopy showed the protein to be distributed throughout the body of the centromere beneath the kinetochore (Cooke et al., 1990). In addition the protein is present at both functional and non-functional centromeres on at least two human stable dicentric chromosomes   (Earnshaw et al., 1989; Earnshaw, unpublished). Thus, the presence of CENP-B is not necessarily indicative of centromere activity.

Consistent to its localization to the centromeric heterochromatin is the finding that CENP-B binds to a specific 17 base pair sequence (the CENP-B box) in centromeric repetitive DNAs from human and mouse (Masumoto et al., 1989). CENP-B is targeted to the centromere in vivo apparently solely as a function of its recognition of the CENP-B box: the amino-terminal 131 amino acids of the protein will target to centromeres in vivo, and also bind to satellite DNA containing the CENP-B box in vitro (Pluta et al., 1992). Thus CENP-B appears to be a structural protein of the centromeric heterochromatin. If human α-satellite chromatin exhibits a preferred nucleosome phasing similar to that exhibited by green monkey α-satellite chromatin, then the CENP-B box is likely to be located in the interior of the solenoidal helix (Widom and Klug, 1985). Based on this assumption, we have proposed a model where the binding of CENP-B with its highly acidic domains might have a profound influence on the structure of chromatin at the centromere (Pluta et al., 1992).

Interestingly, the CENP-B box is present not only in human α-satellite DNA, but also in the minor satellite of the mouse (which is otherwise unrelated to α-satellite DNA). Furthermore, it has thus far been impossible to detect association of CENP-B with the Y chromosome of either the human or mouse (WCE, unpublished; Peter Moens personal communication). In the human, the CENP-B box has yet to be found in any α-satellite arrays cloned from the Y chromosome. Thus, in considering the role of this protein in vivo, it is necessary to take into account its absence (or presence at very low levels) at the Y centromere. We currently imagine that there may be homologue of CENP-B that is found on the Y chromosome and is not detected with available CENP-B-specific reagents.

A protein comigrating with CENP-D was recently purified from human cells, and when its partial amino acid sequence was obtained it was found to be identical to a known cell cycle control protein, RCC1 (Regulator of Chromosome Condensation 1 - Ohtsubo et al., 1989; Bischoff et al., 1990). This result was very surprising for three reasons. Firstly, it seems strange that a cell cycle regulator would be concentrated at the centromere. Secondly, when antibodies have been made to cloned RCC1 protein, these antibodies do not preferentially bind to centromeres: instead they coat the entire chromosome (Ohtsubo et al., 1989). Thirdly, the RCC1 protein is extremely abundant, being present in >10^6 copies per cell. It is hard to imagine that the punctate staining observed at centromeres could account for all of this protein.

At present, the solution to these dilemmas is unknown. One possibility is that the autoantibodies to CENP-D may only recognize a subset of the cellular RCC1 molecules. For example, it is possible that RCC1 protein at centromeres carries a specific modification that is recognized by the autoantibodies. Earlier studies from our laboratory have implied the existence of several such epitopes on CENP-B and CENP-C (Earnshaw et al., 1987). The notion of a structure-specific posttranslational modification is not un-known. Nuclear pores contain a family of at least eight distinct polypeptides (the nucleo-porins (Miller et al., 1991)) that share a common epitope consisting of an O-linked N-acetylglucosamine linked to serine residues. This epitope is specific for nuclear pores, and the entire family of nucleoporins is recognized by several different monoclonal anti-bodies. The nature of the centromere-specific posttranslational modification (if one ex-ists) is unknown at the present time.

Until very recently CENP-C was the least well known of the centromeric autoanti-gens. However, with the availability of cDNA clones encoding this protein, rapid progress has been made in understanding its role in the centromere. The present report summarizes our studies on the isolation and characterization of cDNA clones encoding CENP-C. We then introduce evidence that CENP-C is encoded by a single-copy gene in humans. Our results lead to the suggestion that CENP-C is a component of the inner kinetochore plate that is required for kinetochore function during mitosis.

2. Experimental Procedures
With the following exceptions, all methods are described in Saitoh et al. (1992).

2.1 Northern blot Analysis
Poly(A)-enriched RNA was isolated from HeLa cells using the FastTrack mRNA Isolation Kit (Invitrogen, San Diego, CA). The RNA samples were electrophoresed through a 1% agarose/formaldehyde gel and transferred to Zeta Probe Blotting Membrane (Bio-Rad, Richmond, CA) in 20x SSC. Filters were hybridized at 65 °C in 6x SSC, 25 mM KPO$_4$, 5x Denhardt's, 0.1% SDS containing 1x10^6 cpm/ml of ^{32}P-random priming labeled probe, and were washed in 0.3x SSC, 0.1% SDS at 65 °C. The cDNA inserts of CNPC2, CNPC3 and CNPC7 were used as probes.

2.2 Southern Analysis of Genomic DNA
HeLa cell DNA was digested with the appropriate restriction enzymes, separated on 0.7 % agarose gel and transferred to a 0.45 µm nitrocellulose membrane (Schleicher and Schuell, Keene, NH). The filter was hybridized and washed using the same conditions as for Northern blots. The 369 bp *Hinf*1-*Alu*I fragment of the CNPC3 cDNA insert (nucleotides 921-1290 in Figure 1), was used as a probe.

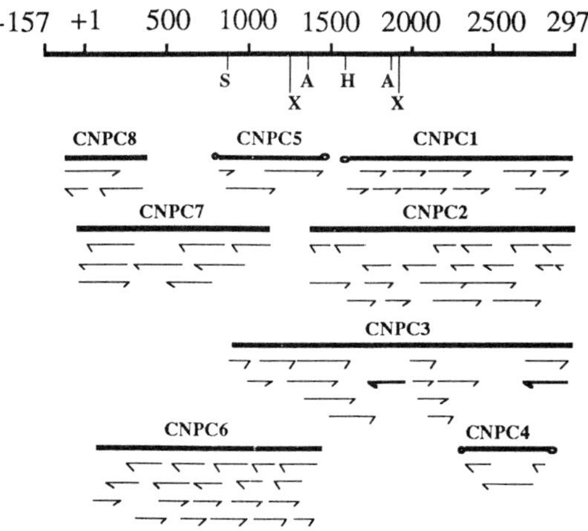

Figure 1: Scheme for determining the DNA sequence of overlapping CENP-C cDNAs. The position of each cDNA is indicated relative to the full length cDNA contig. Positive numbering begins at the adenine of the initiator methionine codon. Arrows indicate the regions and strands subjected to dideoxy chain termination DNA sequencing. Open circles indicate a juncture of CENP-C cDNA sequences with unrelated sequence (i.e. cloning artifacts). The positions of restriction enzyme recognition sites used in cloning or in making ^{32}P-labeled probes are indicated below the bar representing the sequence : S = *Sph* I, X= *Xba* I, A = *Alu* I, H = *Hinf* I. See Saitoh et al. (1992).

3. Results

These experiments have previously been described in Saitoh et al. (1992).

3.1 cDNA cloning of human CENP-C

Three cDNA clones expressing epitopes present on CENP-C were obtained by screening a human placenta λgt11 cDNA library with serum from a patient with scleroderma. Prior to screening the library, this serum, which contains antibodies against CENPs A-C was depleted for antibodies to CENP-B by absorption with a CENP-B fusion protein. Confirmation that the clones express epitopes on CENP-C was obtained by using nitrocellulose lifts of phage plaques to affinity-purify antibodies from autoimmune serum. Antibodies obtained in this way recognized CENP-C on immunoblots of human chromosomal proteins, and bound to centromeres on mitotic spreads of intact human chromosomes (Earnshaw et al., 1989). Characterization of these clones by restriction mapping and DNA sequencing revealed that they share 1.4 kb of sequence identity in their 3' regions (CNPC1, CNPC2 and CNPC3, Figure 1).

Confirmation that the clones encode CENP-C was obtained by raising polyclonal antibodies to fusion proteins encoded by non-overlapping regions of the longest clone, CNPC3. These antibodies recognize a single polypeptide of 140 kDa in immunoblots of chromosomal proteins and recognize centromeres in indirect immunofluorescence of human chromosomes (Figure 2). In this way, we were able to confirm that the cloned proteins share at least three independent epitopes with chromosomal CENP-C.

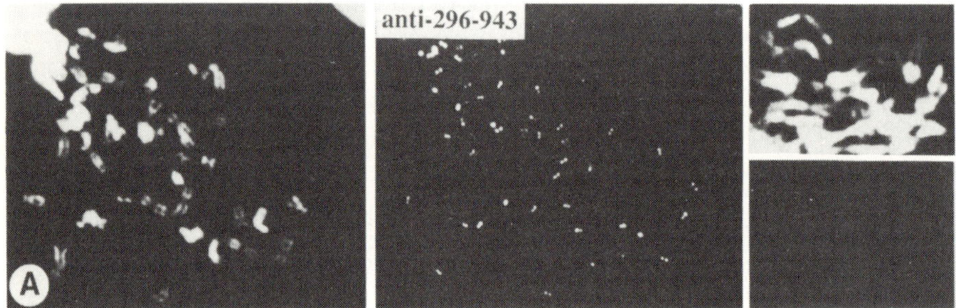

Figure 2: Indirect immunofluorescence of HeLa cell chromosomes with CENP-C-specific antibody. The left panel shows DAPI staining of DNA, the middle panel shows staining with the indicated anti-CENP-C antiserum. The right panel shows DAPI staining (upper) and staining with the corresponding pre-immune serum (lower).

```
1
MAASGLDHLK NGYRRRFCRP SRARDINTEQ GQNVLEILQD CFEEKSLAND FSTNSTKSVP NSTRKIKDTC
IQSPSKECQK SHPKSVPVSS KKKEASLQFV VEPSEATNRS VQAHEVHQKI LATDVSSKNT PDSKKISSRN
INDHHSEADE EFYLSVGSPS VLLDAKTSVS QNVIPSSAKK RETYTFENSV NMLPSSTEVS VKTKKRLNFD
DKVMLKKIEI DNKVSDEEDK TSEGQERKPS GSSQNRIRDS EYEIQRQAKK SFSTLFLETV KRKSESSPIV
RHAATAPPHS CPPDDTKLIE DEFIIDESDQ SFASRSWITI PRKAGSLKQR TISPAESTAL FQGRKSREKH
HNILPKTLAN DKHSHKPHPV ETSQPSDKTV LDTSYALIDE TVNNYRSTKY EMYSKNAEKP SRSKRTIKQK
QRRKFMAKPA EEQLDVGQSK DENIHTSHIT QDEFQRNSDR NMEEHEEMGN DCVSKKQMPP VGSKKSSTRK
DKEESKKKRF SSESKNKLVP EEVTSTVTKS RRISRRPSDW WVVKSEESPV YSNSSVRNEL PMHHNSSRKS
TKKTNQSSKN IRKKTIPLKR QKTATKGNQR VQKFLNAEGS GGIVGHDEIS RCSLSEPLES DEADLAKKKN
LDCSRSTRSS KNEDNIMTAQ NVPLKPQTSG YTCNIPTESN LDSGEHKTSV LEESGPSRLN NNYLMSGKND
VDDEEVHGSS DDSKQSKVIP KNRIHHKLVL PSNTPNVRRT KRTRLKPLEY WRGERIDYQG RPSGGFVISG
VLSPDTISSK RKAKENIGKV NKKSNKKRIC LDNDERKTNL MVNLGIPLGD PLQPTRVKDP ETREIILMDL
VRPQDTYQFF VKHGELKVYK TLDTPFFSTG KLILGPQEEK GKQHVGQDIL VFYVNFGDLL CTLHETPYIL
STGDSFYVPG NYYNIKNLRN EESVLLFTQI KR
                                                                    943
```

Figure 3: Deduced amino acid sequence of the human CENP-C cDNA. The 2,957 bp cDNA sequence compiled from overlapping cDNA clones has here been translated, yielding a 943 residue open reading frame. The putative p34CDC2 phosphorylation site is underlined. See Saitoh et al. (1992).

The complete CENP-C cDNA, was obtained by extensive screening of three additional human cDNA libraries by DNA plaque hybridization using a 5' portions of CNPC3 as well as several oligonucleotide probes (Saitoh et al., 1992). In this way, five additional clones were selected (Figure 1) and their DNA sequences determined . A 3113 bp contiguous sequence corresponding to the CENP-C message was constructed from these overlapping cDNAs. This sequence possesses a single long open reading frame, which begins 158 bp from the 5' end and extends for 2829 bp, encoding a protein of deduced molecular weight 107 kDa (Figure 3).

We used two criteria to determine that the AUG codon 158 bp from the 5' end of the sequence is likely to be the *bona fide* translational initiation codon for CENP-C. Firstly, this is the first AUG to occur within the single long open reading frame, and it matches the Kozak consensus sequence at positions -3 and +4, the two positions deemed most critical for function (Kozak, 1987).

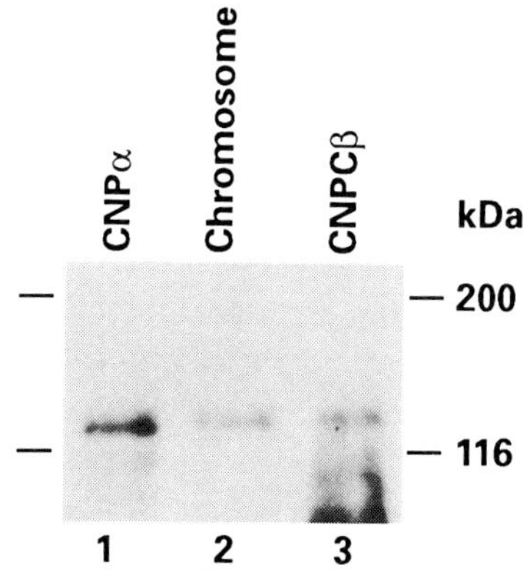

Figure 4: Immunoblot analysis. Fusion protein CNPCα (lane 1), HeLa cell chromosomal CENP-C (lane 2) and bacterially expressed fusion protein CNPCβ (lane 3) were subjected to SDS-PAGE, blotted to nitrocellulose, and probed with GS serum (ACA serum containing high titer of anti-CENP-C antibodies). Positions of protein molecular markers are indicated on the left.

Secondly, we demonstrated that a protein encoded from this start codon roughly comigrates with CENP-C in SDS-PAGE. This was done by comparison of the migration of two bacterially expressed fusion proteins predicted to bracket the *bona fide* protein in size with that of CENP-C from isolated HeLa chromosomes (Figure 4). Fusion protein CNPCα was produced from a cDNA construct containing sequences from 72 bp <u>downstream</u> from the presumed initiator codon to the 3' end of the CENP-C ORF. This protein includes a 15 aa fusion moiety, but is missing 24 aa from the presumed N-terminus. Thus it is predicted to be 9 aa shorter than *bona fide* CENP-C if our assignment of initiator methionine is correct. A second fusion protein, CNPCβ, was produced from a second cDNA construct starting 120 bp <u>upstream</u> from the presumed initiator codon and including the entire CENP-C ORF. As this protein contains an 11 aa fusion moiety, it was predicted to be 45 aa longer than *bona fide* CENP-C.

HeLa chromosomal CENP-C was observed to migrate on SDS-PAGE at a position intermediate between the positions of the shorter and longer fusion proteins (Figure 4). This is consistent with our identification of the initiator AUG based on DNA sequence context and suggests that the sequence reported in Figure 3 includes the entire open reading frame encoding full length CENP-C.

3.2 CENP-C is encoded by a single copy gene that produces one major transcript

Southern (DNA) hybridization to HeLa genomic DNA digested with six different enzymes was performed using a ^{32}P-labeled *Hinf* I-*Alu* I fragment of the CENP-C cDNA as a probe. In three of the six digests the probe hybridized to a single band, indicating that CENP-C is encoded by a single copy gene (Figure 5A). Identical results were obtained when DNA from a non-transformed 46XY cell line obtained in chorionic villus sampling was digested with *Bam* HI, *Bgl* II, *Eco* RI and *Hind* III and probed with the same fragment (data not shown).

Northern (RNA) hybridization analysis of poly-(A) enriched cytoplasmic HeLa RNA was performed using the 2.1 kb CNPC3 cDNA as a probe (Figure 5B). The probe hybridized to a single band of approximately 3.6 kb. This suggests that the full length message encoding CENP-C contains approximately 500 bp of upstream non-translated sequences and/or poly-(A) tail in addition to the sequences we have cloned. Identical results were obtained when non-overlapping portions of clones CNPC2 and CNPC7 were used as probes (data not shown). This is consistent with the hypothesis that the overlapping collection of cDNAs is derived from a single mRNA species.

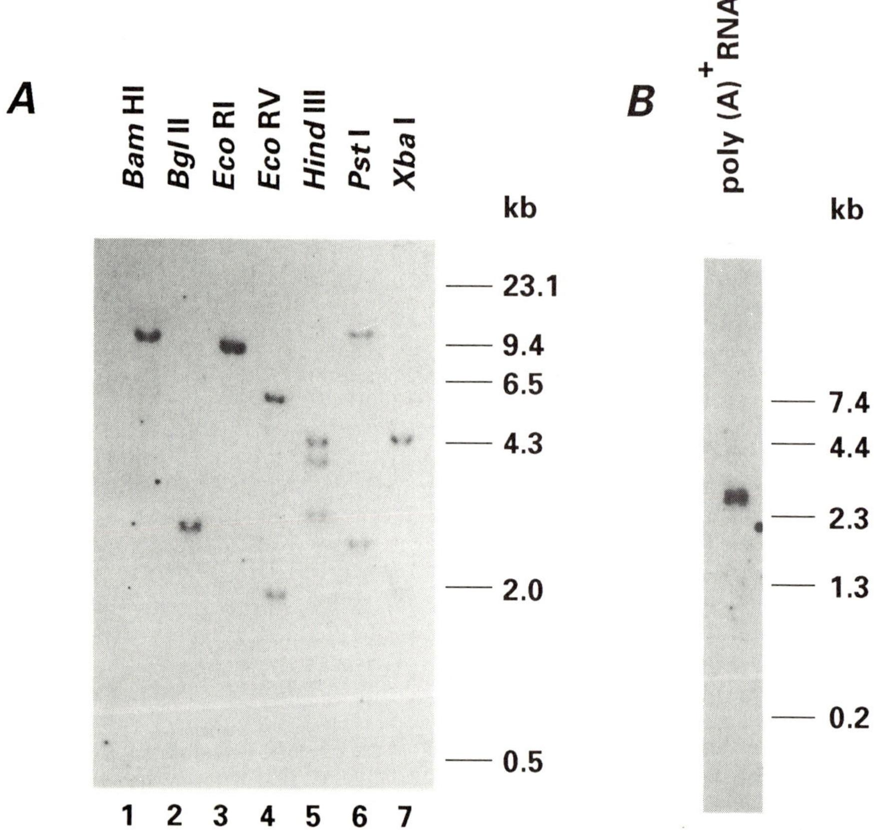

Figure 5: CENP-C is encoded by a single-copy human gene that is transcribed into a 3.6 kb mature mRNA. (A) Southern (DNA) blot analysis. HeLa cell genomic DNA (30 µg/lane) digested by six restriction endonucleases and blotted to nitrocellulose after electrophoresis and denaturation. Fragments homologous to CENP-C were identified by autoradiography after hybridization of a ^{32}P-labeled probe constructed from the 369 bp *Hinf* I-*Alu* I fragment of CNPC3 . Numbers to the right indicate the size, in kilobases, of DNA markers run in an adjacent lane and visualized by ethidium bromide staining. The restriction enzyme used is indicated above each lane. (B) Poly(A)-enriched HeLa cell RNA (3 µg/lane) was

analyzed by RNA blotting after electrophoresis on a denaturing formaldehyde agarose gel. CENP-C mRNA was detected by autoradiography of a blot probed with a ^{32}P-labeled probe from CNPC3. RNA size was determined by comparison to migration of known RNA size markers (Bethesda Research Laboratories) run in an adjacent lane and visualized by staining with ethidium bromide. Migration positions of markers in kb are indicated on the right. Identical results were obtained using CNPC2 or CNPC7 as a probe (data not shown).

3.3 CENP-C is located in the inner kinetochore plate

We have determined the distribution of CENP-C within human mitotic chromosomes by immunoelectron microscopy. These experiments (shown in Figure 6) involved the use of pre-embedding techniques, and the bound antibody was visualized using a 1 nm colloidal gold probe followed by silver enhancement. The gold particles were highly concentrated in a single sharp band immediately beneath the outer kinetochore plate. In many cases they appear to contact the under surface of the plate directly, but we observed only a background labeling of the plate itself or of the fibrous corona.

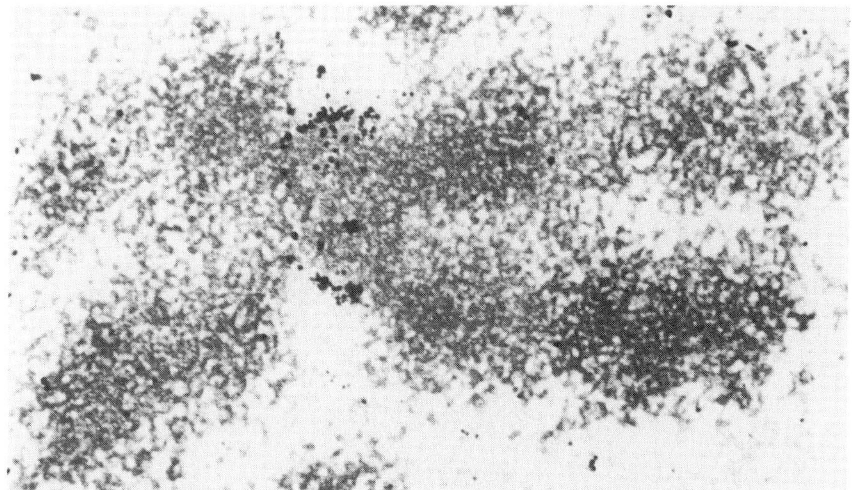

Figure 6: Localization of CENP-C in mitotic chromosomes in situ. A mitotic chromosome from a mitotic HeLa cell (unblocked) was stained with rabbit antiserum anti-g296-943 followed by anti-rabbit IgG conjugated to 1 nm colloidal gold. See Saitoh et al. (1992).

We conclude that CENP-C is a component of the inner kinetochore plate. We cannot, however, exclude the possibility that a significant amount of CENP-C may also be present in the middle zone between the plates.

3.4 Characterization of antibodies used in prior microinjection experiments using CENP-C fusion proteins

We noted in previous microinjection studies that certain purified autoantibodies that inhibited mitosis appeared to lack anti-CENP-C as detected by immunoblotting (Bernat et al., 1990). This led to the suggestion that the antibody responsible for mitotic disruption in those experiments might be solely anti-CENP-B. We could not exclude, however, that other inhibitory antibodies were present, but not detected under our immunoblotting conditions. To partly address this issue, we have now performed an immunoblotting analysis using one of these IgGs purified from a patient serum that does not appear to recognize CENP-C in blots of chromosomes (Figure 7, lane 3). Surprisingly, IgGs purified from this serum do bind to a CENP-C fusion protein (Figure 7, lane 5).

We conclude from this that low levels of anti-CENP-C antibodies may be present even in sera that do not appear to recognize this relatively non-abundant antigen in chromosomes. Thus, it is possible that anti-CENP-C could have contributed significantly to the mitotic inhibition observed in our earlier experiments.

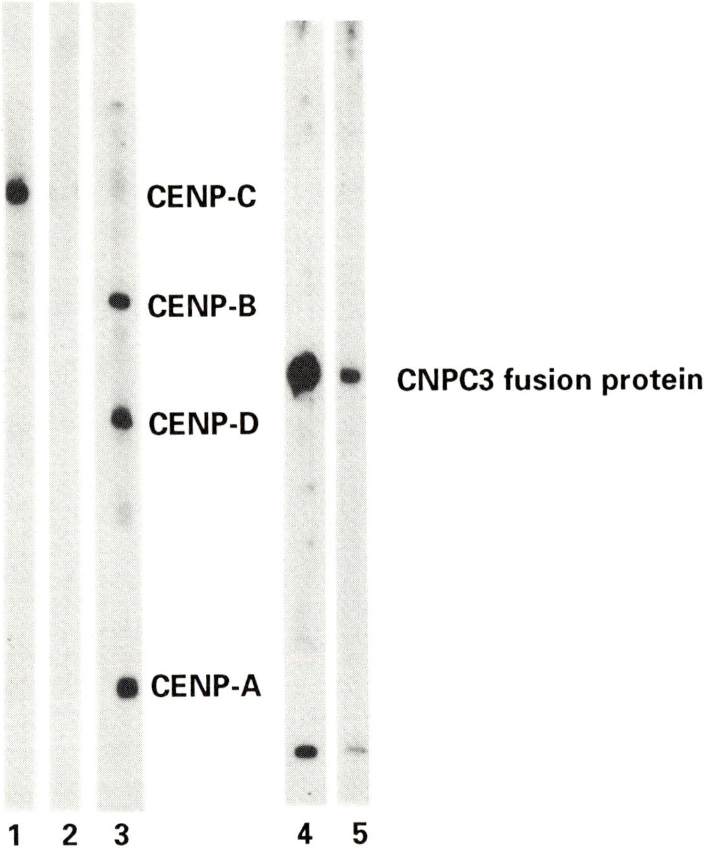

Figure 7: Characterization of antibodies by immunoblotting against chromosomes and the CNPC3 fusion protein. Lanes 1-3: immunoblot of human mitotic chromosomes. Lane 1 - anti-CENP-C IgG (anti-296-943) 1:1000. Lane 2 - Preimmune IgG from this rabbit, 1:100. Lane 3 - IgG purified from sera of patient AP (used extensively in our earlier microinjection studies) 1:1000. Lanes 4, 5: immunoblot of the CNPC3 fusion protein with IgG prepared from anti-CENP-C and autoimmune serum AP. Lane 4 - anti-CENP-C IgG (anti-296-943) 1:1000; Lane 5 - IgG from patient AP, 1:1000. All lanes shown come from a single autoradiographic exposure.

Figure 7 also shows an immunoblot with the IgG preparations derived from one rabbit immunized with a CENP-C fusion protein. This IgG strongly recognizes CENP-C, with a very faint cross-reaction against two other proteins.

4. Discussion
4.1 Molecular cloning of CENP-C

Our cDNA cloning experiments have revealed that CENP-C is a highly basic 107 kDa protein. Chromosomal CENP-C migrates anomalously in SDS-PAGE with M_r 140 kDa.

Comparison of the deduced sequence of Figure 3 with the data bases has been unrevealing: we have failed to discover any significant homology to known proteins. A more detailed examination of the sequence reveals the presence of several potential nuclear localization motifs, as well as the presence of one potential phosphorylation site for the p34^{CDC2} kinase.

4.2 Localization of CENP-C in the centromere

Previous results from our laboratory were consistent with the notion that CENP-C might be localized in the kinetochore. Firstly, when mitotic chromosomes were stained with a CENP-C-specific antibody, all chromosomes were labeled approximately equally, suggesting that roughly equivalent levels of CENP-C are present on all chromosomes.

In a second study, we found that CENP-C appears to be present in nearly uniform amounts, even on centromeres that contain widely variable amounts of α-satellite DNA (Wevrick et al., 1990). In that study, we examined a chromosomal rearrangement in which approximately 80% of the α-satellite DNA array from a chromosome 17 was excised, resulting in production of a 17p$^-$ chromosome and a small marker chromosome (Wevrick et al., 1990). Both of these marker chromosomes segregate normally in mitosis. We showed that, about 80% of the CENP-B label was transferred to the smaller marker chromosome, while staining of the 17p$^-$ chromosome was significantly reduced. In contrast, the marker chromosome appeared to contain a roughly normal level of the CENP-C antigen. Although we could not conclusively identify the 17p$^-$ chromosome in spreads stained with anti-CENP-C, we noted that all chromosomes in the mitotic spreads prepared from these cells appeared to have roughly equivalent levels of CENP-C.

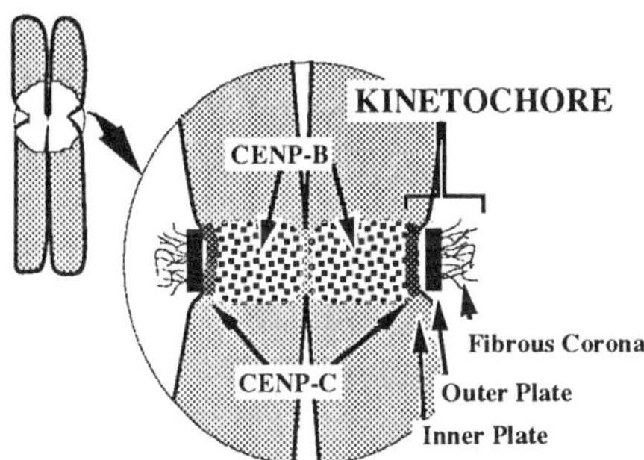

Figure 8: Schematic diagram of the centromere of a metaphase chromosome, indicating the structural domains of the kinetochore and their associated functions. The localization of CENP-B to the centromeric heterochromatin (Cooke et al., 1990) and CENP-C to the inner kinetochore plate are indicated by the stippled and striped patterns, respectively.

Immunoelectron microscopy shows CENP-C to be concentrated in the centromere immediately below the outer kinetochore plate. Although the concentrated distribution of gold particles in this region obscures the underlying chromosomal structure, we infer from the labeling pattern that CENP-C is concentrated in the region of the inner kinetochore plate. Thus CENP-C is the first component of either kinetochore plate to be identified. The relative distributions of CENP-C and CENP-B in the human centromere are shown in Figure 8.

4.3 Function of CENP-C in the kinetochore

The inner plate is the least well characterized region of the kinetochore. In fact, this plate is not even detected as a discrete structure in chromosomes from colcemid-arrested cells, where the heterochromatin in the kinetochore region adopts a uniformly condensed morphology. The function of the inner plate is similarly obscure. It is well known that the majority of the kinetochore-associated microtubules terminate in the outer plate structure. In addition, the fibrous corona appears to contain mechanochemical motor components that are responsible for early prometaphase movements at the very least, and may be responsible for chromosomal movements at all stages of mitosis.

With microtubule attachment and chromosome movement explained as properties of the outer plate and fibrous corona, there would appear to be no functions required of the kinetochore that are left unaccounted for. It might therefore seem perplexing that the presence or absence of a component of the inner kinetochore plate would show an absolute correlation with the ability of the kinetochore to function. Nonetheless, this was the result obtained when a stable dicentric chromosome was stained for the presence of CENP-C.

In a previous study from our laboratory, CENP-C-specific antibodies were found to react only with the active centromere of a mitotically stable dicentric chromosome 13. This chromosome is composed of two identical copies of the maternal homologue linked at or near the ends of their q arms (Rijhsinghani et al., 1988). Despite the presence of two centromere regions, confirmed by C-banding, the chromosome appears to be completely stable in mitosis. Such stability could only arise if one centromere were inactive in binding to the spindle. CENP-C was detected only at the active centromere of this chromosome in indirect immunofluorescence analysis of chromosome spreads (Earnshaw et al., 1989). CENP-B was present at both centromeres. These observations suggested that the presence of CENP-C might be linked with the ability of the centromere to interact with the spindle, while CENP-B might play a structural role common to both active and inactive centromeres.

More direct evidence concerning the role of CENP-C in kinetochore function has come from studies in which purified anticentromere antibodies from autoimmune patients (ACA) were injected into cultured cells (Simerly et al., 1990; Bernat et al., 1990). From an analysis of the effects of antibody injection at different times during the cell cycle, we concluded that the injected antibodies must interfere with events that occur during interphase in order to disrupt the subsequent mitosis (Bernat et al., 1991). Several of the IgG preparations injected in those experiments lacked detectable antibodies to either CENP-C or CENP-A, or both, and we argued that the mitotic inhibition could be due primarily to antibodies recognizing CENP-B, although the involvement of other antibodies was not excluded. We have now shown that anti-CENP-C antibodies are detectable in at least one of these sera, provided that a CENP-C fusion protein is used as a substrate for the immunoblotting experiment. This raises the possibility that antibodies to CENP-C may have contributed substantially to the mitotic inhibition observed in the earlier experiments.

This suggestion is supported by the observation that the levels of CENP-C (detected by indirect immunofluorescence using the rabbit antibodies to CENP-C) were visibly reduced at kinetochores in ACA injected cells (R. Bernat, unpublished). This reduction in staining did not appear to be a consequence of competition for binding to CENP-C by the injected ACA and rabbit antibodies. In control experiments we observed that staining of kinetochores with anti-CENP-C antibodies was not reduced in fixed uninjected cells that had been pre-incubated with ACA. Similar experiments revealed no significant differences in the level of CENP-B at the centromere in injected and uninjected cells. These observations suggest that the disruption of kinetochore function resulting from injection of ACA may be due, at least in part, to an inhibition of CENP-C assembly at the kinetochore.

In our original microinjection study, we noted that introduction of ACA into HeLa cell nuclei during the 2 - 3 hours prior to mitosis resulted in a metaphase arrest (Bernat et al., 1990). A subsequent ultrastructural analysis of the metaphase-arrested cells revealed that this phenotype is correlated with the assembly of a fragile kinetochore. Trilaminar kinetochores are never observed if the injected cells enter mitosis with microtubules present, but are seen at normal frequency if the cells enter mitosis in the absence of microtubules. Electron microscopy of the latter cells revealed that the chromatin in the vicinity of the inner kinetochore plate was less condensed than normal. Our immunoelectron microscopy results with antibodies raised to a CENP-C fusion protein now reveal that this region of less condensed chromatin corresponds exactly to the location of CENP-C (Bernat et al., 1991). This raises the possibility that CENP-C may be involved in stabilization of the outer kinetochore plate in the presence of microtubules.

In addition to the correlative evidence suggesting that CENP-C may be a target of the inhibition of mitosis by ACA, we more recently obtained direct evidence suggesting that CENP-C is required for mitotic events. We have found that injection of antibodies affinity-purified from the serum of a rabbit injected with a CENP-C fusion protein disrupts mitotic events in HeLa cells (Tomkiel et al., in preparation). Thus CENP-C is an essential protein for mitotic progression in human cells.

Our experiments now begin to define a function for the inner kinetochore plate. The assembly of CENP-C into this structure appears to provide a solid substratum for the kinetochore. Without the presence of CENP-C, the kinetochore is structurally weak: it becomes deformed once microtubules attach to it and begin to exert force. Interestingly, this structural deformation is correlated with an inability of the microinjected cells to execute a timely metaphase: anaphase transition. It will remain for future experiments to determine whether CENP-C plays a direct role in the regulation of this transition, or whether the effect of the antibody injections is indirect. In the latter case, CENP-C may be simply one component required to assemble a structurally stable kinetochore, and it may be kinetochore stability *per se* that is required for a timely metaphase:anaphase transition.

5. Acknowledgments

We thank Gail Stetten and Hunt Willard for their past collaborative efforts; S. Tabor and E. Wood for providing the T7 vectors, and N. Saitoh for engineering the modifications to the T7 vectors used here. We are grateful to A. Pluta her helpful comments on the manuscript. We thank C.L.H. Earnshaw for use of his DNA. This work was supported by NIH grant GM35212 to W.C.E. and NIH grant AR37986 to N.F.R., an American Cancer Society postdoctoral grant to J.T. and a postdoctoral grant from the Toyobo Biotechnology Foundation and the Japan Society for the Promotion of Science to H.S.

6. References

Bernat, R. L. et al. (1990). Injection of anticentromere antibodies in interphase disrupts events required for chromosome movement at mitosis. J. Cell Biol. *111*, 1519-1533.

Bernat, R. L. et al. (1991). Disruption of centromere assembly during interphase inhibits kinetochore morphogenesis and function in mitosis. Cell. *66*, 1229-1238.

Bischoff, F. R. et al. (1990). A 47-kDa human nuclear protein recognized by antikinetochore autoimmune sera is homologous with the protein encoded by RCC1, a gene implicated in onset of chromosome condensation. Proc. Nat. Acad. Sci. (USA). *87*, 8617-8621.

Cooke, C. A., Bernat, R. L. and Earnshaw, W. C. (1990). CENP-B: A major human centromere protein located beneath the kinetochore. J. Cell Biol. *110*, 1475-1488.

Earnshaw, W. C. et al. (1986). Three human chromosomal autoantigens are recognized by sera from patients with anti-centromere antibodies. J. Clin. Invest. *77*, 426-430.

Earnshaw, W. C. et al. (1987). Analysis of anti-centromere autoantibodies using cloned autoantigen CENP-B. Proc. Nat. Acad. Sci. (USA). *84*, 4979-4983.

Earnshaw, W. C., Ratrie, H. and Stetten, G. (1989). Visualization of centromere proteins CENP-B and CENP-C on a stable dicentric chromosome in cytological spreads. Chromosoma (Berl.). *98*, 1-12.

Earnshaw, W. C. and Rothfield, N. (1985). Identification of a family of human centromere proteins using autoimmune sera from patients with scleroderma. Chromosoma (Berl.). *91*, 313-321.

Earnshaw, W. C. et al. (1987). Molecular cloning of cDNA for CENP-B, the major human centromere autoantigen. J.Cell Biol. *104*, 817-829.

Kingwell, B. and Rattner, J. B. (1987). Mammalian kinetochore/centromere composition: A 50 kDa antigen is present in the mammalian kinetochore/centromere. Chromosoma (Berl.). *95*, 403-407.

Kozak, M. (1987). An analysis of 5'-noncoding sequences from 699 vertebrate messenger RNAs. Nucleic Acids Res. *15*, 8125-8148.

Masumoto, H. et al. (1989). A human centromere antigen (CENP-B) interacts with a short specific sequence in alphoid DNA, a human centromeric satellite. J. Cell Biol. *109*, 1963-1973.

Miller, M., Park, M. K. and Hanover, J. A. (1991). Nuclear pore complex: Structure, function and regulation. Physiol. Revs. *71*, 909-949.

Moroi, Y. et al. (1980). Autoantibody to centromere (kinetochore) in scleroderma sera. Proc. Nat. Acad. Sci. (USA). *77*, 1627-1631.

Ohtsubo, M., Okazaki, H. and Nishimoto, T. (1989). The RCC1 protein, a regulator for the onset of chromosome condensation locates in the nucleus and binds to DNA. J. Cell Biol. *109*, 1389-1397.

Palmer, D. K. and Margolis, R. L. (1987). A 17-kD centromere protein (CENP-A) copurifies with nucleosome core particles and with histones. J. Cell. Biol. *104*, 805-815.

Palmer, D. K. et al. (1991). Purification of the centromeric protein CENP-A and demonstration that it is a centromere specific histone. Proc. Nat. Acad. Sci. (USA). *88*, 3734-3738.

Pfarr, C. M. et al. (1990). Cytoplasmic dynein is localized to kinetochores during mitosis. Nature. *345*, 263-265.

Pluta, A. F. et al. (1992). Identification of a subdomain of CENP-B that is necessary and sufficient for targeting to the human centromere. J. Cell Biol. *116*, 1081-1093.

Rieder, C. L. (1982). The formation, structure and composition of the mammalian kinetochore and kinetochore fiber. Int. Rev. Cytol. *79*, 1-58.

Rijhsinghani, A. G., Hruban, R. H. and Stetten, G. (1988). Fetal anomalies associated with an inversion duplication 13 chromosome. Obstet Gynecol. *71*, 991-994.

Saitoh, H. et al. (1992). CENP-C, an autoantigen in scleroderma, is a component of the human inner kinetochore plate. Cell. *70*, 115-125.

Sawin, K. E., Mitchison, T. J. and Wordeman, L. G. (1992). Evidence for kinesin-related proteins in the mitotic apparatus using peptide antibodies. J. Cell Sci. *101*, 303-313.

Simerly, C. et al. (1990). Microinjected kinetochore antibodies interfere with chromosome movement in meiotic and mitotic mouse embryos. J. Cell Biol. *111*, 1491-1504.

Steuer, E. R. et al. (1990). Localization of cytoplasmic dynein to mitotic spindles and kinetochores. Nature. *345*, 266-268.

Wevrick, R. et al. (1990). Partial deletion of alpha satellite DNA associated with reduced amounts of the centromere protein CENP-B in a mitotically stable human chromosome. Mol. Cell. Biol. *10*, 6348-6355.

Widom, J. and Klug, A. (1985). Structure of the 300Å chromatin filament: X-ray diffraction from oriented samples. Cell. *43*, 207-213.

Wordeman, L. et al. (1991). Chemical Subdomains within the Kinetochore domain of isolated CHO mitotic Chromosomes. J. Cell Biol. *114*, 285-294.

3 Characterization of the telomeric region of human chromosome 16p

D.R. HIGGS, A.O.M. WILKIE, P. VYAS, M.A. VICKERS, V.J. BUCKLE and P.C. HARRIS

Institute of Molecular Medicine, Oxford, UK

1   Introduction

The terminal regions of chromosomes have specialised molecular and cytological properties, which are reflected in their DNA sequence organisation. Telomeric DNA, the conserved sequence at the extreme ends of chromosomes, comprises in mammals a variable length (2-20 kb in humans; 15-150 kb in mouse) of the simple sequence $(TTAGGG)_n$ orientated 5'->3' toward the chromosome end. It is likely that this repeat motif, together with associated protein components can explain many of the properties of human telomeres; for example, replication of the terminal DNA, protection against exonucleolytic degradation, and the prevention of chromosome fusion (reviewed in Blackburn, 1991).

To extend the analysis of human telomeres, several groups have isolated DNA sequences adjacent to the $(TTAGGG)_n$ repeat. Whereas some of these "telomere-associated sequences" are restricted to a single subtelomeric region (Riethman et al., 1989; Weber et al., 1990) others appear to be more widely, but not universally distributed (Riethman et al., 1989; Cheng et al., 1989; de Lange et al., 1990; Brown et al., 1990; Cross et al., 1990; Weber et al., 1991). At present, no function has been attributed to such sequences in humans.

To date only the most distal 10-20 kb of human chromosomal DNA has been systematically studied. There exists no detailed description of the long range structure around any single end of a particular chromosome, of the extent of the telomere associated DNA and the degree of polymorphism or of the position of genes with respect to the telomere. More complete structural information is required to understand the function of these regions.

In an attempt to develop an integrated structural picture of a human telomere, we have studied the terminal region of the short (p) arm of human chromosome 16 which corresponds to the Giemsa light band (R-band) 16p13.3. Part of this region has been well characterised because it contains the α-globin (HbA) gene cluster

Chromosomes Today   Volume 11. Edited by A.T. Sumner and A.C. Chandley. Published in 1993 by Chapman & Hall, London. ISBN 0 412 47670 3

(reviewed in Higgs et al., 1989) which is the most distal locus described on 16p, both genetically (Donis-Keller et al., 1987; Germino et al., 1990) and physically (Harris et al., 1990).

2.1 Physical linkage of α-globin to the 16p telomere

Previous long-range restriction mapping of band 16p13.3 identified a narrow region distal to the α-globin locus that was apparently readily digested by many rare cutting enzymes (Harris et al., 1990). Further analysis showed that restriction fragments extending to this region reproducibly gave broader bands on pulsed-field gels than were usually observed, indicating slight variability in the size of the fragments detected. Together these results suggested that this region might be the telomere of 16p. To test this hypothesis we analysed mouse/human hybrids containing chromosome 16 using the human-specific telomere associated repeat TelBam3.4 which had previously been shown to be present at the 16p telomere (Brown et al., 1990).

In these hybrids TelBam3.4 identified fragments that were the same size as those detected by the probe α-globin 5' HVR, which lies 90 kb upstream of the α-globin genes (see figure 1).

Further detailed analysis confirmed that TelBam3.4 was linked to the α-globin locus and a map of the entire region from the telomere to approximately 40 kb proximal to the α cluster was constructed. This demonstrated that the α cluster is orientated with its embryonic gene ($\zeta 2$) at the distal (telomeric) end and the fetal/adult genes ($\alpha 2$ and $\alpha 1$) at the proximal (centromeric) end, the α genes being located approximately 170 kb from the end of 16p (figure 2).

To our surprise, on extending this analysis to approximately 50 chromosomes we found that only about two thirds of them had this structure (designated allele A); three other, different 16p subtelomeric structures (designated alleles B, C and D) accounted for the remainder (Wilkie et al., 1991; Harris and Thomas, 1992), many individuals being compound heterozygotes.

2.2 Polymorphism of the subterminal region of 16p

Using restriction enzymes (MluI, PvuI) that cut proximal to the α cluster, but not between the α cluster (coordinate O) and the telomere four different sized terminal regions were identified in which the α genes lie 170, 245, 350 or 430 kb from the telomere; these alleles designated A, D, B and C respectively are shown in figures 2 and 3.

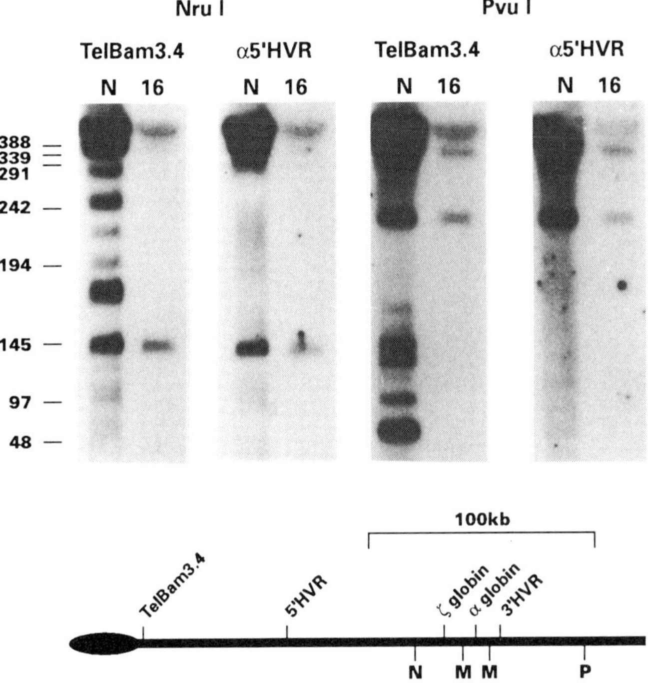

Fig. 1. Physical linkage of α-globin to the 16p telomere. Above, NruI and PvuI-digested DNA run on a CHEF gel and hybridised with α-globin 5'HVR or TelBam3.4. Lane N normal lymphoblastoid cell line; Lane 16, normal chromosome 16-containing hybrid. Size markers are λ DNA concatamers. Below, a map of the end of chromosome 16 showing the relative positions of α-globin 3'HVR, α-globin, α-globin 5'HVR and TelBam3.4. The telomere is indicated by a filled oval. Restriction sites for NruI (N), MluI (M) and PvuI (P) are shown.

Rehybridisation of pulsed-field gels with subtelomeric probes showed that the type A allele was always associated with TelBam3.4 and type B with another human subtelomeric probe, TelBam11. The one chromosome with a type C allele was associated with TelBam3.4, but two chromosomes with type D alleles did not hybridise to any of the human subtelomeric sequences that were tested (Brown et al., 1990; Wilkie et al., 1991; Harris and Thomas, 1992). These observations that different length variants possessed different telomere associated sequences suggested that the length variation had arisen by subtelomeric translocation; to elucidate this further we characterised the divergence point between the alleles.

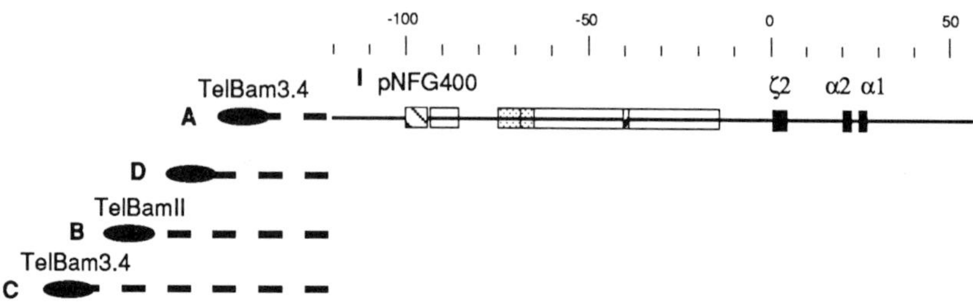

Fig. 2.   A map of the end of chromosome 16.   The α like genes are shown on the right side of the diagram.   Coordinate 0 corresponds to the ζ2 mRNA cap site and sizes are in kilobases.   The telomere is shown as a filled oval and the subtelomeric regions (not to scale) corresponding to alleles A-D as dashed lines.   The boxes in between α-globin and the telomere represent four widely expressed genes. The horizontally striped box (coordinates -93 to -99) indicates the maximum genomic extent of the corresponding gene, the cDNA has not yet been isolated.   The shaded box (coordinates -67.5 to -74) represents the gene encoding methyladenine glycosylase (Vickers et al, in preparation).   The diagonally striped box at coordinate -40 represents the α-globin regulatory elements (HS-40).   The probe pNFG400 is shown as a thin black box.

2.3 Identification of the divergence point between alleles A and B

Since a large segment of DNA around the α cluster had been cloned, probes were available to compare restriction maps of A- and B-type chromosomes using genomic DNA from previously characterised hybrids.   Southern blots generally gave restriction fragments of identical size.   However, exceptions to this were found when a probe (pNFG400) located around coordinate -112 (coordinate 0 being the ζ2 mRNA cap site) was hybridised to digests yielding large restriction fragments (EcoRI and BglII) extending in a distal direction.   Two interesting features were noted: first, different chromosome 16 hybrids gave different fragment sizes, and second, a number of additional fragments were present in genomic DNA that did not map to chromosome 16.   We chose the BglII data for further analysis, because the polymorphic chromosome 16 bands were clearly separated from the non-16 bands (figure 4).

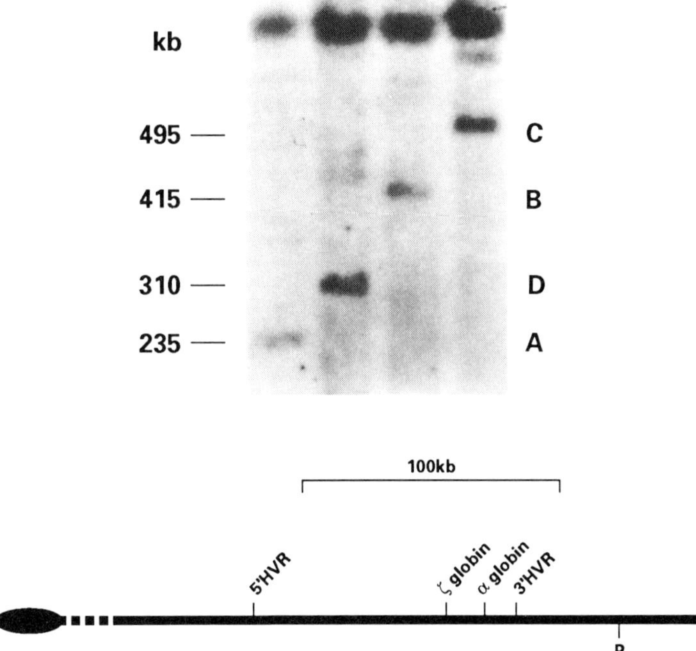

Fig. 3.   The subtelomeric region of chromosome 16 is variable in length.   Above, PvuI-digested DNA from hybrids each containing only one copy of chromosome 16 corresponding to alleles A, B, C and D.   DNA samples were run on a CHEF gel and hybridised to α-globin 5'HVR.   Below, a map of the end of chromosome 16 showing the variable PvuI fragment demonstrated in the autoradiograph.

Analysis of hybrids containing one of the alleles A, B or C showed that the chromosome 16 specific BglII bands associated with each allele differ in size, the bands were named a, b, and c (figure 4).   We have shown complete concordance between the two polymorphic systems, the 16 pter length alleles A, B and C corresponding perfectly with the pNFG400 BglII alleles a, b and c respectively: two type D chromosomes were both associated with type b alleles.   This result implied either that the pNFG400/BglII alleles spanned the breakpoint between the chromosome length alleles or that they were in complete disequilibrium with it.
        To distinguish these alternatives, we cloned a pNFG400/BglII allele of the longest and most frequent (a) type and compared it with the map of a b-B hybrid.   This demonstrated that the polymorphic BglII fragment does indeed straddle the breakpoint of the A and B alleles because all cloned sequences from the a-A allele distal to coordinate -123 (145 kb upstream of the α-globin genes) are absent in the b-B allele.
        Since the A and B subtelomeric regions were non-homologous and yet probes from these regions hybridised to other segments of

the genome we next characterised these cross-hybridising loci. In summary, loci belong to two subtelomeric families related to alleles A or B: those related to allele A were found on Xqter and Yqter, while those similar to allele B were on 9qter, 10qter and 18pter (Wilkie et al., 1991).

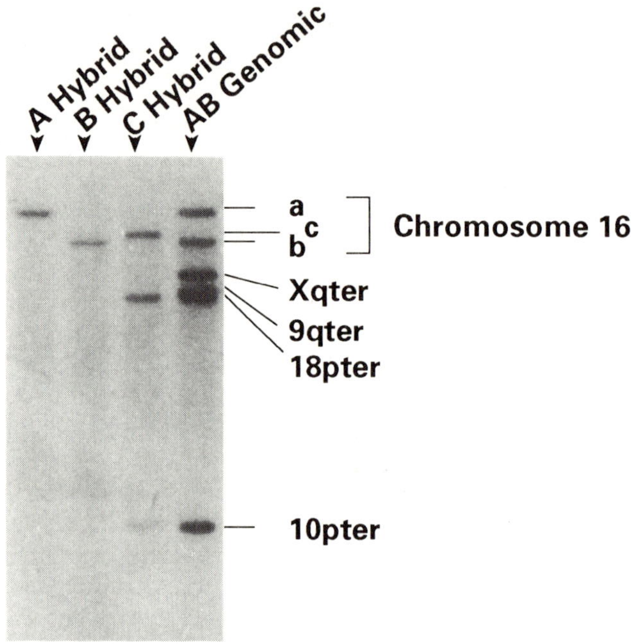

Fig. 4. The subtelomeric probe pNFG400 identifies three alleles on chromosome 16 and four non-16 homologues. A BglII digest run on a conventional agarose gel and hybridised with pNFG400 (see figure 2 for location of the probe). Hybrids containing copies of chromosome 16 with an A, B, C (plus other chromosomes) and a genomic DNA sample from an AB heterozygote are shown.

Together with the observation that different telomere associated sequences lie at the end of the A and B alleles, these data suggest that the A/B length polymorphism has arisen by translocation of subtelomeric sequences present on different chromosomes.

2.4 Further characterisation of the divergence point between alleles A and B

The breakpoint of the exchange event leading to A and B type alleles is unexpectedly complex in structure. Figure 5 summarises the detailed comparison of the A and B 16 pter alleles in a 3 kb segment around the -120 divergence region.   Starting from coordinate -118 and moving toward the telomere six distinct zones (1-6) were identified.

1 As far as -118.9 the maps of the A and B alleles are identical.
2 From -118.9 to the start of the large CA repeat multiple restriction site differences are observed between alleles A and B although some homology is still present.
3 This segment comprises a large CA repeat the length of which differs between A (~850 or 570 bp), B (~60 bp) and C (~80 bp) alleles.   Surprisingly little microheterogeneity was seen within each class, although the two different sized repeats associated with A alleles allowed us to subtype these as A1 and A2.
4 Between coordinates -120.3 and -120.9, the A and B alleles show close homology: 94% at the nucleic acid sequence level.
5 At coordinate -120.9 there is abrupt loss of homology between the A and B alleles.
6 A probe from the B allele hybridises to the A allele indicating that a local insertion/deletion event has occurred to produce segment 5.

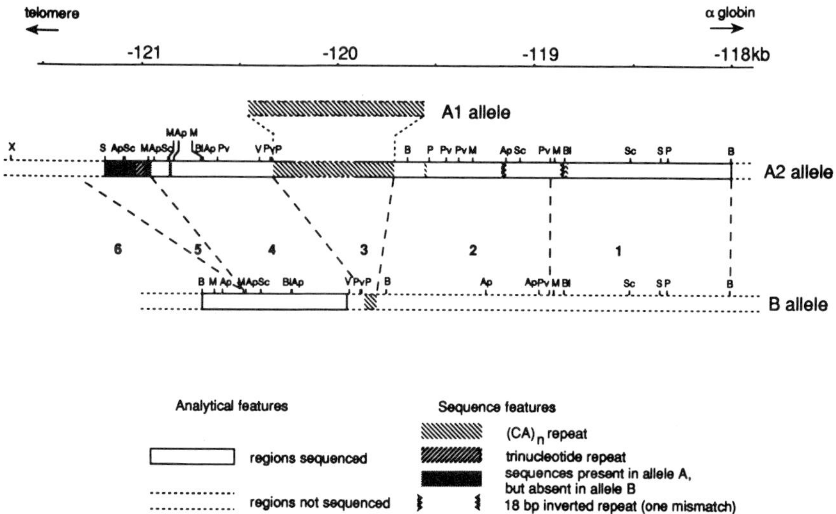

Fig. 5.   Comparative map of the divergence region between the subtelomeric regions of alleles A and B, as described in the text. Ap, ApaI; B, BamHI; Bl, BglI; M, MspI; P, PstI; Pv, PvuII; S, StuI; Sc, SacI; V, EcoRV; X, XbaI.

Final loss of homology between the A and B alleles occurs around coordinate -122.7. These data therefore indicate that the transition from the homologous to the non-homologous region is neither sudden nor gradual but there is an intermediate segment of ~4 kb in which apparently random insertions, deletions and point mutations result in a complex pattern of zones of greater or lesser homology (Wilkie and Higgs, 1992).

2.5 Characterisation of genes within the terminal region of 16p13.3

In the commonest (A) allele the reiterated telomeric repeat sequence $(TTAGGG)_n$ lies approximately 25 kb distal to the region of divergence described above (Wilkie et al., 1991). The structural genes of the α-globin cluster lie 120-145 kb proximal to the divergence region and we have shown that critical *cis*-acting sequences located 40 kb upstream of the α cluster in the direction of the telomere, are required for their regulation (Higgs et al., 1990; figure 2). We have now surveyed the entire region of cloned DNA from coordinate -130 kb to +30 kb for the presence of expressed sequences other than the erythroid-specific α-globin genes.

To identify new genes within this region we have exploited the fact that all housekeeping genes transcribed by RNA polymerase II and approximately 50% of tissue-specific genes are associated with short stretches (0.5-3 kb) of unmethylated CpG-rich DNA (Bird, 1986; Gardiner-Garden and Frommer, 1987) which can readily be detected using methylation sensitive restriction enzymes such as SstII. Using this approach, in addition to the previously described CpG islands associated with the $\zeta 2$, $\psi\zeta 1$, $\alpha 2$, $\alpha 1$ and $\theta 1$ globin genes (Bird, 1986; Fischel-Ghodsian et al., 1987), we identified five unmethylated SstII sites at coordinates -99, -89/-91, -80, -74 and -14 that could represent CpG islands and hence locate previously unidentified genes. Further characterisation proved that by comparison with previously defined criteria (Bird, 1986; Gardiner-Garden and Frommer, 1987) the regions around -99, -74 and -14 appear to be *bona fide* CpG islands and therefore have a high probability of being associated with genes. The region around -89/-91 also shares many of the features associated with CpG islands, but the pattern of methylation is unlike that of previously described islands. The region corresponding to the Sst II site at -80 was not analysed further.

We then used DNA fragments associated with each island to search for mRNA transcripts on Northern blots that included RNA from a variety of erythroid and non-erythroid cell lines. Each of these fragments detected different mRNA transcripts in a wide variety of cell lines; cDNAs corresponding to three transcripts were subsequently isolated. Each cDNA revealed a high degree of evolutionary conservation as judged by "zoo blot" analysis. The genomic extent of each cDNA is shown in figure 2. Although we have not identified a cDNA corresponding to the CpG island at -99, a poly(A) enriched mRNA transcript is identified by a fragment spanning coordinates -93 to -99, indicating that the most distal gene lies remarkably close to the subtelomeric region of this chromosome and only ~20 kb from the A/B divergence region.

Thus we have shown that the telomeric region of chromosome 16 contains a high density of constitutively expressed and tissue-specific globin genes, with at least eight genes located in a region of 180 kb (Vyas et al., 1992). For two of the genes, the genomic DNA encoding their primary RNA transcripts overlaps, a situation that also occurs in other regions of densely packed genes. Surprisingly, the critical *cis*-acting regulatory sequence (HS-40, see figure 2) controlling expression of the tissue-specific α-globin genes lies within the intron of a widely expressed gene associated with the CpG island at -14. Detailed analysis of HS-40 has shown that it binds lineage restricted and ubiquitous *trans*-acting factors which, once bound to HS-40 may facilitate interactions with the promoters of the α and ζ globin genes (Jarman et al., 1991). Similar elements exist upstream of the β-globin locus (referred to as the β-globin locus control region or β-LCR) (Grosveld et al., 1987). It has been shown that such elements act as transcriptional enhancers and, for the β-LCR, have an effect on long range chromatin structure (Forrester et al., 1990). Since unlike the β-LCR, HS-40 lies in a region of constitutively open chromatin (Vyas et al., 1992), it may be required only for its enhancer effect.

Provisional analysis of the three megabases of DNA lying proximal to the α cluster in 16p indicates that it too contains many CpG islands suggesting that the entire segment of DNA in this region may contain a high density of genes. (Harris et al., 1990; Germino et al., 1992).

2.6 Naturally occurring deletions of the 16p telomere

The proximity of the α-globin genes to the 16p telomere, together with the relative ease with which hemizygous deletions (--/$\alpha\alpha$) can be identified (producing the haematological phenotype of α thalassaemia; reviewed in Higgs et al., 1989), provides a useful model system for studying the mechanism and consequences of terminal chromosome deletions. Many small (3-30 kb) interstitial deletions involving the α-globin cluster are known (Higgs et al., 1989) and we have described translocations involving 16p13.3 that, when inherited in an unbalanced manner, give rise to α thalassaemia (for example see Lamb et al., 1989).

In addition we have observed an α thalassaemia mutation due to truncation of the short arm of chromosome 16 which deletes HS-40, and we have shown that $(TTAGGG)_n$ has been added directly to the site of the break. The mutation is stably inherited, proving that telomeric DNA alone is sufficient to stabilize the broken chromosomal end (Wilkie et al., 1990).

3   Discussion

We have cloned and characterised the terminal 200 kb of the short arm of chromosome 16 and this may provide a useful model for understanding the general structure of telomeric regions of human chromosomes. This region is highly polymorphic, containing many single base variations, insertion/deletion polymorphisms and at least five hypervariable minisatellites. In addition we have identified a major length polymorphism in the terminal region of this chromosome that has probably arisen by occasional exchanges between the subtelomeric regions of non-homologous chromosomes. Such polymorphic variation may be a general feature of telomeres. For example it has been previously shown that such regions may contain a high density of minisatellites (Royle et al., 1988; Wells et al., 1989) and it is of interest that similar long-range polymorphism has been shown for the human chromosome 16q telomere (Harris and Thomas, 1992).

These findings are of interest since it has been suggested that early synaptonemal complex formation may involve repeated trials of homology checking between pairs of (frequently non-homologous) chromosomes before the correct pairing is found (Carpenter, 1987; Chandley, 1989). This would be most favoured at telomeres because they are mechanically tethered and aggregated (de Lange, 1992); however it is not excluded at other parts of the chromosome. The likely genetic importance of this process may be reflected in the bias of both chiasmata (Laurie and Hulten, 1985) and recombination events (Donis-Keller et al., 1987) toward telomeric regions. The structure of 16pter and its non-16 homologues then appear to pose two difficulties for this mechanism: the two common 16pter alleles are non homologous in their subtelomeric region and each has closer homologues on other chromosomes.

Assuming that heterozygotes for the 16 pter polymorphism are of normal fertility, chromosome synapsis must usually proceed normally despite these potential problems; therefore, homology of subtelomeric regions is not a prerequisite for synoptonemal complex formation. In this situation the specific recognition of homologous chromosomes presumably initiates either just proximal to the non-homologous region or at an interstitial site(s). The non-homologous sequences may remain unpaired throughout synapsis because recombination appears suppressed between the proximal and distal limits of the non-homologous section. Circumstantial evidence suggests that recombination may be relatively frequent in the 150 kb segment immediately proximal to this section in that the pNFG400/BglII polymorphism shows no significant disequilibrium with the α-globin haplotype. It is of interest that a chromosome $(\alpha\alpha)^{TI}$, in which the entire subtelomeric sequence from 16p has been deleted, appeared to be inherited and propagated in a stable manner (Wilkie et al., 1990).

The high density of genes and CpG islands that we have identified may be typical of telomeric regions. Recent work has shown that the human DNA fraction richest in G+C nucleotides (equivalent to the isochore family H3) corresponds to chromosomal T bands (thermal denaturation resistant telomeric bands) and chromomycin bands which are predominantly located at telomeres (Saccone et al., 1992). The H3 fraction represents only 3.5% of the genome but may contain at least a third of human genes, including the α genes, suggesting that many human genes lie close to the telomeres. In many respects the α genes appear to be typical of genes found in such areas in that they are GC rich (>65% G+C), associated with CpG islands, replicate early in the cell cycle and are located in a relatively open chromatin conformation. Although it has been suggested that housekeeping genes are especially concentrated in the H3 fraction, the α-globin genes are expressed in a tissue-specific manner and as discussed previously this may have important implications for their hierarchical control (Vyas et al., 1992).

4   Acknowledgements

This work was supported by the MRC, The Wellcome Trust and Action Research for the Crippled Child. We are grateful to Professor D.J. Weatherall for his support and encouragement.

5   References

Bird, A.P. (1986) CpG-rich islands and the function of DNA methylation. **Nature**, 321, 209-213.

Blackburn, E.H. (1991) Structure and function of telomeres. **Nature**, 350, 569-573.

Brown, W.R.A. et al. (1990) Structure and polymorphism of human telomere-associated DNA. **Cell**, 63, 119-132.

Carpenter, A.T.C. (1987) Gene conversion, recombination nodules, and the initiation of meiotic synapsis. **BioEssays**, 6, 232-236.

Chandley, A.C. (1989) Asymmetry in chromosome pairing: a major factor in de novo mutation and the production of genetic disease in man. **J Med Genet**, 26, 546-552.

Cheng, J.-F., Smith, C.L. and Cantor, C.R. (1989) Isolation and characterization of a human telomere. **Nucl Acids Res**, 17, 6109-6127.

Cross, S. et al. (1990) The structure of a subterminal repeated sequence present on many human chromosomes. **Nucl Acids Res**, 18, 6649-6657.

de Lange, T. (1992) Human telomeres are attached to the nuclear matrix. **EMBO J**, 11, 717-724.

de Lange, T. et al. (1990) Structure and variability of human chromosome ends. **Mol Cell Biol**, 10, 518-527.

Donis-Keller, H. et al. (1987) A genetic linkage map of the human genome. **Cell**, 51, 319-337.

Fischel-Ghodsian, N., Nicholls, R.D. and Higgs, D.R. (1987) Long range genome structure around the human α-globin complex analysed by PFGE. **Nucl Acids Res**, 15, 6197-6207.

Forrester, W.C. et al. (1990) A deletion of the human β-globin locus activation region causes a major alteration in chromatin structure and replication across the entire β-globin locus. **Genes Dev**, 4, 1637-1649.

Gardiner-Garden, M. and Frommer, M. (1987) CpG islands in vertebrate genomes. **J Mol Biol**, 196, 261-282.

Germino, G.G. et al. (1990) Identification of a locus which shows no genetic recombination with the autosomal dominant polycystic kidney disease gene on chromosome 16. **Am J Hum Genet**, 46, 925-933.

Germino, G.G. et al. (1992) The gene for autosomal dominant polycystic kidney disease lies in a 750-kb CpG-rich region. **Genomics**, 13, 144-151.

Grosveld, F. et al. (1987) Position-independent, high-level expression of the human β-globin gene in transgenic mice. **Cell**, 51, 975-985.

Harris, P. and Thomas, S. (1992) Length polymorphism of the subtelomeric regions of both the short (p) and long arms (q) of chromosome 16. **Cytogenet Cell Genet**, In press,

Harris, P.C. et al. (1990) A long range restriction map between the α-globin complex and a marker closely linked to the polycystic kidney disease I (PKDI) locus. **Genomics**, 7, 195-206.

Higgs, D.R. et al. (1989) A review of the molecular genetics of the human α-globin gene cluster. **Blood**, 73, 1081-1104.

Higgs, D.R. et al. (1990) A major positive regulatory region located far upstream of the human α-globin gene locus. **Genes Dev**, 4, 1588-1601.

Jarman, A.P. et al. (1991) Characterization of the major regulatory element upstream of the human α-globin gene cluster. **Mol Cell Biol**, 11, 4679-4689.

Lamb, J. et al. (1989) Detection of breakpoints in submicroscopic chromosomal translocation, illustrating an important mechanism for genetic disease. **Lancet**, ii, 819-824.

Laurie, D.A. and Hulten, M.A. (1985) Further studies on chiasma distribution and interference in the human male. **Ann Hum Genet (Lond)**, 49, 203-214.

Riethman, H.C. et al. (1989) Cloning human telomeric DNA fragments into *Saccharomyces cerevisiae* using a yeast-artificial-chromosome vector. **Proc Natl Acad Sci USA**, 86, 6240-6244.

Royle, N.J. et al. (1988) Clustering of hypervariable minisatellites in the proterminal regions of human autosomes. **Genomics**, 3, 352-360.

Saccone, S. et al. (1992) The highest gene concentrations in the human genome are in telomeric bands of metaphase chromosomes. **Proc Natl Acad Sci USA**, 89, 4913-4917.

Vyas, P. et al. (1992) Cis-acting sequences regulating expression of the human α globin cluster lie within constitutively open chromatin. **Cell**, 69, 781-793.

Weber, B. et al. (1991) A low-copy repeat located in subtelomeric regions of 14 different human chromosomal termini. **Cytogenet Cell Genet**, 57, 179-183.

Weber, B. et al. (1990) Characterization and organization of DNA sequences adjacent to the human telomere associated repeat $(TTAGGG)_n$. **Nucl Acids Res**, 18, 3353-3361.

Wells, R.A., Green, P. and Reeders, S.T. (1989) Simultaneous genetic mapping of multiple human minisatellite sequences using DNA fingerprinting. **Genomics**, 5, 761-772.

Wilkie, A.O.M. and Higgs, D.R. (1992) An unusually large $(CA)_n$ repeat in the region of divergence between subtelomeric alleles of human chromosome 16p. **Genomics**, 13, 81-88.

Wilkie, A.O.M. et al. (1991) Stable length polymorphism of up to 260 kb at the tip of the short arm of human chromosome 16. **Cell**, 64, 595-606.

Wilkie, A.O.M. et al. (1990) A truncated human chromosome 16 associated with α thalassaemia is stabilized by addition of telomeric repeat $(TTAGGG)_n$. **Nature**, 346, 868-871.

4 The vertebrate genome: isochores and chromosomal bands

G. BERNARDI

Institut Jacques Monod, France

1 Introduction

Vertebrate genomes are mosaics of *isochores*, namely of
long, compositionally homogeneous DNA segments which can be
subdivided into a small number of families characterized by
different GC levels.In the human genome,which is
representative of a number of mammalian genomes and, more
broadly, of the genomes of warm-blooded vertebrates,the
compositional spectrum of isochores ranges between 30% and
60% GC, and five families of isochores have been identified,
two GC-poor families, L1 and L2, representing together 62% of
the genome, and three GC-rich families, H1, H2 and H3,
representing 22%, 9% and 3%, respectively. The remaining 4%
of the genome consists of satellite and ribosomal DNAs, which
can also be visualized as isochores, because of their
homogeneous base composition (Bernardi, 1989).

2 Compositional patterns and compositional correlations

The *compositional distribution* of large (*ca.* 100 Kb) DNA
fragments (such as those forming current DNA preparations)
represents a *compositional pattern* that reflects the
isochore pattern. Other compositional patterns are
represented by the compositional distributions of exons (and
of their codon positions) and of introns.
These compositional patterns characterize *genome phenotypes*
(Bernardi and Bernardi, 1986), which are very different in
cold- and warm-blooded vertebrates. The main differences are
that the former are much less heterogeneous in composition,
are generally GC-poorer, and never attain very high GC
levels compared to the latter (Bernardi and Bernardi,
1990a,b; 1991). Smaller compositional differences exist
among the genomes of either cold-blooded or warm-blooded
vertebrates.

Chromosomes Today Volume 11. Edited by A.T. Sumner and A.C. Chandley. Published in
1993 by Chapman & Hall, London. ISBN 0 412 47670 3

G. Bernardi

Compositional correlations hold between exons (and their codon positions) and the isochores in which they are embedded, as well as between exons and the corresponding introns (Bernardi et al., 1985; 1988; Bernardi and Bernardi, 1985; 1986; Bernardi, 1989; Aïssani et al., 1991; D'Onofrio et al., 1991; Mouchiroud et al., 1991). These compositional correlations link, in a linear fashion, the coding sequences and the non-coding sequences which surround them, or are contained in them.

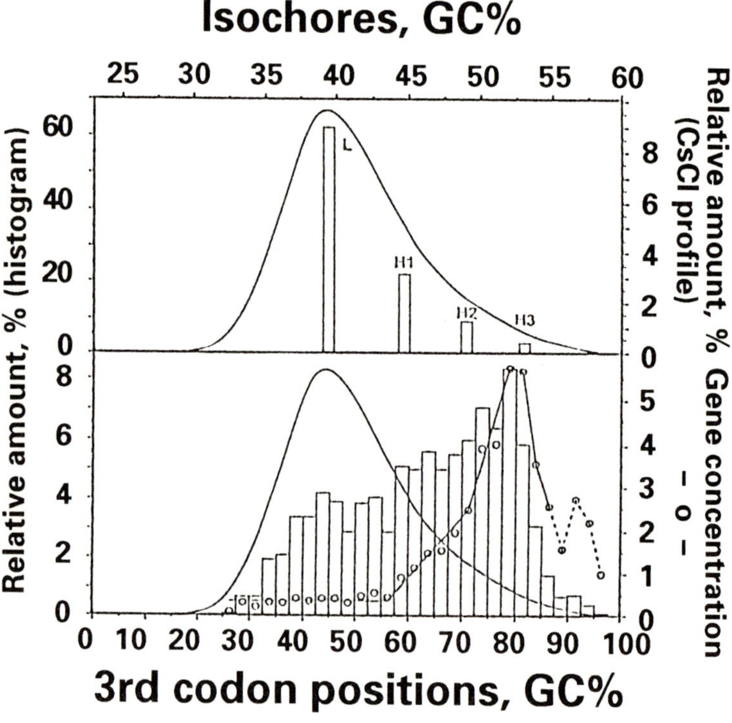

Fig. 1. Top. A histogram of relative amounts of isochore families (or "major DNA components") L (L1 + L2) H1, H2, H3 from the human genome. The upper scale concerns the GC levels of the isochores, as well as those of a CsCl profile of human DNA.

Bottom. Histogram of relative amounts of human genes divided in classes according to GC levels of third codon positions. The profile of gene concentration in the human genome is also shown. The profile was obtained by dividing the relative amounts of genes in each 2.5% GC interval of the histogram by the corresponding relative amounts of DNA, as deduced from the CsCl profile. (Modified from Mouchiroud et al., 1991).

3 The gene distribution in the human genome

The compositional correlation which links GC levels of third codon positions of human genes with the GC levels of the extended sequences in which the genes are located can be used in order to assess gene distribution in the different isochore families and to quantify the finding (Bernardi et al., 1985) that gene distribution in the human genome is strikingly non-uniform. This approach (Fig. 1) has shown that, while 34% of all genes currently present in gene banks are contained in isochore families L1 and L2, 38% are contained in H1 and H2, and 28% in H3 (Mouchiroud et al., 1991). If the gene sample used is representative of all human genes, and if account is taken of the different relative amounts of isochore families (see Introduction), gene concentration in H3 would be 16 times higher than in L1+L2 and 8 times higher than in H1+H2. These ratios are, however, probably underestimated, because housekeeping genes, which are likely to be more abundant in H3 than in other isochore families (see next section), are currently very much under-represented in gene banks (Mouchiroud et al., 1991).

The results just described indicate that increasing gene concentrations are accompanied by increasing GC levels in the genome of warm-blooded vertebrates; the evolutionary process underlying this phenomenon will be discussed later. Very interestingly, the gradient of gene concentration is paralleled by a series of changes in a number of properties which have functional significance. This will be illustrated by describing the extreme situation found in the GC-richest isochore family H3.

4 The human genome core

The isochore family H3 corresponds to a genome compartment endowed with very remarkable properties for which the name *genome core* is proposed (see also next section). This family has not only the highest GC level and the highest gene concentrations, but also the highest concentrations of CpG doublets (Bernardi, 1985), the only potential sites of methylation in vertebrates, and the highest concentrations of CpG islands (Aïssani and Bernardi, 1991a,b), which are very GC-rich sequences characterized by abundant, unmethylated CpG doublets (Bird, 1986). Since CpG islands, which are located in the 5'flanking sequences of genes, are preferentially associated with housekeeping genes (Gardiner-Garden and Frommer, 1987), the latter should be more abundant in H3 than in the other isochore families.

The coding sequences of the H3 isochores are much higher in GC level than their genomic environment, compared with those from other isochore families, especially from GC-poor isochores (Aïssani et al., 1991). Moreover, these genes and

their associated CpG islands are characterized by a
particular chromatin structure, with nucleosome-free
regions, absence or scarcity of histone H1, and acetylation
of histones H3 and H4 (Tazi and Bird, 1990; see also Aïssani
and Bernardi, 1991a,b). These properties make these
chromosomal regions more "open", as also indicated by their
sensitivity to nuclease attack (Kerem et al., 1984).

The H3 isochore family presumably has the highest level
of transcription because of its very high concentration of
genes, and especially of housekeeping genes. It also has the
highest recombination rate, possibly because of its "open"
chromatin structure and to the abundance of repetitive
sequences, like Alu sequences and minisatellites. The very
high recombination rate of H3 isochores may also be largely
responsible for the much higher rate of karyotypic
rearrangements (and speciation) shown by warm-blooded
vertebrates compared to cold-blooded vertebrates (Bernardi,
1992).Indications exist that the H3 isochores may be the
main integration regions for the majority of (GC-rich)
retroviral sequences (see Rynditch et al., 1991; Zoubak et
al., 1992).

The H3 isochore family has an extremely biased codon
usage, a number of codons being absent or very scarce
because of the very high GC levels in third codon positions,
and an extreme amino acid utilization, which favors
aminoacids corresponding to codons having only G and/or C in
the first two codon positions (D'Onofrio et al., 1991),
namely arginine (quartet codons), alanine, glycine and
proline, rather than those corresponding to codons with only
A and/or T in those positions, like lysine, or those
corresponding to codons having both G/C and A/T in first and
second codon positions, like serine. Finally, the location
of the H3 family of isochores in metaphase chromosomes will
be discussed in section 6.

5 The evolutionary origin of the genome core

In terms of base composition, the genome of warm-blooded
vertebrates appears to comprise a *paleogenome*, characterized
by GC-poor isochores which have not changed in composition
relative to the corresponding isochores of cold-blooded
vertebrates, and a *neogenome*, characterized by isochores
which have become GC-rich (Bernardi, 1989; see Fig. 2). Very
interestingly, the GC increase in the genome of warm-blooded
vertebrates concerns only a minority of the genomes, about
one third of it, which contains, however, the majority of
the genes, at least two thirds of them. Indeed, as already
mentioned, GC increases parallel gene concentration (Fig.
1).

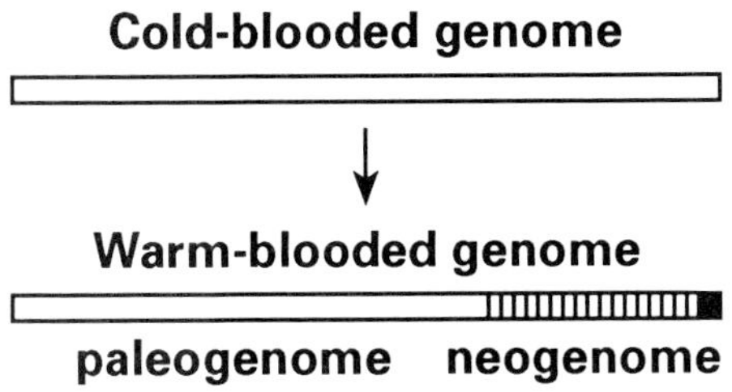

Cold-blooded genome

Warm-blooded genome

paleogenome   neogenome

Fig. 2. Scheme of the compositional genome transition
accompanying the emergence of warm-blooded from cold-blooded
vertebrates. The compositionally homogeneous, GC-poor
genomes of cold-blooded vertebrates are changed into the
compositionally heterogeneous genomes of warm-blooded
vertebrates. The latter comprise a paleogenome
(corresponding to about two thirds of the genome) which did
not undergo any large compositional change and a neogenome
(corresponding to the remaining one third of the genome,
with the GC-richest part only representing 3% of the
genome). In the scheme, the mosaic structure of the warm-
blooded vertebrate genome is neglected; GC-poor isochores
(open bar), GC-rich isochores (hatched bar) and GC-richest
isochores (black bar) are represented as three contiguous
regions. Gene concentration increases from GC-poor to GC-
rich to GC-richest isochores·(From Bernardi, 1992).

   The main reason for proposing the name of *genome core*
for the GC-richest isochore family of the human genome is
that the strikingly non-uniform gene distribution observed
for the human genome and, in particular, the existence of
isochores with very high gene concentrations appear to be
shared by all warm-blooded vertebrates and, very probably,
by all vertebrates.
   Indeed, several data point to the fact that the
formation of GC-rich and GC-richest isochores is due to
events (consisting in the fixation of biased mutations)
superimposed on a gene concentration pattern already present
in cold-blooded vertebrates. First of all, in spite of
significant, yet relatively minor differences, the genomes
of mammals are very similar in their compositional
organization. This was shown by the analysis of DNA (Sabeur
et al., paper in preparation) which indicated fundamentally
similar isochore patterns, and by the analysis of homologous
coding sequences (Mouchiroud et al., paper in preparation)
which showed very similar GC levels in all codon positions
and particularly in third codon positions. If one recalls
that compositional correlations exist between coding
sequences (and their codon positions) and the large DNA
fragments in which they are located, the two facts mentioned
above indicate that the compositional patterns of mammals,

and, therefore, the corresponding gene concentration patterns, are at least largely conserved. The comparison of mammalian (human) and avian (chicken) genomes and of their homologous coding sequences also indicate a large similarity although, expectedly, a lesser one. This similarity suggests, in turn, that the reptiles from which both classes of warm-blooded vertebrates derived, also showed comparable gene concentration patterns. The same conclusion could also be reached on another ground, namely that, if such was not the case, a tremendous reshuffling of gene distribution should have occurred at the transition between reptiles and warm-blooded vertebrates (Bernardi and Bernardi, 1990b).

Second, in some genomes from cold-blooded vertebrates a certain compositional heterogeneity exists. In these cases, a plot of GC levels of third codon positions of genes against GC levels of the corresponding genomes, as derived from modal buoyant densities, shows a deviation towards lower values (Bernardi and Bernardi, 1991). This is consistent with the fact that these genes are mainly present in the GC-richest compartments of the corresponding genomes. In other words, this finding suggests that the gene concentration pattern is already similar to that of warm-blooded vertebrates, even if the actual GC levels are quite different, compositional differences being modest in cold-blooded vertebrates.

Third, preliminary experiments (paper in preparation) have shown that the H3 isochore family from man cross-hybridizes with the GC-richest fractions not only from other mammals, but also from cold-blooded vertebrates.

An implication of the evolutionary conservation of the gene concentration pattern and especially of the genome core, is that the increase in genome size from fishes to mammals (polyploidy being neglected here) should largely involve an expansion of intergenic sequences located in gene-poor regions, but not of those located in the genome core. Indeed, if such expansion involved the genome core, it would strongly decrease its gene concentration.

6 Isochores and chromosomal bands

A number of findings indicate that GC-poor isochores are located in G(iemsa) bands, whereas GC-rich isochores are located in R(everse) bands of human metaphase chromosomes (see Bernardi, 1989, and Bickmore and Sumner, 1989, for reviews; see also Medrano et al. 1988, and Schmid and Guttenbach, 1988). In the latter case, at least, the correspondence cannot be a direct one, for the simple reason that GC-rich and GC-poor isochores are in a 1:2 ratio, whereas R- and G-bands are in a 1:1 ratio (Gardiner et al., 1990). This may mean (i) that standard R-bands comprise more "thin" G-bands (only detected at high resolution) than

standard G-bands comprise "thin" R-bands (as suggested by
Ikemura and Aota, 1988); and/or (ii) that DNA compaction is
higher in G-bands than in R-bands. Moreover, while GC-poor
isochores, which are present in G-bands, differ very little
from each other in composition and represent 62% of the
genome, GC-rich isochores, which are located in R-bands,
encompass a wide GC range and represent only 22%, 9% and 3%,
respectively , of the genome (Bernardi et al., 1985;
Bernardi, 1989). This situation should lead by itself to
inter and/or intra-band compositional heterogeneity in R-
bands.

An approach developed in order to solve the problem of
the correlations between isochores and chromosomal bands is
compositional mapping (Bernardi, 1989). This consists in
hybridizing probes corresponding to landmarks that are
localized on a physical map, or on a chromosome band map, to
compositional DNA fractions. This approach allows to assess
the GC level of about 200 Kb around the landmarks

Compositional mapping, as applied to the long arm of
human chromosome 21 (Gardiner et al., 1990), showed that
practically all probes for loci present in G-bands
hybridized to GC-poor isochores, whereas probes located in
R-bands hybridized to either GC-poor or GC-rich isochores.
In other words, G-bands are GC-poor and at least largely
homogeneous in base composition, whereas R-bands are
compositionally heterogeneous. The GC-richest region of the
long arm of the human chromosome 21, hybridizing on isochore
family H3, was shown to correspond to the telomeric band,
which is a T-band (Dutrillaux, 1973), namely one of the 20
or so most heat-denaturation-resistant R-bands (Fig. 3).

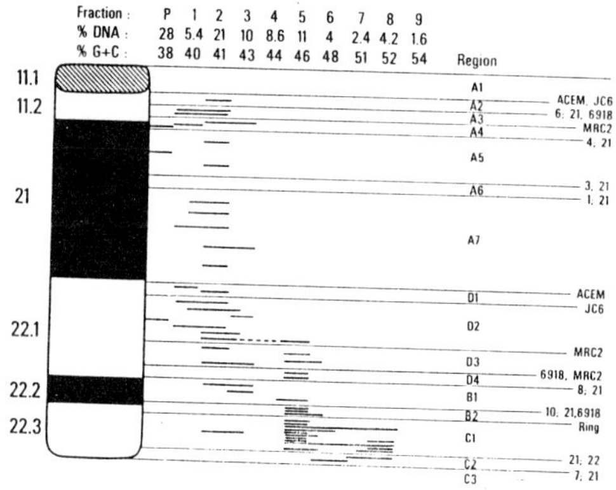

Fig. 3. Compositional map of the long arm of human
chromosome 21. Long horizontal lines indicate positions of
the breakpoints associated with the rearranged chromosomes
listed at the right of the figure. Short horizontal lines
indicate the compositional fractions to which probes located
in different chromosomal regions hybridized. (From Gardiner
et al., 1990).

The meaning of T-bands was not clear for twenty years. Indeed, the high resistance to temperature denaturation of these bands could be due to high GC levels in the corresponding DNA, or to particular chromatin structures or to other reasons. The work of Ambros and Sumner (1987) showed that T-bands are characterized by DNA having high GC levels. While this result pointed to the involvement of DNA, it still did not say, however, whether these GC-rich regions corresponded to repeated or single-copy DNA.

Recent investigations suggested a preferential location of very GC-rich genes in T-bands (Bernardi, 1989; Gardiner et al., 1990; Ikemura and Wada, 1991; De Sario et al., 1991). Moreover, while the telomeric repeats common to all chromosomes hybridized on isochore families H1, H2 and H3, telomere probes corresponding to T-bands hybridized on the GC-richest isochore family, H3, whereas telomeric probes corresponding to R-bands hybridized on GC-rich families H1 and H2 (De Sario et al., 1991). A clear solution of the problem was obtained, however, when *in situ* suppression hybridization on human metaphase chromosomes of the single-copy sequences from the H3 isochore family showed (Fig. 4) that these sequences were precisely located at T-bands (Saccone et al., 1992).

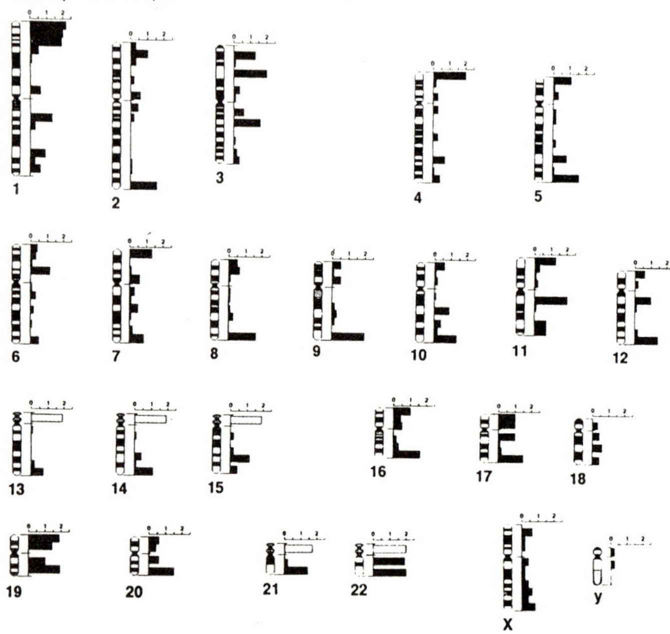

Fig. 4. Histogram (solid bars) showing the distribution of sequences hybridizing with the H3 DNA fraction on human G-banded chromosomes. Open bars correspond to rDNA. Scales are percentage of total number of signals. (From Saccone et al., 1992).

It should be noted that high gene concentrations in some specific T-bands had been occasionally reported. For instance, most of the small nuclear DNA genes, a family of 30-125 genes per haploid genome, were localized at 1p36.3 (Naylor et al., 1984) which is a T-band. Along the same line, one could quote the telomeric localization of ribosomal genes on the short arms of chromosomes 13,14,15,21 and 22. These bands could, in fact, be considered as a sub-class of T-bands. It should also be noted that evidence exists at least for some of the intercalary T-bands to be the result of telomere fusion. For instance, the intercalary T-band 11q13 is the result of a pericentric inversion juxtaposing the telomeric region to the centromere (Dutrillaux, 1979).

Very recent compositional mapping results on the Xq26-Xqter region (Pilia et al., 1992, submitted) confirmed the GC-poorness and compositional homogeneity of G-bands, as well as the heterogeneity of R-bands, and revealed isochores of the H3 family even in chromosomal regions, like the telomeric Xq28 band, which did not correspond to T-bands. This might be due to the fact that the GC-richest region of Xq is narrower than usual T-bands. It is, therefore, possible that a number of T-bands are still escaping cytogenetic detection.

If several T-bands have not yet been detected, the high concentration of genes in telomeres indicated by *in situ* hybridization (Saccone et al., 1992) becomes an even more conspicuous property of the vertebrate genome, whose implications certainly deserve further investigations. Indeed, human telomeres are tightly associated with the nuclear matrix, via their TTAGGG repeats forming the terminal 2-30 Kb of chromosomes (De Lange, 1992), as well as with the nuclear envelope (Henderson and Larson, 1991).

In some cases, both telomeric and intercalary T-bands make up a large proportion of chromosomes. This is true of chromosome 20,21 and, more so, of chromosomes 19 and 22 (in the case of chromosomes 21 and 22, the T-bands on the short arm are ribosomal bands; see above). This raises two questions. The first one is whether these high concentrations of H3 isochores lead to overall increased GC contents of those chromosomes. This appears to be true since, as shown by Korenberg and Engels (1978), these four chromosomes are the richest in GC. The second question, is whether a high gene density can be detected in chromosomes which have high densities of H3 isochores. This also appears to be the case since gene densities are the highest for chromosomes 11,17,19,22 and X (McKusick, 1991) all of which are very high in H3 (except for X in which the high gene density is, in all likelihood, the result of investigations concentrating on this chromosome).

7 References

Aissani,B. and Bernardi, G. (1991a) CpG islands : features
    and distribution in the genome of vertebrates.
    Gene, 106, 173-183.
Aissani,B. and Bernardi, G. (1991b) CpG islands, genes and
    isochores in the genome of vertebrates.
    Gene, 106,185-195.
Aissani, B. D'Onofrio,G. Mouchiroud, D. Gardiner, K.
    Gautier, C. and Bernarid, G. (1991) The compositional
    properties of human genes. **J. Mol. Evol.**, 32,497-503.
Ambros, P.F. and Sumner, A.T. (1987) Correlation of
    pachytene chromomeres and metaphase bands of human
    chromosomes, anddistinctive properties of telomeric
    regions. **Cytogen. Cell Genet.**, 44,223-238.
Bernardi, G. (1985) The organization of the vertebrate
    genome and the problem of the CpG shortage.
    in **Chemistry, Biochemistry and Biology of DNA
    Methylation** (eds G.L.Cantoni and A. Razin), Alan
    Liss, New York, N.Y., pp. 3-10.
Bernardi, G. Olofsson,B. Filipski,J. Zerial,M. Salinas,J.
    Cuny,G. Meunier-Rotival, M. and Rodier, F. (1985)
    The mosaic genome of warm-blooded vertebrates.
    Science, 228,953-958.
Bernardi, G. and Bernardi, G. (1985) Codon usage and genome
    composition. **J. Mol. Evol.**, 22,363-365.
Bernardi, G. and Bernardi, G. (1986) Compositional
    constraints and genome evolution.
    J. Mol. Evol., 24,1-11.
Bernardi, G. Mouchiroud, D. Gautier, C. and Bernardi, G.
    (1988) Compositional patterns in vertebrate genomes :
    conservation and change in evolution.
    J. Mol. Evol., 28,7-18.
Bernardi, G. (1989) The isochore organization of the human
    genome. **Ann. Rev. Genet.**, 23,637-661.
Bernardi, G. and Bernardi G. (1990) Compositional patterns
    in the nuclear genomes of cold-blooded vertebrates.
    J. Mol. Evol., 31,265-281.
Bernardi, G. and Bernardi, G. (1990b) Compositional
    transitions in the nuclear genomes of cold-blooded
    vertebrates. **J. Mol. Evol.**, 31,282-293.
Bernardi, G and Bernardi G. (1991) Compositional properties
    of nuclear genes from cold-blooded vertebrates.
    J. Mol. Evol., 33,57-67.
Bernardi, G. (1992) Genome organization and species
    formation in vertebrates. **J. Mol. Evol.**, in press.
Bickmore, W.A. and Sumner A.T.(1989) Mammalian chromosome
    banding - an expression of genome organization.
    TIG, 5,144-148.
Bird, A. (1986) CpG-rich islands and the function of DNA
    methylation. **Nature**, 321,209-203.

De Lange, T. (1992) Human telomores are attached to the
    nuclear matrix. **EMBO J.**, 11,717-724.
De Sario, A. Aissani, B. and Bernardi, G. (1991)
    Compositional properties of telomeric regions from
    human chromosomes. **FEBS Letters**, 295,22-26.
D'Onofrio, G. Mouchiroud, D. Aissani, B. Gautier, C. and
    Bernardi, G. (1991) Correlations between the
    compositional properties of human genes, codon usage and
    aminoacid composition of proteins.
    J. Mol. Evol., 32,504-510.
Dutrillaux, B. (1973) Nouveau système de marquage
    chromosomique : les bandes T. **Chromosoma**, 41,395-402.
Dutrillaux,B.(1979) Chromosomal evolution in primates :
    tentative phylogeny from *Microcebus murinus* (Prosimian)
    to man. **Hum. Genet.**, 48,251-314.
Gardiner,K. Aissani, B. and Bernardi, G. (1990b)
    A compositional map of human chromosome 21.
    EMBO J., 9,1853-1858.
Gardiner-Garden, M. and Frommer, M. (1987) CpG islands in
    vertebtate genomes. **J. Mol. Biol.**, 196,261-282.
Henderson, E.R. and Larson, D.D.(1991) Telomeres - what's
    new at the end? **Curr. Opin. Genet. & Develop.**,
    1,538-543.
Ikemura, T.and Wada, K. (1991) Evident diversity of codon
    usage patterns of human genes with respect to chromosome
    banding patterns and chromosome numbers,relation between
    nucleotide sequence data and cytogenetic data.
    Nucl. Acids Res., 13,1915-1922.
Kerem, B.-S. Goiten,R. Diamond,G. Cedar, H. and Marcus, M.
    (1984) Mapping of DNAase I sensitive regions of mitotic
    chromosomes. **Cell**, 38,493-499.
Korenberg, J.R. and Engels, W.R.(1978) Base ratio, DNA
    content, and quinacrine-brightness of human chromosomes.
    Proc. Natl. Acad. Sci. USA, 75,3382-3386.
Medrano,L. Bernardi,G. Couturier,J. Dutrillaux,B. and
    Bernardi, G. (1988) Chromosome banding and genome
    compartmentalization in fishes.
    Chromosoma, 96,178-183.
McKusick, V.A. (1991) Genomic mapping and how it has
    progressed. **FASEB J.**, 26,50-64.
Mouchiroud, D. D'Onofrio,G. Aissani,B. Macaya,G.
    Gautier, C. and Bernardi, G. (1991) The distribution
    of genes in the human genome. **Gene**, 100,181-187.
Naylor, S.L. Zabel,B.U. Manser,T. Gesteland, R. and
    Sakaguchi, A.Y. (1984) Localization of human U1 small
    nuclear RNA genes to band p36.3 of chromosome 1 by *in
    situ* hybridization. **Somat.Cell & Mol.Genet.**,
    10,307-313.
Pilia, G. Little,R.D. Aissani,B. Bernardi, G. and
    Schlessinger, D.(1992) Isochores and CpG islands in YAC
    contigs covering the q26.1-qter region of the human X
    chromosome. Submitted for publication.

Rynditch, A. Kadi,F. Geryk,J. Zoubak,S. Svoboda, J. and
    Bernardi, G. (1991) The isopycnic, compartmentalized
    integration of Rous sarcoma virus sequences.
    Gene, 106,165-172.
Saccone,S. De Sario,A. Della Valle, G. and Bernardi, G.
    (1992) The highest gene concentrations in the human
    genome are inT-bands of metaphase chromosomes.
    Proc. Natl. Acad. Sci. USA, 89,4913-4917.
Schmid, M. and Guttenbach, M. (1988) Evolutionary diversity
    of reverse (R) fluorescent chromosome bands in
    vertebrates. **Chromosoma**, 97,101-114.
Tazi, J. and Bird, A.P. (1990) Alternative chromatin
    structure at CpG islands. **Cell**, 60,909-920.
Zoubak, S. Rynditch, A. and Bernardi, G. (1992) Compositional
    bimodality and evolution of retroviral genomes.
    Gene, in press.

The nucleolus

5 Ribosomal genes and nucleolar morphology

F. WACHTLER, W. MOSGÖLLER, C. SCHÖFER
University of Vienna, Austria

J. SYLVESTER
Hahnemann University, USA

P. HOZAK
Czechoslovak Academy of Sciences, Czechoslovakia

M. DERENZINI
University of Bologna, Italy

and A. STAHL
Medical Faculty, Marseille, France

1 Introduction

Nucleoli were noted by biologists very early (Fontana, 1781) and a lot of data has been accumulated on the structure of nucleoli in the last decades. Due to the work of Perry (1962; Perry et al., 1961) it is known that the nucleolus is the site of ribosome biogenesis. Its visibility both in the light- and the electron microscope is due to the fact that the genes for ribosomal RNA are not only transcribed but also processed and complexed with ribosomal proteins in nucleoli, rendering them morphologically different from the rest of the karyoplasm.

The last few years have led to an essentially consistent view on the organization and transcription of rRNA-genes and ribosome biogenesis. Currently new insights on the regulation of the transcription of these genes, the processing of rRNA and interactions between nucleic acids and specific nucleolar and ribosomal proteins accumulate at an enormous speed (Reeder, 1990; Sollner-Webb and Mougey, 1991; Hernandez-Verdun, 1991).

Interestingly, it has not been possible so far to correlate what is known about the function of ribosomal genes to the morphology of nucleoli, more specifically to answer the question where in the nucleolus ribosomal genes are located and transcribed. Understanding nucleolar structure and function is not only of interest to basic research but is needed also as a basis for using nucleolar morphology as a diagnostic tool in pathology (Egan and Crocker, 1992).

In the following we shall take a rather broad approach to answer these questions. We shall first discuss the arrangement of ribosomal genes, the morphology of nucleoli, the site of DNA in nucleoli and shall attempt to locate the process of transcription in nucleoli by locating the uptake of uridine and enzymes required for the transcriptional process.

Chromosomes Today   Volume 11. Edited by A.T. Sumner and A.C. Chandley. Published in 1993 by Chapman & Hall, London. ISBN 0 412 47670 3

We shall first present findings from our own group. Subsequently, we shall point out the differences between our findings and those from other workers in the field. We shall try to point out reasons for divergent conclusions and shall finally attempt to suggest a consensus interpretation the can accommodate most of the published findings.

2 Ribosomal Genes

In humans, where most of our data have been obtained, the genes for 28 S, 5.8 S and 18 S ribosomal RNA are located at the secondary constrictions ("nucleolus organizer regions") of the five pairs of acrocentric chromosomes (Evans et al., 1974). The genes for the 5 S rRNA are located in chromosome 1 and do not participate in the formation of the nucleolus. The three sequences for 28 S, 5.8 S and 18 S rRNA are transcribed as a single 45 S rRNA primary transcript. The sequences for the said three rRNA molecules are separated by "transcribed spacer sequences". Several (a few dozens) of such transcription units are located at any given NOR. The transcription units are arranged in a tandem way and are interspersed by "intergenic spacer sequences". The rRNA genes are transcribed by polymerase I (or A), an enzyme engaged only in the transcription of the 45 S rRNA sequence.

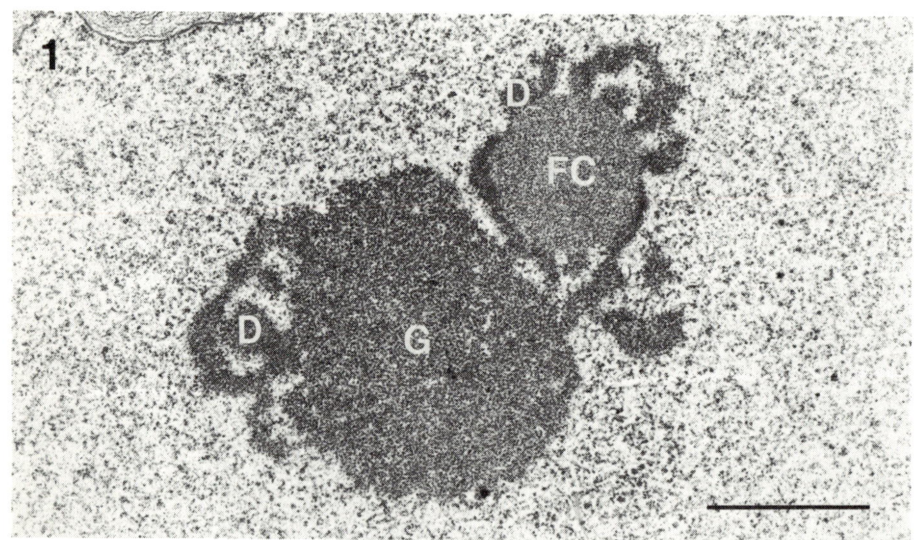

fig. 1. Nucleolus of a human Sertoli cell. FC...fibrillar centre, D...dense fibrillar component, G...granular component; bar: 1 µm.

3 Nucleolar morphology (see fig. 1)

3.1 Nucleolar components
Nucleoli of mammalian cells typically display three components in routine electron micrographs (Goessens, 1984, see fig. 1).

3.1.1 fibrillar centres (Recher et al., 1969)
These are mostly spherical structures of various size and number in nucleoli. Fibrillar centres are thought to consist mainly of proteins although the occurrence of nucleic acids has been reported within them. Fibrillar centres can be stained with silver.

3.1.2 dense fibrillar component
This component is composed of electron-opaque fibres that can be arranged in a three dimensional array of sheets or strands ("nucleolonema"). It can likewise be stained with silver, but not all areas can be stained with the same intensity. The dense fibrillar component is thought to consist of proteins and of RNA. RNA processing is thought to occur there (Puvion-Dutilleul et al., 1991).

3.1.3 granular component
This nucleolar material displays a more granular structure. Transitions between the dense fibrillar component and the granular component are not infrequent. It seems to be generally accepted, that the granular component consists of pre-ribosomes and corresponds to a later step in ribosome biogenesis. The amount of granular component in different nucleoli can be very different.

3.2 Arrangement of nucleolar components in nucleoli
Fibrillar centres are always surrounded by the dense fibrillar component, the latter can, however project away from the vicinity of fibrillar centres. The granular component is generally found at the periphery of nucleoli. Despite of the seemingly uniform role of nucleoli in all cells (the production of ribosomes), nucleoli vary greatly in number and morphology in different cells. Nucleoli in most cells display various amounts of the above mentioned nucleolar components intermingled in a somewhat irregular way.

On the other hand, nucleoli with very unique and spatially ordered arrangement of nucleolar components can be observed. An example for such a regular arrangement of nucleolar components are the "ring-shaped" nucleoli seen in lymphocytes and monocytes, where the nucleolar components are arranged in a somewhat concentric way: A single large fibrillar centre is surrounded by a shell of dense fibrillar component that merges into a peripheral layer of granular material.

Nucleoli with morphologies so typical, that the cell type and the species can be diagnosed from the nucleolus exist likewise. The nucleolus of human Sertoli cells is an example of such a peculiar and unique nucleolar structure (Cataldo et al., 1988, see fig. 1). These nucleoli contain mostly only one large spherical fibrillar centre that is surrounded by dense fibrillar component. Adjacent to this, a homogeneous block of granular material can be seen, that is partly surrounded by dense fibrillar component. Thus, in Sertoli cells, the nucleolar components are spatially separated.

In the following, we shall mainly present data on Sertoli cells, because of the above-mentioned particular nucleolar morphology that lends itself particularly well to the type of investigation we have performed. Other types of cells gave similar results in our hands.

4 DNA at and in nucleoli
Condensed chromatin can sometimes be seen at the periphery of nucleoli ("nucleolus associated heterochromatin") in conventional electron micrographs and occasionally even in the light microscope after suitable preparation. Vacuoles apparently identical to karyoplasm can frequently be seen in nucleoli. It can be assumed that these vacuoles contain decondensed DNA.

The question, whether and where DNA can be found within the above-mentioned nucleolar components has been investigated with several procedures.

4.1 The "Feulgen-like Osmium-Ammine" procedure
This method (Cogliati and Gautier, 1973) was used to visualize DNA in Sertoli cell nucleoli (Schöfer et al., 1992; see fig. 2). With this procedure, DNA can be seen inside Sertoli cell nucleoli. The major part of DNA can be seen in nucleolar vacuoles and lining the strands of the dense fibrillar component. It seems particularly noteworthy that some DNA penetrates from the periphery of the dense fibrillar component into the strands of dense fibrillar component in a highly dispersed form.

There appear to be areas of the dense fibrillar component where DNA resides only at the periphery or the surface of the latter and other areas that are penetrated by DNA in a decondensed form. It therefore appears, that the dense fibrillar component is not homogeneous with respect to DNA content.

A minor amount of DNA seems to be present at the periphery of fibrillar centres and may extend into the fibrillar centres.

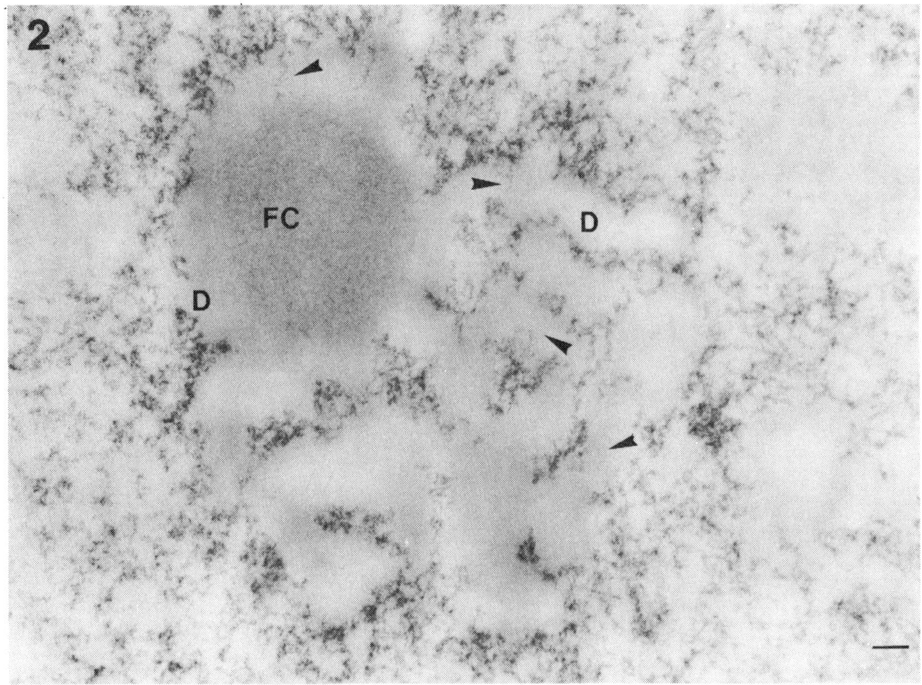

fig. 2. Osmium-ammine staining of a human Sertoli cell nucleolus. FC...fibrillar centre, D...dense fibrillar component, arrowheads point to DNA unravelling into the dense fibrillar component; bar: 100 nm

4.2 In situ hybridization using total human genomic DNA

This procedure (Schöfer et al., 1992; see fig. 3) resulted in a distribution of the label similar to the Feulgen like osmium-ammine staining procedure, revealing most of the DNA associated with the dense fibrillar component. No signal above background could be detected in the fibrillar centres.

4.3 Immunolocalization of DNA using anti-DNA antibodies

An antibody against DNA revealed DNA over the condensed chromatin associated with the nucleolus, over nucleolar vacuoles and over the dense fibrillar component, but no significant signal over the fibrillar centres could be seen (Schöfer et al., 1992; see fig. 4).

The distribution of the label over the dense fibrillar component is not homogeneous, but appears to be concentrated in clusters, thus corroborating the observed inhomogeneity of the dense fibrillar component.

The site of early replicating DNA within nucleoli was found to be the dense fibrillar component (Ghosh et al., 1990).

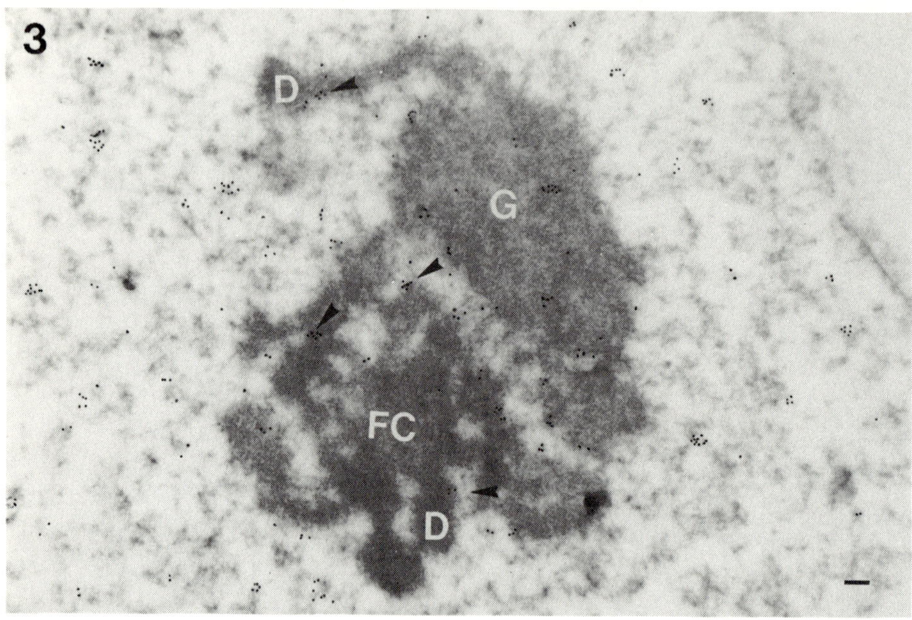

fig. 3. Human Sertoli cell nucleolus after in situ
hybridization with human total genomic DNA. Hybrids are
visualized by gold grains (arrowheads). In this figure and
in figs. 5, 7, 8 and 9 the visibility of the gold particles
has been enhanced using India ink. FC...fibrillar centres,
D...dense fibrillar component, G...granular component; bar:
100 nm.

4.4 In situ hybridization using probes for various parts of the ribosomal gene

Probes for the transcribed part of the ribosomal gene
repeat, the external transcribed spacer and for the
intergenic spacer were used at the light- and electron
microscopic level. These investigations showed, that the
ribosomal gene repeat is present inside nucleoli in a
dispersed form, often as delicate reticules (Wachtler et
al., 1986; 1989, see fig. 5).

Double labelling at the light microscopic level showed,
that there is no compartmentalization of different parts of
the ribosomal gene repeat into different nucleolar
compartments (Wachtler et al., 1991).

In situ hybridization at the electron microscopic level
showed that the rRNA gene is predominantly associated with
the dense fibrillar component (see fig. 5). The fibrillar
centres show no label that would be significantly different
from back ground values (Wachtler et al., 1992).

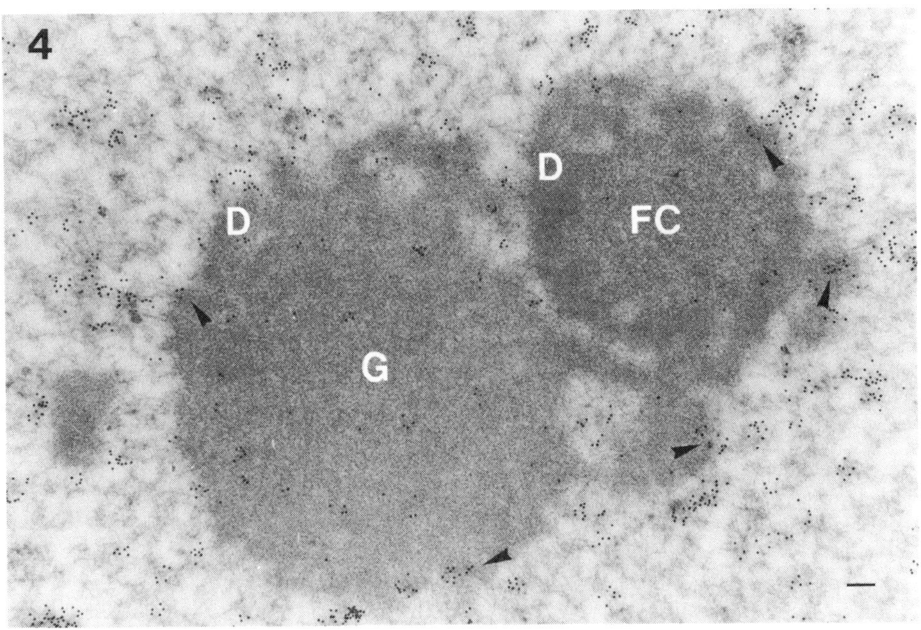

fig. 4. Human Sertoli cell nucleolus after immunogold detection of DNA (arrowheads). FC...fibrillar centres, D...dense fibrillar component, G...granular component; bar: 100 nm.

4.5 In situ hybridization with DNA from nucleolar extracts

By biochemical fractionation procedures it is possible to isolate fibrillar centres with the adjacent dense fibrillar component ("fibrillar complex"). DNA from such preparations can be isolated, characterized, labelled and used for in situ hybridization at the light- and electron microscopic level. Filter hybridization of DNA from such extracts probed for the presence of rRNA genes showed that these genes are highly enriched in such preparations.
No evidence for the presence of other genes in these preparations could be obtained so far.

In situ hybridizations at the light microscopic level showed that fibrillar complex DNA hybridized to the secondary constrictions of chromosomes and to nucleoli only. In situ hybridization at the EM level showed signal over the dense fibrillar component, but not over the fibrillar centres.

4.6 In situ hybridization with specific non-ribosomal DNA-probes

Hybridization with probes from chromosomal regions neighbouring the ribosomal genes showed, that those are

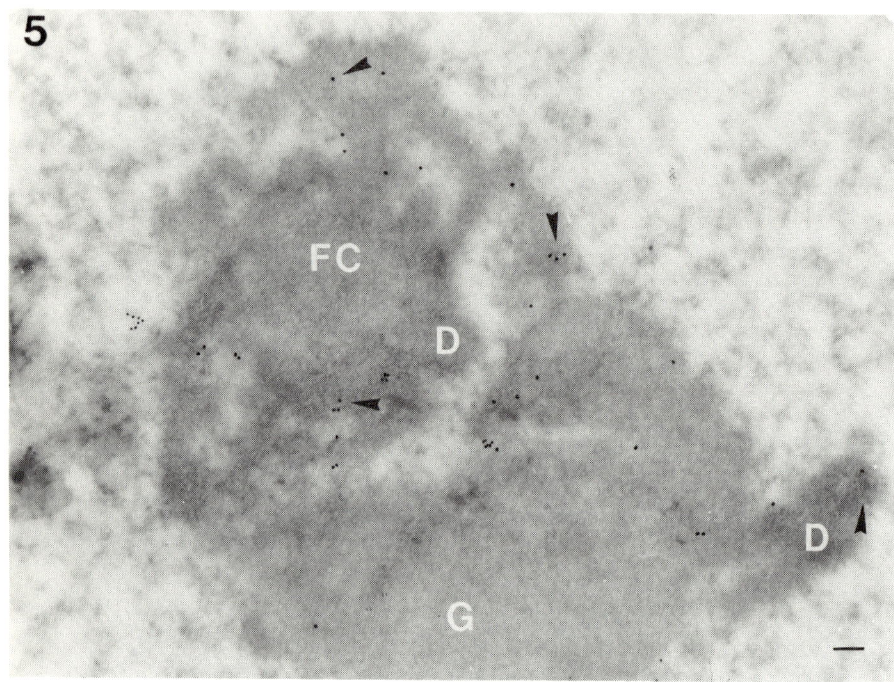

fig. 5. Nucleolus of a human Sertoli cell after in situ hybridization with a probe for the transcribed part of the human ribosomal gene repeat. Hybrids are visualized by immunogold (arrowheads). FC...fibrillar centre, D...dense fibrillar component, G...granular component; bar: 100nm.

located at the nucleolar periphery, but not inside the nucleolus (see fig. 6). Probes with repetitive DNA that is not present in the ribosomal gene repeat indicates, that non-ribosomal DNA is present in the nucleolus (Sylvester et al., 1991).

fig. 6. Nucleus of a human Sertoli cell after simultaneous in-situ-hybridization with a probe for the transcribed part of the human ribosomal gene repeat (A) and a probe for chromosomal segment proximal to the nucleolus organizer region of the short arm of chromosome 15 (B). An arrow points to the fibrillar centre,

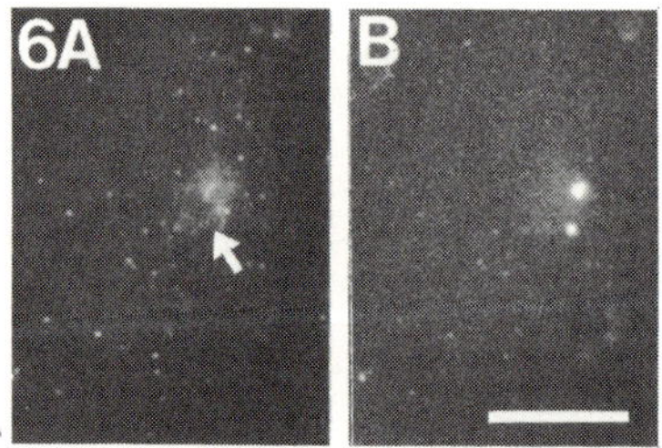

that can be seen as a dark area within the nucleolus; bar: 10 µm.

5 Localization of uridine uptake in nucleoli

5.1 Autoradiography

The uptake of tritiated uridine and subsequent visualization of the radioactive nucleotide by autoradiography have been performed by several groups. In the majority of reports the signal was found over the dense fibrillar component. This holds also true for human Sertoli cells (Wachtler et al., 1989).

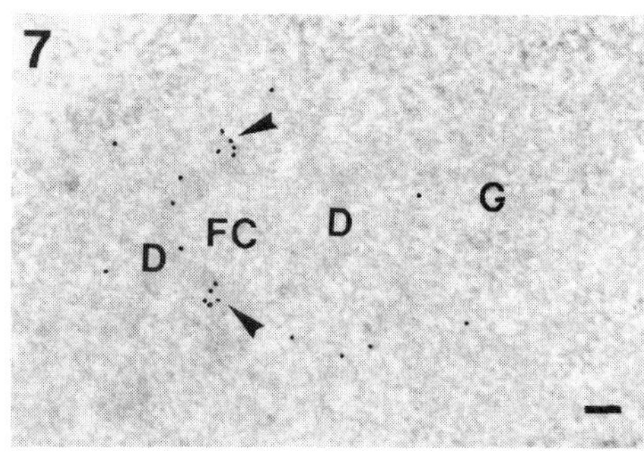

fig. 7. Detail from a horse melanoma cell nucleolus. Incorporated digoxigenin-labelled uridine is visualized by immunogold detection (arrowheads).
FC...fibrillar centre,
D...dense fibrillar component, G...granular component; bar: 100 nm.

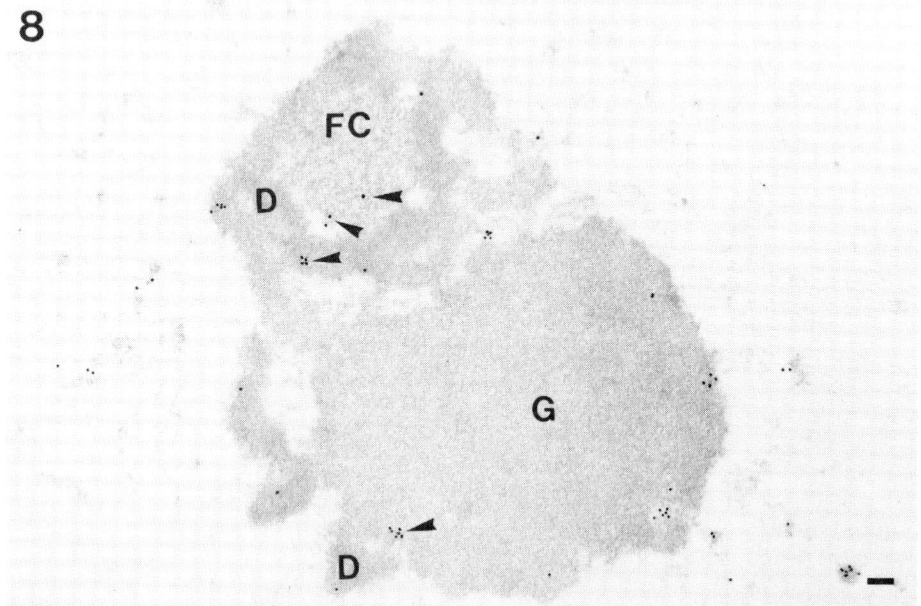

fig. 8. Immunogold detection of topoisomerase I in a human Sertoli cell nucleolus (arrowheads). FC...fibrillar centres, D...dense fibrillar component, G...granular component; bar: 100 nm

5.2 Visualization of uridine uptake without autoradiography

The low spatial resolution inherent to autoradiography does, however, generally not allow to distinguish between the periphery of fibrillar centres and the adjacent dense fibrillar component as the source of radiation. Furthermore, it can be argued that the chase usually following the radioactive pulse would result in a shift of the material from the putative site of transcription to the site of processing of rRNA. To avoid these shortcomings we used a non-autoradiographic procedure to visualize the uptake of uridine, namely the detection of incorporated digoxigenin-uridine by antibodies. Filter hybridization of RNA extracted from cells showed the incorporation of this compound into RNA.

Visualization of digoxigenin-uridine after incorporation for five minutes without chase showed that uridine is taken up into the dense fibrillar component but not into the fibrillar centres (see fig. 7).

6 Localization of enzymes required for rRNA gene transcription

Topoisomerase I and polymerase I have been localized with the help of specific antibodies at the EM level. Within nucleoli, significant levels of signal indicating the presence of topoisomerase I were found mostly in the dense fibrillar component (see fig. 8).

Polymerase I, on the other hand, was found not only in the dense fibrillar component but also in the fibrillar centres (Mosgöller et al., 1992, see fig. 9). The activity rather than the presence of polymerase I has been investigated by Gruca et al. (1978) and was likewise seen in the dense fibrillar component.

7 Active vs. inactive ribosomal genes

So far, we have dealt with ribosomal genes within nucleoli that are active in the process of rRNA-gene transcription. Cells with low amounts of rRNA gene transcription exist, e. g. lymphocytes and monocytes from peripheral blood. In these cells, rRNA genes can be found in condensed chromatin outside the nucleoli and at a distance therefrom (Wachtler et al., 1986).

Changes in transcriptional activity can be induced by drugs like actinomycin D, but occur also spontaneously, e. g. during the pachytene stage in meiosis (Tres, 1975). In these cases, the ribosomal genes are retracted from nucleoli into condensed chromatin (Stahl et al., 1991, see fig. 10).

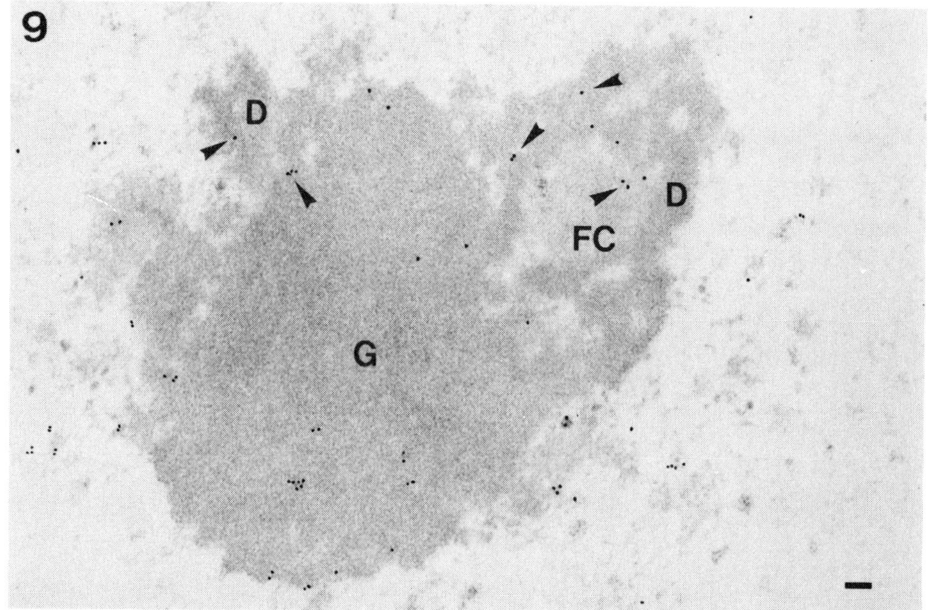

fig. 9. Immunogold detection of polymerase I in a human Sertoli cell nucleolus (arrowheads). FC...fibrillar centres, D...dense fibrillar component, G...granular component; bar: 100 nm

8   Discussion

8.1   Location of rRNA gene transcription is a controversial subject

Two main propositions have been put forward with respect to the location and site of transcription of ribosomal genes within nucleoli (for a review see Jordan, 1991): One group of researchers feels that the fibrillar centres are the site of transcription (Thiry and Thiry-Blaise, 1989; Puvion-Dutilleul et al., 1991) whereas others (including ourselves) assume the location and transcription to be associated with the dense fibrillar component (Ghosh and Paweletz, 1990; Wachtler et al., 1992).

Both propositions basically rely on the application of similar techniques. How can these discrepancies be explained ? We think that there are three main reasons:

1) The choice of cells: If the question is asked whether the dense fibrillar component or the fibrillar centres are the site of rDNA location and transcription, nucleoli of most cells are not ideal for this investigation because nucleolar components are intermingled in most cells, rendering it sometimes difficult to attribute the signal to

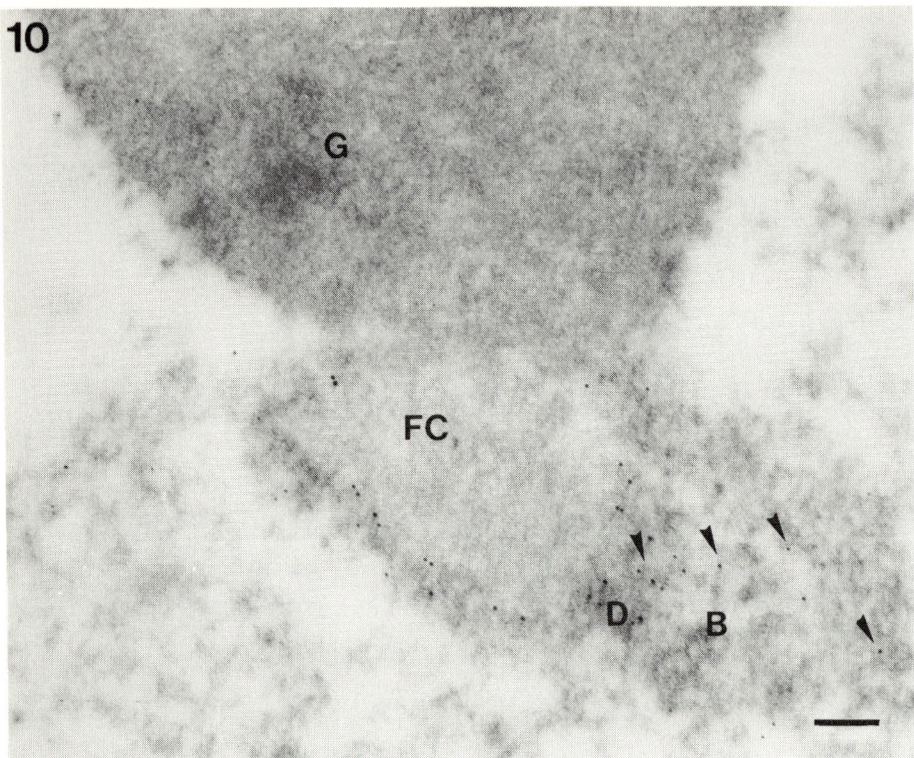

fig. 10. Nucleolus of a human spermatocyte in the pachytene
stage after in situ hybridization with a probe for the
transcribed part of the human ribosomal rRNA gene. Hybrids
are visualized by immunogold detection. B...bivalent
(meiotic chromosome), FC...fibrillar centre, D...dense
fibrillar component, G...granular component; bar: 100 nm.

any particular nucleolar component. Also, in most cells the
dense fibrillar component is found exclusively in the
vicinity of the fibrillar centres, making the distinction
between the periphery of the fibrillar centres and the part
of the dense fibrillar component facing fibrillar centres
difficult.

2) Sensitivity of the immunogold detection of hybrids.
Both the sensitivity and the signal/background ratio of the
immunogold detection are low when compared to other
cytochemical procedures. It therefore can be very
misleading to draw conclusions from visual inspection of
electron micrographs only. A way to overcome this problem
is to perform stereometric analysis based on a larger
number of cells. For the same reason, small amounts of an
epitope or a highly dispersed epitope may not be

detectable. Applied to the visualization of ribosomal genes in nucleolar components minor amounts of ribosomal genes may very well go by undetected.

3) Differences between different types of cells in different species.

Whereas it is evident that nucleoli in plants (Deltour and Motte, 1990), insects (Knibiehler et al., 1982), amphibians and so forth are organized in a somewhat different way from that in mammals, it is generally assumed, that the relationship between nucleolar structure and nucleolar function is the same in all types of cells. This need not necessarily be so and divergent results could be accounted for by differences in nucleolar orgnization.

8.2 A consensus interpretation
It has been tacitly assumed so far that nucleolar components defined by their morphology are homogeneous with respect to their composition and function. The observation, that only some parts of the dense fibrillar component are penetrated by DNA and the observation that a silver staining reaction stains only some parts of the dense fibrillar component indicates that the composition of the dense fibrillar component is not homogeneous.

Therefore, it may not be possible at all to answer the question where the ribosomal genes are being located and transcribed by identifying a nucleolar component that corresponds to this process. Our data and most of the published data (not the conclusions) are compatible with the proposition that most of the ribosomal genes within a nucleolus are located at some sites at the periphery or the surface of the strands of the dense fibrillar component.

9 Acknowledgements
This work was supported by "Österreichischer Fonds zur Förderung der wissenschaftlichen Forschung", P 7820 MED

10 References
Cataldo, C., Soucher, C. and Stahl, A. (1988) Three dimensional ultrastucture and quantitative analysis of the human Sertoli cell nucleolus. Biol. Cell, 63, 277-285
Cogliati, R., and Gautier, A. (1973) Mise en evidence de l´ADN et des polysaccharides a l'aide d'un nouveau reactif "de type Schiff". C. R. Acad. Sci. Ser. D, 276, 3041-3044.

Deltour, R., and Motte, P. (1990) The nucleolonema of plant and animal cells. Biol. Cell, 68, 5-11.

Egan, M. J., and Crocker, J. (1992) Nucleolar Organiser Regions in Pathology. Br. J. Cancer, 5, 1-7.

Evans, H. J., Buckland, R. A. and Pardue, M. L. (1974) Location of genes coding for 18S and 28S ribosomal RNA in the human genom. Chromosoma, 48, 405-426.

Fontana, F. (1781) Traite sur le venin de la vipere, avec des observations sur la structure primitive du corps animale. Florence.

Ghosh, S. et al (1990) Timings of DNA synthesis in the nucleoli of Allium cepa L. Cell Biol. Int. Rep., 14, 173-177.

Ghosh, S. and Paweletz, N. (1990) Localization of ribosomal cistrons at the ultrastructural level by in situ hybridization technique. Cell Biol. Int. Rep., 14, 521-525.

Goessens, G. (1984) Nucleolar structure. Int. Rev. Cytol., 87, 107-158.

Gruca, S.et al (1978) Intranucleolar localization of the RNA-polymerase A activity in isolated nuclei of rat liver. Exp. Cell Res., 114, 462-467.

Hernandez-Verdun, D. (1991) The Nucleolus Today. J. Cell Sci., 99, 465-471.

Jordan, E. G. (1991) Interpreting nucleolar structure: where are the transcribing genes?. J. Cell Sci., 98, 437-442.

Knibiehler, B., Mirre, C. and Rosset, R. (1982) Nucleolar organizer structure and activity in a nucleolus without fibrillar centres: The nucleolus in an established Drosophila cell line. J. Cell Sci., 57, 351-364.

Mosgöller, W.et al (1992) Localization of polymerase I and its activity in human nucleoli. Europ. J. Cell Biol. (Suppl. 36), 57, 56.

Perry, R. P. (1962) The cellular sites of synthesis of ribosomal and 4 S RNA. Proc. Natl. Acad. Sci. USA, 48, 2179-2186.

Perry, R. P., Hell, A. and Errera, M. (1961) The role of the nucleolus in ribonucleic acid and protein synthesis. I Incorporation of cytidine into normal and nucleolar inactivated Hela cells. Biochim. Biophys. Acta, 49, 47-57.

Puvion-Dutilleul, F., Bachellerie, J. P., and Puvion, E. (1991) Nucleolar Organization of HeLa Cells As Studied by Insitu Hybridization. Chromosoma, 100, 395-409.

Recher, L., Whitescarver, J. and Briggs, L. (1969) Fine structure of a nucleolar constituent. J. Ultrastuct. Res., 29, 1-14.

Reeder, R. H. (1990) rRNA synthesis in the nucleolus. TIG, 6, 390-395.

Schöfer, C.et al (1992) Localization of DNA in the nucleolus of human Sertoli cells. Europ. J. Cell Biol. (Suppl.36), 57, 70.

Sollner-Webb, B., and Mougey, E. B. (1991) News from the
    nucleolus: rRNA gene expression. TIBS, 16, 58-62.
Stahl, A.et al (1991) Nucleoli, nucleolar
    chromosomes and ribosomal genes in the human
    spermatocyte. Chromosoma, 101, 231-244.
Sylvester, J. E.et al (1991)
    Nucleolar organizing regions within interphase nuclei.
    Am. J. Hum. Genet. (Suppl.), 49, 308.
Thiry, M., and Thiry-Blaise, L. (1989) In situ
    hybridization at the electron microscopic level: an
    improved method for the precise localization of
    ribosomal DNA and RNA. Europ. J. Cell Biol., 50,
    235-243.
Tres, L., L. (1975) Nucleolar RNA synthesis of meiotic
    prophase spermatocytes · in the human testis.
    Chromosoma 53, 141-151.
Wachtler, F.et al (1989) Ribosomal DNA
    is located and transcribed in the dense fibrillar
    component of human Sertoli cell nucleoli. Exp. Cell
    Res., 184, 61-71.
Wachtler, F.et al (1986) On the position of nucleolus
    organizer regions (NORs) in interphase nuclei. Studies
    with a new, non-autoradiographic in situ hybridization
    method. Exp. Cell Res., 167, 227-240.
Wachtler, F.et al (1992) Human ribosomal
    RNA gene repeats are localized in the dense fibrillar
    component of nucleoli. Light- and electron microscopic
    in-situ-hybridization in human Sertoli cells.
    Exp. Cell Res., 198, 135-143.
Wachtler, F.et al (1991) Transcribed and
    nontranscribed parts of the human ribosomal gene repeat
    show a similar pattern of distribution in nucleoli.
    Cytogenet. Cell Genet., 57, 175-178.

6 Nucleolar proteins during mitosis

D. HERNANDEZ-VERDUN, P. ROUSSEL and T. GAUTIER

Institut Jacques Monod, France

1 Introduction

Nucleoli are the sites of ribosomal gene (rDNA) transcription, and also the assembly and processing of eukaryotic ribosomal subunits Hadjiolov (1985). Specific proteins associated with rDNA transcription machinery and pre-ribosomal subunits are sorted and assembled in these nuclear territories. In addition proteins that are thought to provide structural support for the spatial arrangement of the rDNA and to play a role in the ordered translocation of pre-ribosomal subunits from their primary sites of assembly to their sites of translocation.

The different steps of ribosome biogenesis correspond to three nucleolar domains that can be identified by their morphology: the fibrillar centers (FCs), the dense fibrillar component (DFC) and the granular component (GC). The two first domains correspond to the transcription steps and the third to maturation and storage of the ribosomal subunits, for review see Fischer et al. (1991), Hernandez-Verdun (1986,1991), Jordan (1991).

During the cell cycle rDNA transcription starts in telophase and stops in late G_2 phase when the chromatin condenses into chromosomes. The entry into mitotic phase (M phase) is correlated with specific phosphorylation of major nucleolar proteins by cdc2 kinase (Peter et al. 1990). It has been suggested that this phosphorylation controls mitotic changes in nucleolar protein localization and activity (Peter et al. 1990). During prophase nucleoli disintegrate and their proteins become differently distributed within the dividing cell, for review see Sommerville (1986). Some proteins remain associated with rDNA in the nucleolar organizer region (NOR); some nucleolar proteins disperse at the periphery of all chromosomes while others are scattered throughout the cytoplasm without any detectable sites of accumulation. During telophase these disparate nucleolar constituents rapidly reassemble in an apparently coordinate fashion and

Chromosomes Today Volume 11. Edited by A.T. Sumner and A.C. Chandley. Published in 1993 by Chapman & Hall, London. ISBN 0 412 47670 3

accumulate in the NORs, contributing to the reformation of functioning nucleoli.

To analyse the assembly and the deassembly of the nucleoli during the cell cycle, we have studied nucleolar proteins which remain associated with the NOR during mitosis and those forming a perichromosomal layer around all the chromosomes.

2 Nucleolar Organizer regions (NORs)

During mitosis NORs are associated with a subset of proteins. Some of these proteins can be detected by cytochemical reactions first described in the seventies (Goodpasture and Bloom, 1975; Howell and Black, 1980). These cytochemical methods are based on the specific argyrophilic affinity of the NOR-associated proteins, the so-called Ag-NOR proteins. It has been established that only the NORs in which transcription is occurring are associated with the Ag-NOR proteins: inactive NORs contain no Ag-NOR proteins. In human cells, the amount of Ag-NOR protein varies between active NORs. Recently, it has been reported that the amount of Ag-NOR protein has a prognostic value for human cancer, for review see Derenzini and Ploton (1991).

Some NOR-associated nucleolar proteins have also been revealed by specific antibodies. They include part of RNA polymerase I, DNA topoisomerase I and nucleolin molecules (Gas et al. 1985; Ochs et al. 1983; Scheer and Rose 1984) and more recently RNA polymerase I transcription factor, UBF (Chan et al. 1991 and the present work).

Table 1. Proteins associated with active NORs.

Proteins	MW (kDa)	Localization Interphase	Mitose	refs.
RNA pol I		CF	NOR	Scheer and Rose 1984
Nucleolin	100	DFC+GC	NOR	Ochs et al. 1983
p135	135	DFC	NOR	Pfeifle et al. 1986
DNA topo I	95		NOR	Guldner et al. 1986
UBF	97-94	DFC+FC	NOR	Chan et al.1991 and this work
Ag-NOR	X	DFC+FC	NOR	Hernandez-Verdun et al.1980, Ploton et al. 1987 and this work

X: several nucleolar proteins under investigation. RNA pol I: RNA polymerase I, DNA topo I: DNA topoisomerase I, UBF: Upstream binding factor.

2.1 Ag-NOR proteins

The Ag-NOR proteins can be used as markers of "active" ribosomal genes. However they have not been identified to date. To identify the Ag-NOR proteins, we have adapted the Ag-NOR staining procedure to Western blots (Hozak et al. 1992). In cellular, nuclear and nucleolar protein extracts from HeLa cells, Ag-NOR staining reveals only a few proteins as compared to those revealed by general Protogold staining. Thus, detection of Ag-NOR proteins immobilized on nitrocellulose appears to be as selective as in situ methods, for further details see Hozak et al. (1992). On blots from 1D gel electrophoresis, we observed two major positive bands (105 and 38 kDa), and also bands around 40 kDa, 130 kDa and occasionally 180 and 29 kDa in human nuclear extracts (Hozak et al. 1992). On 2D gel electrophoresis Ag-NOR proteins especially the minor proteins were more clearly resolved (Fig. 1C). We think that this approach will help further identification of Ag-NOR proteins associated with different cell activities.

The two major Ag-NOR-stained human cell proteins of 105 kDa (pI 5.3) and 38 kDa (pI 5.4), have the same apparent molecular weights as two major nucleolar proteins, nucleolin and B23. That nucleolin is an Ag-NOR protein was confirmed using crude and purified nucleolin (in preparation). We were also able to define precisely the domain of the protein involved in the staining. Its silver stainability is localized in the amino-terminal acidic domain, and is not dependant on the phosphorylation state of the protein. These results clarify the principle of Ag-NOR staining.

We analyzed the pattern of Ag-NOR silver staining in non-synchronized cells (Hozak et al. 1992), interphasic cells and mitotic cells. In situ, it is obvious that the amount of Ag-NOR proteins is lower in mitosis than in interphase (Fig. 1A). The Western blot patterns of interphase and mitotic HeLa cell extracts were similar: most positive bands were similar but these were additional bands of low molecular mass in mitotic extracts (Fig. 1B). However the identification of mitotic Ag-NOR proteins requires further investigation. The extra bands may correspond to specific mitotic Ag-NOR proteins or to degradation products of the major interphasic Ag-NOR proteins.

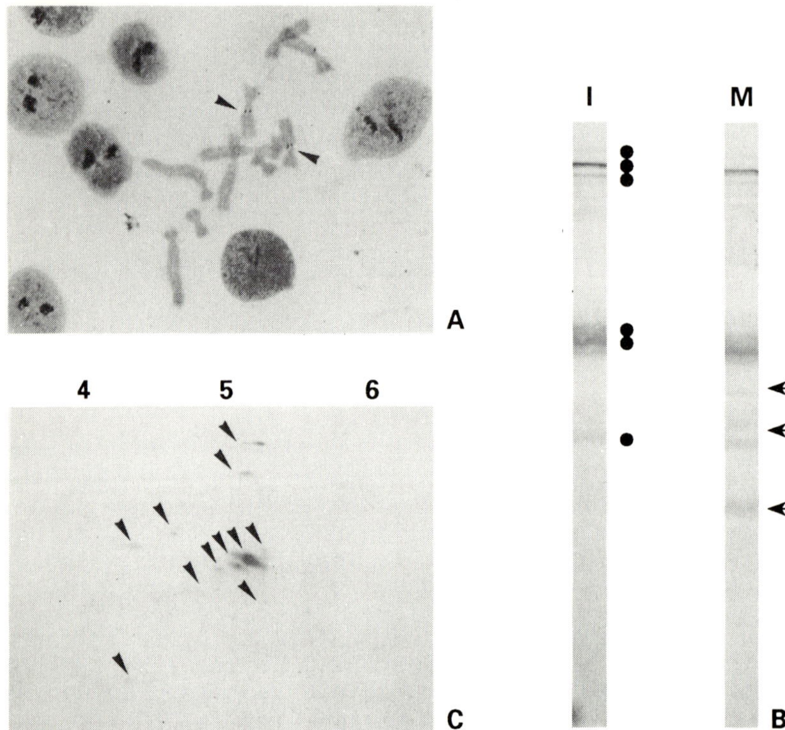

Fig. 1. A- Ag-NOR staining of PtK$_1$ cells in interphase and
mitosis. Note the positive staining of 2 nucleoli in
nuclei and 2 chromosomes during mitosis. In PtK$_1$ cells the
NOR-bearing chromosomes are X chromosomes and the NOR
forms the secondary constriction (arrow heads). B- 1D
Western blot of human nuclear extracts after Ag-NOR
staining. I: interphasic cells; M: mitotic cells;
equivalent numbers of cells per lane. C- 2D Western blots
of human nuclear extracts after Ag-NOR staining. Note that
Ag-NOR proteins have acidic pI (arrow heads).

2.2 UBF

Transcription of rDNA involves several factors beside RNA
polymerase I, for reviews see Reeder (1990) and Sollner-
Webb and Mougey (1991). For accurate in vitro initiation
by RNA polymerase I at least two factors, UBF (upstream
binding factor) and SL1 proteins are required (Bell et al.
1988).
Using anti-UBF sera, we investigated the distribution of
rDNA transcription factor UBF in human cells in
correlation with rDNA transcription activity at various
stages of the cell cycle (Fig. 2A,A'). During interphase

anti UBF labelled the nucleoli forming a chain-like structure composed of beads. At the end of the G2 phase, as rDNA transcription is inactivated, UBF was redistributed and appears associated in differing amounts with chromosomes being condensed (Fig. 2A,A'). UBF remained associated with 6 or 8 NORs among the 10 NOR-bearing chromosomes. Therefore, the 10 human NORs seem not to be equally involved in rDNA transcription. During anaphase, UBF is equally partitioned between the two chromosomal sets. The same number of spots and the same fluorescence intensities were observed during telophase. The activation of rDNA transcription was not correlated with a detectable increase in the amount of UBF. However the presence of the proteins required for transcription, including UBF and as previously described RNA polymerase I close to the rDNA during mitosis may allow the rapid switch-on of rDNA transcription in telophase.

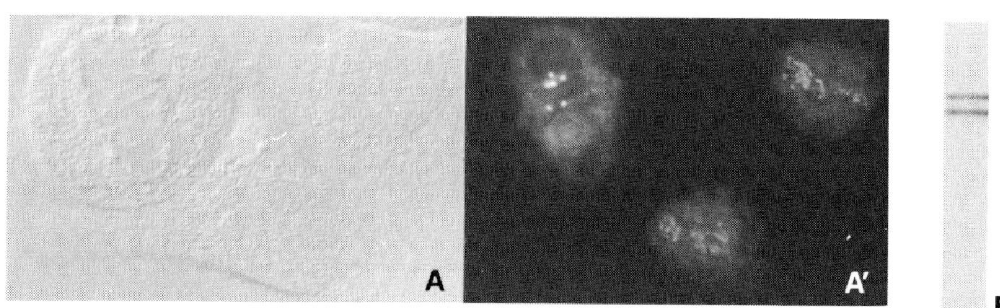

Fig. 2. A- Interphasic and mitotic HEp-2 cells in interference contrast; A'- Same cells labelled with anti-UBF antibodies (photo courtesy of C. Masson). B- Western blot of human nuclear proteins incubated with anti-UBF antibodies. The two polypeptides 97-94 kDa are labelled equally.

3 Chromosome periphery

During mitosis, nuclear proteins become distributed throughout the mitotic cells and adopt a totally different organization. Some remain associated with the chromosome surface (Sommerville, 1986). Among them, we can distinguish those which transiently associate and are later released, therefore named "chromosomal passenger proteins" (Earnshaw and Bernat, 1991), and those which stay in association with the chromosome periphery. Using antibody probes, it has been possible to localize an increasing number of proteins within the chromosome (Rattner 1992): many of those confined to the outer margin of the chromosome have a nucleolar origin (see table 2).

Table 2. Nucleolar proteins associated with the perichromosomal layer.

Proteins	MW (kDa)	Localization Interphase	Mitosis	refs.
PCN	30-36	NP	PC	Shi et al. 1987
Fibrillarin	33	DFC+GC	PC	Ochs et al. 1985
B23	38	DFC+FC	PC	Ochs et al. 1983
Ribocharin	40	GC+NP	PR	Hügle et al. 1985b
S1	45	GC+CP	PR	Hügle et al. 1985a
53	53	GC	PC	Gautier et al. 1992a
66	66	GC	PC	"
103	103	GC+DFC	PC	"
p400+	>400	NP	PC	Dilworth 1991
Ki-67	?	DFC	PC	Verheijen et al. 1989

PC: periphery of the chromosomes, PR: perichromosomal region, NP: nucleoplasm, CP: cytoplasm.

3.1 Characterization of nucleolar RNP in the perichromosomal layer

We tested over 40,000 sera from patients suffering from autoimmune diseases, and found 36 anti-nucleolar sera, 10 of which bound to the chromosome periphery during mitosis. These sera were used to characterize some of the nucleolar antigens accumulated at the chromosome periphery and to study their behaviour during mitosis (see Fig. 3). One dimensional Western blot analysis indicated that 3 kinds of antigens were recognized by the antisera (52, 68, and 103 kDa). One serum precipitated an antigen-associated RNA with the same electrophoretic mobility as the U3 RNA (Gautier et al. 1992b). Presumably, therefore, nucleolar RNPs participate in the formation of perichromosomal layer during mitosis. Immunoelectron microscopy demonstrates that the nucleolar antigens recognized by the ten sera were located in the GC during interphase. These findings that at least one of these antigens was associated with U3, are consistent with the recently identified locations for U3 RNA in DFC and in GC (Puvion-Dutilleul et al. 1991).

3.2 Structure of the perichromosomal layer

The nucleolar antigens can be seen under the electron microscope at the periphery of the chromosomes in closely packed fibrils unevenly distributed around most of the chromosome sections. We investigated the ultratructural organization of this material after quick-freezing in liquid helium. It displayed fibrous structures, as well as dense granules, some of them arranged in a network. The granules

were frequently linked by thin filaments and their diameters
varied from 11 to 16 nm. Their characteristics resemble
those of the interchromatin granules observed in the
interphase nuclei of adjacent cells prepared in the same way
(Gautier et al. 1992b).

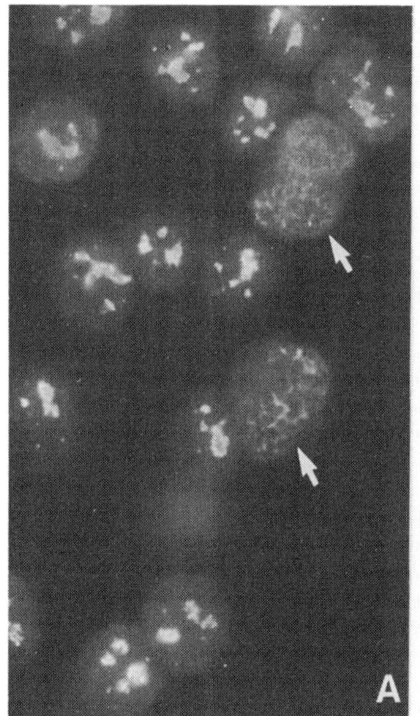
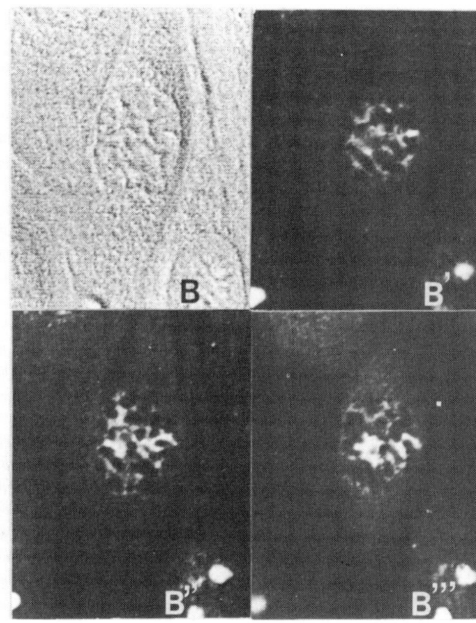

Fig. 3. A- Immunofluorescence of HeLa cells labelled with
one autoimmune serum. Each nucleolus scores positive and the
periphery of the chromosomes is strongly labelled in the
different stages of mitosis (arrows). B- Phase contrast of
PtK$_1$ cell. B'to B'''- Serial confocal optical sections, 1 µm
apart, of immunofluorescence labelling. The labelling,
observed throughout entire nuclear volume, forms a network
that extends around the condensing chromosomes.

3.3 Formation and kinetics of the perichromosomal layer

Three-dimensional reconstructions of the cells labelled with
one serum showed that at the beginning of prophase when the
chromatin condenses, the nucleolar antigens migrate from the
nucleoli to form a network in the nucleoplasm. The antigens
dispersed from two helicoidal structures towards the
periphery of the nucleus creating 3D-star pattern. During
the progression of prophase, an increasing amount of the
antigens became dispersed throughout the nucleus. At the end
of prophase, the labelling formed a sharp outline of even
thickness all along the border of each chromosome. During
metaphase and anaphase, the antigens remained at the borders
of the chromosomes (Gautier et al. 1992c).

In late telophase, chromosomes were still coated with the antigens, and clusters corresponding to the newly reformed nucleoli were detected. The antigens remained associated with the chromosome periphery after hypotonic shock, but were released from the nucleoli by this treatment. It was therefore possible to perform immunofluorescence experiments on isolated chromosomes: we were able to isolate the chromosomes surrounded by the antigens, and found that each chromosome was completely covered except the centromeric regions, where the centromeres were not labelled and appeared as dark zones (Gautier et al. 1992c).

Thus the perichromosomal region appears to be a very complex structure involving nucleolar RNPs. The mechanism by which these proteins are translocated and addressed to the chromosome surface and how this movement is regulated very early in prophase are unknown. The presence of a structure at the chromosome periphery seems to be the result of well coordinated movements in space and time during mitosis in which phosphorylation cycles may play a central role.

4 Conclusions

A general reorganization of the cell's inner architecture takes place when the cell enters mitosis. The reorganization of the nucleolar territory is presented in Fig. 4 which summarizes the data reported in this chapter concerning the proteins associated with the NORs and nucleolar RNP localized at the chromosome surface during mitosis.

The NOR proteins characterized during mitosis are all part of the ribosomal transcription machinery. However we do not know if all the proteins necessary for accurate transcription of rDNA are present in the mitotic NORs since few of them have been identified. Further work is necessary to continue the molecular dissection of the mitotic NORs and investigate the relationship between NOR proteins and chromatin. This knowlegde will help to elucidate the structure and the organization of the secondary constriction of the chromosomes.

The data accumulated on the distribution of nucleolar proteins during mitosis have led to a reevaluation of the role of the chromosome during cell division. There is a growing consensus that the chromosome is not a passive structure but integrated into the process of cell division (Earnshaw and Bernat, 1991). Therefore, the relocation of some nucleolar antigens at the periphery of the chromosomes would suggest a different function for this material. The layer of antigens which seems to be a general feature in higher eukaryotes, is present from prophase to telophase and is strongly bound to the chromosomes, suggesting that it is important for correct processing of mitosis. It has been proposed that proteins of this layer could be involved in chromatin condensation (Dilworth and Dingwall, 1988; McKeon

et al. 1984; Shi et al. 1987). Other proteins including
nucleolin and topoisomerase II have been proposed as being
involved in the condensation/decondensation cycles of the
chromatin (Dilworth and Dingwall, 1988; Earnshaw et al.
1985; Earnshaw and Heck, 1985). Such a sheath around the
chromosomes at the time of lamin dissociation, led to the
idea of a chromosome protective function (Yasuda and Maul,
1990). The layer covering the chromosome may therefore be
composed of various nuclear proteins that specifically
interact with the chromosome surface at a particular time
during the mitotic process. Finally, the distribution of the
antigens on the chromosome surface could have a storage
role, as has been proposed for nucleoplasmin and p400+
(Dilworth, 1991). This region was also demonstrated to be a
transient accumulation site for various proteins including
ribosomal protein S1 (Hügle et al. 1985a), ribocharin (Hügle
et al. 1985a,b), and U-snRNPs recognized by anti-Sm and
anti-fibrillarin sera (Spector and Smith, 1986; Yasuda and
Maul, 1990). In this case, lining the chromosome periphery
could be an efficient mechanism to distribute nuclear and
nucleolar proteins equally between daughter cells.

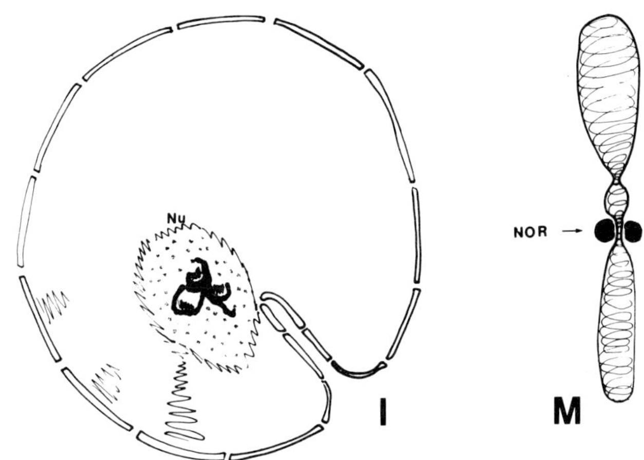

Fig. 4. Nucleolar proteins in PtK1 cells during interphase
and mitosis. I: interphase; M: mitosis; NOR: nucleolar
organizer region; Nu: nucleolus.
In PtK1 cells, the NORs form the secondary constriction of
X chromosomes. The perichromosomal layer covers the whole
chromosome surface including the telomere but excluding
the centomere when the complex structure separating the
cytoplasm and the genome during interphase is
destabilized.

5 References

Bell, S.P.et al. (1988) Functional Cooperativity Between Transcription Factors UBF1 and SL1 Mediates Human Ribosomal RNA Synthesis. **Science**, 241, 1192-1197.

Chan, E.K.L.et al. (1991) Human autoantibody to RNA polymerase I transcription factor hUBF. Molecular identity of nucleolus organizer region autoantigen NOR-90 and ribosomal RNA transcription upstream binding factor. **J. Exp. Med.**, 174, 1239-1244.

Derenzini, M. and Ploton, D. (1991) Interphase nucleolar organizer regions in cancer cells. **Int. Rev. Exp. Pathol.**, 32, 149-191.

Dilworth, S.M. (1991) A perichromosomal region contains proteins phosphorylated during mitosis in *Xenopus-laevis* cells. **J. Cell Sci.**, 98, 309-315.

Dilworth, S.M. and Dingwall, C. (1988) Chromatin assembly *in vitro* and *in vivo*. **BioEssays**, 9, 44-49.

Earnshaw, W.C. and Bernat, R.L. (1991) Chromosomal passengers: toward an integrated view of mitosis. **Chromosoma**, 100, 139-146.

Earnshaw, W.C.et al. (1985) Topoisomerase II is a structural componant of mitotic chromosome scaffolds. **J. Cell Biol.**, 100, 1706-1715.

Earnshaw, W.C. and Heck, M.M.S. (1985) Localization of topoisomerase II in mitotic chromosomes. **J. Cell Biol.**, 100, 1716-1725.

Fischer, D., Weisenberger, D. and Scheer, U. (1991) Assigning functions to nucleolar structures. **Chromosoma**, 101, 133-140.

Gas, N., Escande, M.-L. and Stevens, B.J. (1985) Immunolocalization of the 100 kDa nucleolar protein during the mitotic cycle in CHO cells. **Biol. Cell**, 53, 209-218.

Gautier, T.et al. (1992a) Identification and characterization of a new set of nucleolar ribonucleoproteins which line the chromosomes during mitosis. **Exp. Cell Res.**, 200, 5-15.

Gautier, T.et al. (1992b) The ultrastructure of the chromosome periphery in human cells. An *in situ* study using cryomethods in electron microscopy. **Chromosoma**, 101, 502-510.

Gautier, T.et al. (1992c) Relocation of nucleolar proteins around chromosomes at mitosis- A study by confocal laser scanning microscopy. **J. Cell Sci.**, 102, 729-737.

Goodpasture, C. and Bloom, S.E. (1975) Visualization of nucleolar organizer regions in mammalian chromosomes using silver staining. **Chromosoma**, 53, 37-50.

Guldner, H.-H.et al. (1986) Scl 70 autoantibodies from scleroderma patients recognize a 95 kDa protein identified as DNA topoisomerase I. **Chromosoma**, 94, 132-138.

Hadjiolov, A.A. (1985) **The nucleolus and the ribosome biogenesis.** Springer-Verlag, Wien, New-York.

Hernandez-Verdun, D. (1986) Structural organization of the nucleolus in mammalian cells., in **Methods and Achievements in experimental pathology** (eds G. Jasmin and R. Simard), Karger, S., Basel, pp. 26-62.

Hernandez-Verdun, D. (1991) The nucleolus today. **J. Cell Sci.**, 99, 465-471.

Hernandez-Verdun, D.et al. (1980) Ultrastructural localization of Ag-NOR stained proteins in the nucleolus during the cell cycle and in other nucleolar structures. **Chromosoma**, 79, 349-362.

Howell, W.M. and Black, D.A. (1980) Controlled silver-staining of nucleolus organizer regions with a protective colloidal developer: a 1-step method. **Experientia**, 36, 1014-1015.

Hozak, P., Roussel, P. and Hernandez-Verdun, D. (1992) Procedures for specific detection of silver stained nucleolar proteins on Western blots. **J. Histochem. Cytochem.**, 40: 1089-1096.

Hügle, B.et al. (1985a) Localization of ribosomal protein S1 in the granular component of the interphase nucleolus and its distribution during mitosis. **J. Cell Biol.**, 100, 873-886.

Hügle, B., Scheer, U. and Franke, W.W. (1985b) Ribocharin: a nuclear M_r 40,000 protein specific to precursor particles of the large ribosomal subunit. **Cell**, 41, 615-627.

Jordan, E.G. (1991) Interpreting nucleolar structure; where are the transcribing genes. **J. Cell Sci.**, 98, 437-449.

McKeon, F.D.et al. (1984) The redistribution of a conserved nuclear envelope protein during the cell cycle suggests a pathway for chromosome condensation. **Cell**, 36, 83-92.

Ochs, R.et al. (1983) Localization of nucleolar phosphoproteins B23 and C23 during mitosis. **Exp. Cell Res.**, 146, 139-149.

Ochs, R.L.et al. (1985) Fibrillarin: a new protein of the nucleolus identified by autoimmune sera. **Biol. Cell**, 54, 123-134.

Peter, M.et al. (1990) Identification of major nucleolar proteins as candidate mitotic substrates of Cdc2 kinase. **Cell**, 60, 791-801.

Pfeifle, J., Boller, K. and Anderer, F.A. (1986) Phosphoprotein pp135 is an essential component of the nucleolus organizer region (NOR). **Exp. Cell Res.**, 162, 11-22.

Ploton, D.et al. (1987) Behaviour of nucleolus during mitosis. A comparative ultrastructural study of various cancerous cell lines using the Ag-NOR staining procedure. **Chromosoma**, 95, 95-107.

Puvion-Dutilleul, F.et al. (1991) Localization of U3 RNA molecules in nucleoli of HeLa and mouse 3T3 cells by high resolution in situ hybridization. **Eur. J. Cell Biol.**, 56, 178-186.

Rattner, J.B. (1992) Integrating chromosome structure with function. **Chromosoma**, 101, 259-264.

Reeder, R.H. (1990) rRNA synthesis in the nucleolus. **Trends Genet.**, 6, 390-395.

Scheer, U. and Rose, K.M. (1984) Localization of RNA polymerase I in interphase cells and mitotic chromosomes by light and electron microscopic immunocytochemistry. **Proc. Natl. Acad. Sci.**, 81, 1431-1435.

Shi, L.et al. (1987) Involvement of a nucleolar component, perichromonucleolin, in the condensation and decondensation of chromosomes. **Proc. Natl. Acad. Sci. USA**, 84, 7953-7956.

Sollner-Webb, B. and Mougey, E.B. (1991) News from the nucleolus: rRNA gene expression. **TIBS**, 16, 58-62.

Sommerville, J. (1986) Nucleolar structure and ribosome biogenesis. **Trends Biochem. Sci.**, 11, 438-442.

Spector, D.L. and Smith, H.C. (1986) Redistribution of U-snRNPs during mitosis. **Exp. Cell Res.**, 163, 87-94.

Verheijen, R.et al. (1989) Ki-67 detects a nuclear matrix-associated proliferation-related antigen II. Localization in mitotic cells and association with chromosomes. **J. Cell Sci.**, 92, 531-540.

Yasuda, Y. and Maul, G.G. (1990) A nucleolar auto-antigen is part of a major chromosomal surface component. **Chromosoma**, 99, 152-160.

7 Ribosomal RNA gene expression and localization in cereals

A.R. LEITCH
Queen Mary and Westfield College, UK
and J.S. HESLOP-HARRISON
John Innes Centre, UK

Abstract
The rRNA genes are the most accessible genes for the study
of gene expression at a molecular, structural, and
ultrastructural level. This paper describes the occurrence
and distribution of rRNA genes in wheat, barley and rye.
These cereal species are closely related and
intercrossable but the rRNA genes show only a very loose
conservation to chromosomes of the same homoeologous group
which is surprising considering the close co-linearity of
single copy markers. The cereals have an apparent excess
in the number of genes required to sustain ribosome
synthesis. Some of the "redundant" genes may occur in
inactive rDNA loci that are not associated with a
nucleolus. Factors affecting locus suppression are
reviewed. Loci involved in rDNA transcription include
regions that contain actively transcribing genes and
regions that contain inactive genes. The active genes are
probably associated with the dense fibrillar component of
the nucleolus, while the inactive genes remain condensed.
In rye, the condensed, inactive genes occur in
perinucleolar chromatin and the active genes are
decondensed at the distal end of the rDNA locus. In wheat
condensed genes occur inside and outside the nucleolus
with each active locus showing several sites with
decondensed genes separated by condensed, intranucleolar
presumably inactive genes. In rye and wheat active rDNA
loci extend through the nucleolus and active rDNA
sequences do not loop out from perinucleolar chromatin.
Keywords: rDNA, rRNA genes, Nucleolus, Wheat, Rye, Barley,
Gene expression

1 Introduction

The genes specifying the ribosomal RNA (rRNA) genes can be
present in many thousands of copies in plant genomes. Each
rDNA unit contains one copy of the genes encoding the 18S,
5.8S and 26S ribosomal subunits and an extended intergenic
region (Fig. 1). The rDNA units occur in tandem arrays at

Chromosomes Today   Volume 11. Edited by A.T. Sumner and A.C. Chandley. Published in
1993 by Chapman & Hall, London. ISBN 0 412 47670 3

one or more chromosomal loci which may also be called the nucleolar organizing region (NOR), although some rDNA loci may not organize a nucleolus in some or all tissues.

The gene regions of the rDNA unit mostly show substantial sequence homology between plant species, presumably to conserve ribosome structure. Even between prokaryotes and eukaryotes there are extensive regions of sequence homology (see Appels and Honeycut 1986).

In contrast the intergenic region shows considerable sequence divergence between plants (Dvorak and Appels 1982). The intergenic region contains subrepeat sequences in a tandem array, which in wheat (<u>Triticum aestivum</u> L.) are about 135 bp long (see Flavell 1989). The repeat length, methylation status, sequence composition and copy number can vary both within a locus and between species, the variation giving rise to different lengths of the rDNA unit itself. In wheat each rDNA unit is about 9 kb long (Gerlach and Bedbrook 1979). The intergenic region contains the presumed gene promoter and gene terminator sequences (Flavell 1989).

In wheat cultivars, and within individual plants, there is variable methylation of the intergenic repeats within individual loci (Flavell et al. 1988). Active rDNA loci have a lower proportion of rRNA genes with specific CpG residues methylated. When such loci are transferred to a new genetic background where the locus activity is suppressed, then the proportion of genes within the locus that are methylated at these cytosines increases (Flavell 1986). Thus, methylation of these specific CpG residues is probably involved in the regulation of rRNA gene transcription (Flavell et al. 1988).

rDNA unit

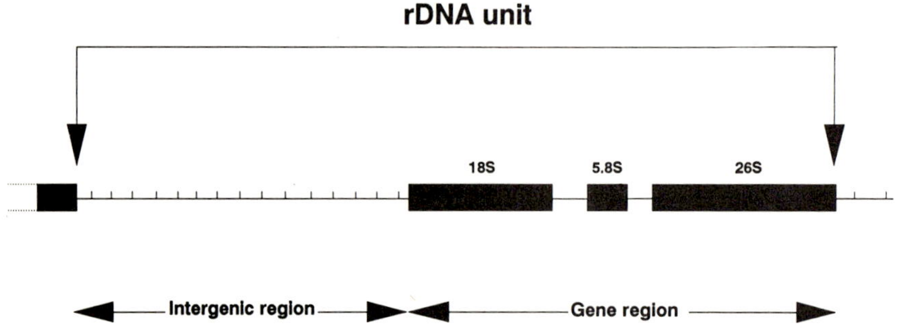

Fig. 1 The rDNA repeat unit

2 Occurrence of the rDNA repeat unit

The number of rDNA repeat units can vary in number both between and within a species. Aneuploid derivatives of the wheat variety "Chinese Spring" contain between

approximately 1,900 (haploid copy number of a nullisomic
6B tetrasomic 6A wheat line) and 7,300 (haploid copy
number of a tetrasomic 6B wheat line) rRNA genes per cell
(Flavell et al. 1988). Wheat has about 17,000 Mbp of DNA,
and therefore rDNA constitutes between 0.1% and 0.4% of
the wheat genome. In a long term wheat suspension culture,
a reduction of rRNA genes to about 1000 copies was
observed (Leitch et al. unpublished) suggesting that some
or all gene redundancy has been lost during the period of
culture. The high variability in gene copy number is also
found in maize, where a five-fold difference does not
appear to affect the amount of rRNA produced or the
general phenotype of the plants (Buescher et al. 1984).
These wide variations in rDNA content demonstrate loose
constraints on copy number although there is presumably a
minimum number to sustain ribosome synthesis and a maximum
rDNA content such that the cell cycle time and cell
resources are not adversely affected.

Gene dosage may also vary within and between cereal
tissues. The gametophyte generation has half the rDNA
content of diploid cells and cereal antipodal cells are
endopolyploid such that the nucleus in barley (<u>Hordeum
vulgare</u> L.) is 512C (Bennett 1984) and hence contains in
excess of 1,500,000 rRNA genes if there is no selective
loss of rDNA sequences. Rogers and Bendich (1987a) found
that in <u>Vicia</u> <u>faba</u> different tissues (e.g. leaves and
roots) can also have different numbers of rRNA genes. The
mechanisms or reasons for altered gene dosage in different
tissues are unknown.

The best way to determine the distribution of rDNA in a
genome is by using <u>in situ</u> hybridization. Mukai et al.
(1991) found rDNA sequences on chromosomes 1B, 6B, 5D, 1A
and 7D of wheat. Rye (<u>Secale</u> <u>cereale</u> L.) has around 3,000
rRNA genes (haploid copy number, see Rogers and Bendich
1987b), accounting for 0.3% of the genome, but only one
rDNA locus occurring on chromosome 1R (Appels et al.
1980). In barley, there are between 2,900 and 4,200 rRNA
genes (haploid copy number, see Rogers and Bendich 1987b)
with two major sites on chromosomes 6 (6I) and 7 (5I) with
approximately twice as many genes on chromosome 6 (Appels
et al. 1980), and additional minor sites (Leitch and
Heslop-Harrison 1992) on chromosome 5 (1I, 50-100 copies),
chromosome 1 (7I, 5-10 copies) and chromosome 2 (2I, 5-10
copies).

A comparison of the chromosomal distributions of the
genes between homoeologous chromosomes in cereals shows
that there are only weak similarities between different
genomes. This is surprising, especially considering that
there is near complete co-linearity of low copy/single
sequence to homoeologous chromosomes of the wheat and rye
genomes (Devos et al. 1992). An absence of conservation of
rDNA to homoeologous chromosomes may reflect different

genome organization constraints imposed on rRNA genes to those of low copy sequences.

The evolutionary constraints leading to the distributions and variabilities of tandem arrays of rRNA genes is unknown. Unequal crossing over at meiosis or unequal somatic recombination may play a role in changing the gene dosage at a particular locus. Chromosome organization is different in different tissues (e.g. Manuelidis and Borden 1988) and chromosomes may occupy specific domains of the nucleus when they are transcriptionally active (Bennett 1984, Leitch et al. 1991). An advantage of duplication of gene loci is that different gene sets can be activated in different nuclear organizations.

3 Expression between rDNA loci

The activity of rDNA loci in cereals has been examined using nucleolar volumes and number (Jordan et al. 1982; Martini and Flavell, 1985), silver staining (Moreno et al. 1990, Vieira et al. 1990a,b) and in situ hybridization (Appels et al. 1986a, Gustafson et al. 1988, Mukai et al. 1991, Leitch et al. 1992).

"Chinese Spring" wheat has ten rDNA loci (Mukai et al. 1991), but only chromosomes 6B and 1B are normally actively involved in nucleolus formation (Jordan et al. 1982; Martini and Flavell, 1985), whilst the minor loci on chromosome 5D, 1A and 7D usually show little transcriptional activity. When chromosome 1U from Aegilops umbellulata, which carries an rDNA locus, is added to the chromosome complement of the wheat "Chinese Spring", the wheat rDNA loci are largely suppressed and the 1U locus is the active locus (Martini et al. 1982). Flavell et al. (1988) compared nucleolar volume in various wheat genotypes with gene methylation status and showed that the most active rDNA loci have the highest proportion of genes with non-methylated CCGG sequences and *vice versa*.

Hybrids between wheat and rye often show suppression of activity of the NOR on chromosome 1R of rye when in competition with either of the two major nucleolar organizing chromosomes of wheat origin (Lacadena et al. 1988, Fig. 2b). The genetic control of activity involves the presence of chromosome arms other than the one carrying the nucleolar organizer region (Vieira et al. 1990b). Cytosine methylation correlates with locus activity, since treatment with 5-azacytidine activates the loci of rye origin that are normally quiescent (Vieira et al. 1990a). The organization of the interphase nucleus also correlates with the activity of particular rDNA loci. Nicoloff et al. (1979) showed in barley lines with reciprocal translocations that NOR's translocated to non-NOR bearing chromosomes show suppression of the locus.

This suppression may be due to nuclear architecture
constraints imposed by the abnormal karyotype (Bennett
1984). Bennett (1984) showed in various hybrids between
grass species that the nuclear location of the rRNA genes
in somatic cells correlates with the activity of the locus
such that there was a tendency for the nucleolar
organizers of the central genome to be expressed and the
peripheral genome to be suppressed.

Viegas, Silver and Queiroz (poster presentation at this
symposium) have showed that the expression of rDNA loci is
also cell type dependent. In Triticale somatic cells (root
tip and pre-meiotic), the two major wheat loci are active
whilst the rye rDNA locus is inactive. However at pollen
grain mitosis the rye NOR showed silver staining
indicative of locus activity at the preceding interphase.
These data show that locus inactivity is reversible during
development and that genes may become activated upon
demand for ribosomes.

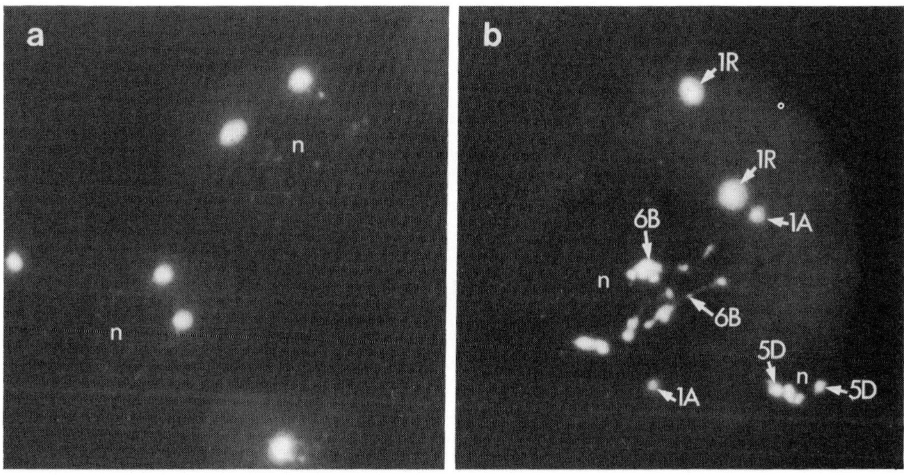

Fig. 2. *In situ* hybridization of rDNA probes to root tip
interphase nuclei of (a) rye variety "Petkus Spring" and
(b) wheat variety "Beaver" which includes the short arm of
chromosome 1R of rye translocated onto the long arm of
chromosome 1B of wheat. In rye most signal is
perinucleolar with diffuse signal within the nucleolus
(n). In wheat the major loci on chromosomes 6B are most
active, with a low level of activity on the chromosome 5A
locus. Condensed rRNA genes with strong *in situ* signal are
found in the nucleoli (n). The rRNA genes of chromosome 1R
origin and on chromosome 1A are inactive in this cell. The
probable identities of rDNA loci were determined from
morphology and expected activity (Leitch et al. 1992,
Martini and Flavell, 1985).

4 Expression within rDNA loci

Nucleoli enlarge and fuse during the cell cycle
demonstrating a dynamic morphology that changes with cell
activity (see Jordan et al. 1982). The nucleolus is
located towards the centre of the nucleus in somatic cells
of different cereal species. However, near to the start of
meiosis the nucleolus moves to a peripheral position,
often in contact with the nuclear envelope (Bennett 1984).
During the different cell stages of meiosis the morphology
and organization of the nucleolus and the activities of
rRNA genes alter (Williams et al. 1973).

In cereals, regulation of gene activity occurs not only
by the suppression of whole loci but also genes within
individual loci (Flavell and O'Dell, 1990, Leitch et al.
1992). rRNA gene expression can not only be studied from a
molecular biologists perspective, but also at the
cytogenetic and ultrastructural level because
transcription occurs within the nucleolus of the
interphase nucleus. There are several recent reviews on
the ultrastructure of nucleoli (Goessens 1984; Risueño and
Medina 1986; Schwarzacher and Wachtler 1986; Deltour and
Motte 1990; see the Wachtler, Mosgöller, Schöfer et al.
chapter) and the distribution of their proteins (e.g.
Hernandez-Verdun 1991; see Hernandez-Verdun, Roussel and
Gautier chapter) and these will not be addressed here.

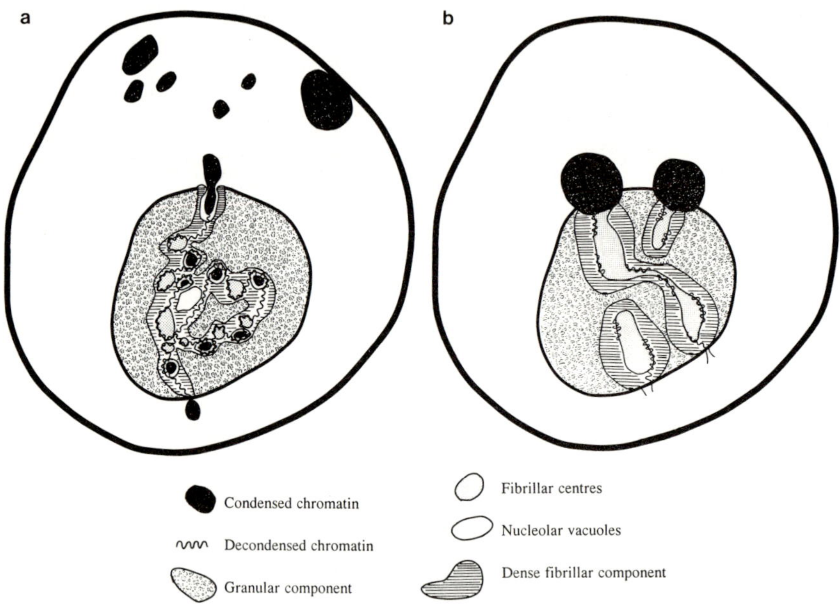

Fig. 3 Diagrammatic representation of the location of rDNA
sequences in interphase nuclei of (a) wheat and (b) rye.
The intranucleolar components are defined in the key.

The detection of rRNA genes using <u>in situ</u> hybridization and the imaging of cells, nuclei and chromatin using light and electron microscopy enables individual features of the active rDNA loci to be examined in wheat and rye. Models for rRNA gene expression patterns within the nucleolus are shown in Fig. 3. In rye, both rDNA loci are usually active in root tip nuclei and the gene units at the distal (telomeric) end of the locus are those most often expressed (Fig. 2a, 3b). These expressed genes are decondensed and found inside the nucleolus, probably associated with the dense fibrillar component. The inactive genes remain condensed and perinucleolar (Leitch et al. 1992). In wheat, activity occurs at decondensed genes that are probably associated with the dense fibrillar component. The active genes are dispersed at a number of sites along each active rDNA locus, an expression pattern that leaves "islands" of condensed genes within the nucleolus (Fig. 2b, 3a). Inactive condensed genes are also found outside the nucleolus (Leitch et al. 1992).

Root tip interphase nuclei of cereals are arranged in a Rabl (1885) configuration with the telomeres clustered at one pole of the nucleus and centromeres at the other (Anamthawat-Jónsson and Heslop-Harrison 1990). Each active locus within wheat and rye follows this Rabl orientation and extends through the nucleolus from perinucleolar condensed chromatin. The same rDNA carrying chromatin axis emerges from the nucleolus at the opposite side to the entry point. Thus the chromosome arm runs through the nucleolus and does not loop back upon itself.

It is suggested that increased rDNA transcriptional activity is correlated with decondensation (and demethylation) of rDNA sequences in the cereals. There may also be increased transcriptional activity at the decondensed sequences, particularly at increased temperatures, as occurs in bacteria (Nomura et al. 1984). In rye, activation of genes would result in genes outside the nucleolus being incorporated into the nucleolus. In wheat, genes from outside the nucleolus may be activated as in rye, but increased activity can also occur by decondensation of intranucleolar chromatin leading to fewer intranucleolar condensed genes.

5 Acknowledgements

We thank BP and Venture Research International for support. We also thank Dr W. Mosgöller and Mrs M. Shi for help with the data presented in this paper.

6 References

Anamthawat-Jónsson, K. and Heslop-Harrison, J.S. (1990) Centromeres, telomeres and chromatin in the interphase nucleus of cereals. **Caryologia** 43, 205-213.

Appels, R., Gerlach, W.L., Dennis, E.S., Swift, H. and Peacock, W.J. (1980) Molecular and chromosomal organization of DNA sequences coding for ribosomal RNAs in cereals. **Chromosoma** 78, 293-311.

Appels, R. and Honeycut, R.L. (1986) rDNA: evolution over a billion years, in **DNA Systematics Volume II: Plants**, (ed Dutta, S.K.), CRC Press, Florida, pp.81-135.

Bennett, M.D. (1984) Nuclear architecture and its manipulation, in **Gene manipulation in plant improvement** (ed Gustafson JP), Plenum Publishing Corporation, New York, pp. 469-502.

Buescher, P.J., Phillips, R.L. and Brambl, R. (1984) Ribosomal RNS contents of maize genotypes with different ribosomal RNA gene numbers. **Biochem. Genet.** 22, 923-930.

Deltour, R. and Motte, P. (1990) The nucleolonema of plant and animal cells: a comparison. **Biol. Cell** 68, 5-11.

Devos, K.M., Atkinson, M.D., Chinoy, C.N., Liu, C.J. and Gale, M.D. (1992) RFLP-based genetic map of the homoeolgous group 3 chromosomes of wheat and rye. **Theor. Appl. Genet.** 83, 931-939.

Dvorák, J. and Appels, R. (1982) Chromosome and nucleotide sequence differentiation in genomes of polyploid <u>Triticum</u> species. **Theor. Appl. Genet.** 63, 349-360

Flavell, R.B. (1986) The structure and control of expression of ribosomal RNA genes. **Oxford Surveys of Plant Mol. and Cell Biol.** 3, 252-274.

Flavell, R.B. (1989) Variation in structure and expression of ribosomal DNA loci in wheat. **Genome** 31, 963-968.

Flavell, R.B. and O'Dell, M. (1990) Variation and inheritance of cytosine methylation patterns in wheat at the high molecular weight glutenin and ribosomal RNA gene loci. **Development** Supplement, 15-20.

Flavell, R.B., O'Dell, M. and Thompson, W.F. (1988) Regulation of cytosine methylation in ribosomal DNA and nucleolus organizer expression in wheat. **J. Mol. Biol.** 204, 523-534.

Gerlach, W.L. and Bedbrook, J.R. (1979) Cloning and characterization of ribosomal RNA genes from wheat and barley. **Nuc. Acids Res.** 7, 1869-1885.

Goessens, G. (1984) Nucleolar structure. **Int. Rev. Cytol.** 87, 107-158.

Gustafson, J.P., Dera, A.R. and Petrovic, S. (1988) Expression of modified rye ribosomal RNA genes in wheat. **Proc. Natl. Acad. Sci.** USA 85, 3943-3945.

Hernandez-Verdun, D. (1991) The nucleolus today. **J. Cell Sci.** 99, 465-471.

Jordan, E.G., Martini, G., Bennett, M.D. and Flavell, R.B. (1982) Nucleolar fusion in wheat. **J. Cell Sci.** 56, 485-495.

Lacadena, J.R., Cremeño, M.C., Orellana, J. and Santos, J.L. (1988) Nucleolar Competion in <u>Triticeae</u>, in **Kew Chromosome Conference III.** (ed. Brandham P.E.), HMSO, London, pp. 151-165.

Leitch, I.J. and Heslop-Harrison, J.S. (1992) Physical mapping of 18S-5.8S-26S rRNA genes in barley by <u>in situ</u> hybridization. **Genome** (in press).

Leitch, A.R., Mosgöller, W., Shi, M. and Heslop-Harrison, J.S. (1992) Different patterns of rDNA organization at interphase in nuclei of wheat and rye. **J. Cell Sci.** 101, 751-757.

Leitch, A.R., Schwarzacher, T., Mosgöller, W., Bennett, M.D. and Heslop-Harrison, J.S. (1991) Parental genomes are separated throughout the cell cycle in a plant hybrid. **Chromosoma** 101, 206-213.

Martini, G., O'Dell, M and Flavell, R.B. (1982) Partial inactivation of wheat nucleolar organisers by the nucleolus organiser chromosomes from <u>Aegilops umbellulata</u>. **Chromosoma** 84, 687-700.

Manuelidis, L. and Borden, J. (1988) Reproducible compartmentalization of individual chromosome domains in human CNS cells revealed by <u>in situ</u> hybridization and three-dimensional reconstruction. **Chromosoma** 96, 397-410.

Martini, G. and Flavell, R.B. (1985) The control of nucleolus volume in wheat, a genetic study at three developmental stages. **Heredity** 54, 111-120.

Nicoloff, H., Anastassova-Kristeva, M., Rieger, R., and Künzel G. (1979) 'Nucleolar dominance' as observed in barley translocation lines with specifically reconstructed SAT chromosomes. **Theor. Appl. Genet.** 55, 247-251.

Moreno, F.J., Rodrigo, R.M. and Garcia-Herdugo, G. (1990) Ag-NOR proteins and rDNA transcriptional activity in plant cells. **J. Histochem. Cytochem.** 38, 1879-1887.

Mukai, Y., Endo, T.R. and Gill, B.S. (1991) Physical mapping of the 18S.26S rRNA multigene family in common wheat: Identification of a new locus. **Chromosoma** 100, 71-78.

Nomura, M., Gourse, R. and Baughman, G. (1984) Regulation of the synthesis of ribosomes and ribosomal components. **Ann. Rev. Biochem.** 53, 75-117.

Rabl, C. (1885) Über Zelltheilung. **Morphol. Jahrb.** 10, 214-330.

Risueño M.C. and Medina F.J. (1986) The nucleolar structure in plant cells. **Cell Biology Reviews (RBC)** 7, 1-154.

Rogers, S.O. and Bendich, A.J. (1987a) Heritability and Variability in ribosomal RNA genes of <u>Vicia faba</u>. **Genetics** 117, 285-295.

Rogers, S.O., and Bendich, A.J. (1987b) Ribosomal RNA genes in plants: variability in copy number and in the intergenic spacer. **Plant Mol. Biol.** 9, 509-520.

Schwarzacher, H.G., and Wachtler, F. (1986) Nucleolus organizer regions and nucleoli: Cytological findings. **Chromosomes Today** 9, 252-260.

Vieira, R. Queiroz, A., Morais, L., Barao, A., Mello-Sampayo, T. and Viegas, W. (1990a) 1R chromosome nucleolus organizer region activation by 5-azacytidine in wheat x rye hybrids. **Genome** 33, 707-712.

Vieira, R. Queiroz, A., Morais, L., Barao, A., Mello-Sampayo, T. and Viegas, W. (1990b) Genetic control of 1R nucleolus organizer region expression in the presence of wheat genomes. **Genome** 33, 713-718.

Williams, E., Heslop-Harrison, J. and Dickinson, H.G. (1973) The activity of the nucleolus organising region and the origin of cytoplasmic nucleoloids in meiocytes of *Lilium*. **Protoplasma** 77, 79-93.

8 Compartmentalization of the cell nucleus: case of the nucleolus

I. RASKA and M. DUNDR

Czechoslovak Academy of Sciences, Czechoslovakia

1 Introduction

The nucleolus is the most prominent nuclear organelle. Only one major
function is ascribed to the nucleolus: here ribosomal RNA (rRNA) is
transcribed, processed and assembled into preribosomal particles
(Reeder, 1990, Warner, 1990). The growing eukaryotic cell contains
thousands of different transcripts, but the nucleolar rRNA synthesis
accounts for nearly half of the cellular transcriptional activity.
Besides tandemly repeated rRNA genes, the presence of more than 100 ribosomal proteins, nucleolar proteins and ribonucleoproteins (RNPs) is
required for the nucleolar function, including RNA polymerase I, DNA
topoisomerases, transcription factors, processing enzymes, 5S rRNA and
U3, U8, U13 and U14 small nucleolar RNAs (snoRNAs) (Reeder, 1990, Warner, 1990).

By means of conventional electron microscopy (EM), three basic nucleolar components are identified: 1) relatively electron lucent fibrillar centers (FCs) which are thought to be interphasic counterparts of
nucleolar organizer regions (NORs) of mitotic chromosomes, 2) dense
fibrillar components (DFCs) surrounding usually FC and apparently containing transcripts of rRNA genes, 3) granular components (GCs) which
are believed to be, at a least in part, preribosomes. Besides these
components, two further intranucleolar components are described: nucleolar interstices of low electron density as well as intranucleolar condensed chromatin structures (e.g. Smetana and Busch, 1974, Goessens,
1984, Derenzini et al., 1990, Jordan, 1991, Hernandez-Verdun, 1991). It
is clear that the identification of just a few structural components is
not sufficient for a description of the detailed nucleolar structure
-function compartmentalization. Nevertheless, there is a general
agreement that FCs, DFCs and GCs correspond in some way to different
(and vectorial) events in the transcription of rRNA and formation of

---

Abbreviations: DFC=Dense fibrillar component, DNase I=Deoxyribonuclease
I, EM=Electron microscopy, EMAC=Ultrastructural immunocytochemistry and
in situ hybridization (ultrastructural affinity cytochemistry),
EMARG=Electron microscopic autoradiography, EMI=Ultrastructural immunocytochemistry, FC=Fibrillar center, GC=Granular component, ISH=*In situ*
hybridization, m$_3$G RNA=RNA containing a trimethylguanosine cap,
NOR=Nucleolar organizer region, rDNA=Ribosomal DNA, RNP=Ribonucleopro-
tein, rRNA=Ribosomal RNA, snoRNA=Small nucleolar RNA, TdT=Terminal deoxynucleotidyl transferase, UBF=Upstream binding factor.

Chromosomes Today   Volume 11. Edited by A.T. Sumner and A.C. Chandley. Published in
1993 by Chapman & Hall, London. ISBN 0 412 47670 3

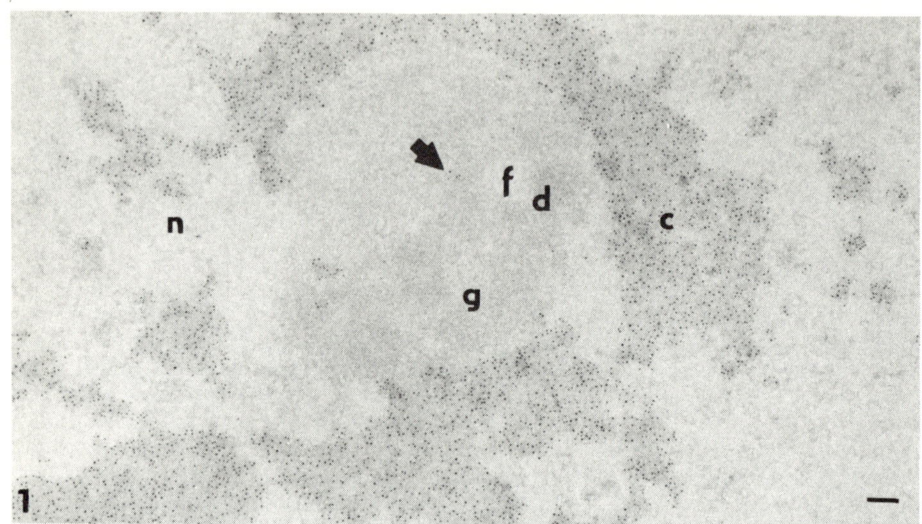

Fig. 1. Rat hepatocyte fixed in aldehyde and embedded in lowicryl K4M
is immunostained with a monoclonal antibody to DNA. The bound antibodies are visualized by the secondary antimouse antibodies coupled to 5
nm gold particles (Amersham Company; secondary antibodies bound to colloidal gold are used in most consecutive Figures, an alternative use of
protein A-gold complexes is specified). Condensed chromatin is heavily
labelled. Note the presence of a few gold particles over DFCs (arrow)
within the nucleolus. Abbreviations used in this and consecutive Figures are: cytoplasm=cy, nucleoplasm=n, FC=f, DFC=d, GC=g, nucleolar in-
terstices=i, chromatin structures=c. The bar in this and all
consecutive Figures correponds to 0.1 micrometer.

preribosomes. This conclusion has been reached on the basis of electron
microscopic autoradiography (EMARG) using tritiated uridine incorporation. A large scale nucleolar compartmentalization had been established
already in the mid sixties. The EMARG signal is first observed over
DFCs and with a prolongation of uridine incorporation, it moves to GCs
as well (Granboulan and Granboulan, 1965). These findings of principle
have been repeatedly confirmed by other groups and in today's EMARG
picture of nucleoli, the autoradiographic grains due to a short uridine
incorporation are located to DFCs or to both DFCs and peripheral part
of FCs (e.g. Fakan and Puvion, 1980, Puvion and Moyne, 1981, Derenzini
et al., 1987, Wachtler et al., 1990, Thiry and Goessens, 1991).
However, the detailed picture of the nucleolar subcompartmentalization
is still controversial, since different groups situate the transcribing
rRNA genes either to the peripheral part of FCs or to DFCs (see also
Jordan (1991) for a discussion of other possibilities). Of primordial
importance are the results of *in situ* hybridization of various ribosomal DNA (rDNA) sequences (see the paper of Wachtler et al. of this volume for a detailed discussion).
   The immunocytochemical (EMI) and *in situ* hybridization (ISH) methods
depicting specific macromolecules and macromolecular complexes repre-

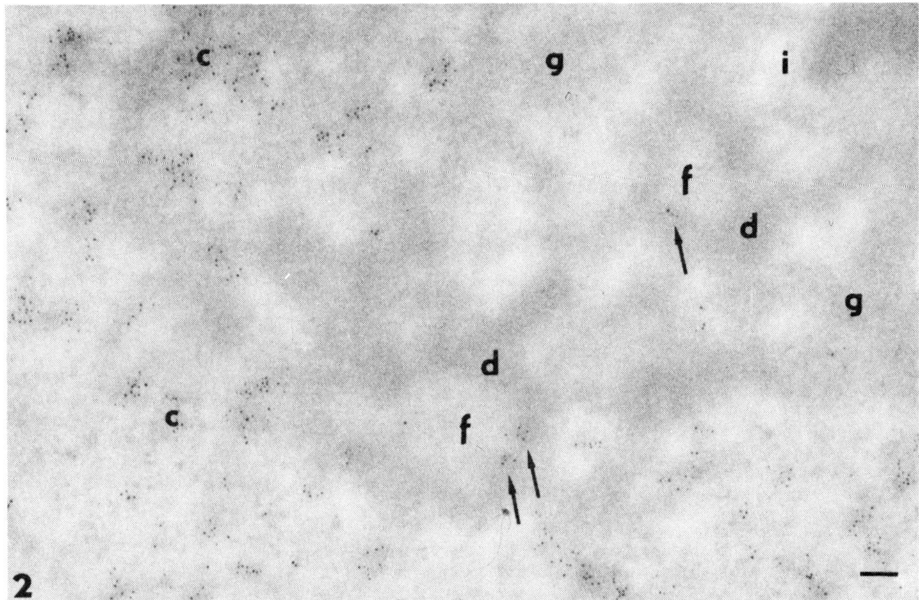

Fig. 2. EMI detection of DNA in thin sectioned HeLa cell embedded in lowicryl. Some of the intranucleolar 5 nm gold particles are clearly associated with DFCs (arrows).

sent in our opinion the most convenient means how to deepen the knowledge about the *in situ* functional morphology of the nucleolus. In the present review we attempt to document the most important colloidal gold affinity cytochemistry (EMAC) data on the nucleolus, particularly that of cycling mammalian cells, with an emphasis on the localization of nucleolar chromatin as well as proteins, RNAs and RNPs which are more or less directly related to the synthesis and processing of rRNA, and the formation of preribosomes. For the localization of rDNA sequences, the behaviour of nucleolar components during mitosis and the nucleolar organization in plants we refer to detailed papers of this volume by Wachtler et al., Hernandez-Verdun et al., and Leitch et al.

2 Localization of nucleolar chromatin

Much endeavour has been put in the visualization of nucleolar chromatin. One reason for it is the fact that the existence of intranucleolar chromatin has been questionable for a long time in the cytological literature (see Busch and Smetana (1970) for a review).

We are of opinion that the localization of nucleolar chromatin by immunocytochemistry has a limited value for the identification of rRNA transcription sites (Raska et al., 1988; see Raska et al., 1990 for a detailed discussion). Ribosomal DNA sequences represent only a small fraction, in order of one or at most a few %, of the total intranucleolar DNA (Bachellerie et al., 1977).

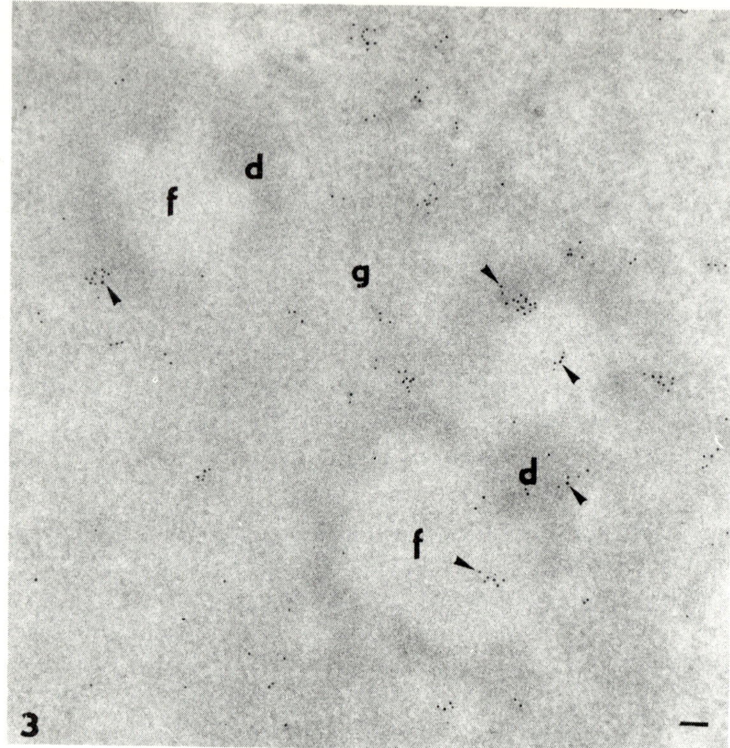

Fig. 3. *In situ* nick-translation performed on thin sectioned HeLa cell embedded in lowicryl. 5 nm gold particles (arrowheads) reveal the presence of incorporated uridine tagged with biotin. Intranucleolar label is associated with both FCs and DFCs.

By means of EMI, we get compatible results with probes to double-stranded DNA, single-stranded DNA, double- and single-stranded DNA, histones, and nucleohistones such as complexes of DNA with histones H2A and H2B (Raska et al., 1988, 1990, 1992, Dundr et al., in preparation). The label is found on condensed intranucleolar chromatin present most frequently in nucleolar interstices at boundaries with FCs, DFCs and GCs (Figs. 1, 2). Even though not every individual FC and/or DFCs is labelled, a low label is also regularly found not only within FCs, but also within DFCs (Figs. 1, 2) of mammalian tissue cells as well as of cultured cells (Raska et al., 1990, 1992, Dundr et al., in preparation).

Previous results have demonstrated the nucleosomal arrangement of rRNA (Reeves, 1988, Cartwright and Elgin, 1988), while other data favour a picture of transcribing rRNA genes being devoid of histones (Scheer and Zentgraf, 1982, Labhart and Koller, 1982, Conconi et al., 1989). One group of EMI results favours the nucleosomal arrangement of ribosomal chromatin (Thiry and Muller, 1989), while other results, supported in addition by cytochemical osmium amine staining of DNA, are in favour of non-nucleosomal arrangement (Derenzini et al., 1987). In our

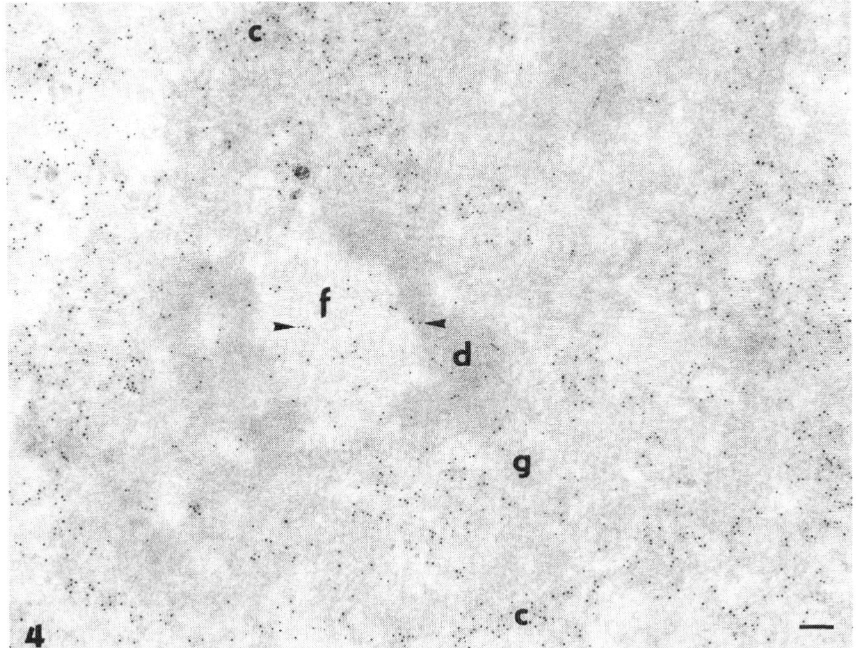

Fig. 4. TdT experiment performed on a thin sectioned HeLa cell embedded in lowicryl. 5 nm gold particles (arrowheads) reveal the presence of bromodeoxyuridine incorporated within DNA. Chromatin structures are stained, within the nucleolus, both FCs and DFC are labelled .

hands, we get a comparable distribution of the label with probes both to DNA and to (nucleo)histones (Raska et al, 1988, 1990, 1992, Dundr et al., in preparation). Our conclusion is, however, that the EMI techniques used do not allow one to claim unequivocally the presence or absence of histones in the (transcribing) ribosomal chromatin (Raska et al., 1990).

Two further immunocytochemical results concerning the visualization of nucleolar DNA, obtained by *in situ* nick-translation and terminal deoxynucleotidyl transferase (TdT) methods applied to thin sectioned material, have to be documented. Deoxyribonuclease I (DNase I) sensitive DNA domains (e.g. Graaf et al., 1990), such as one would expect for active rRNA genes, have been observed in FCs and not in DFCs of Ehrlich ascites cells, by means of nick-translation (Thiry, 1991). This approach employs the exonuclease activity of an exogenous DNA polymerase at DNase I nicks for the incorporation of biotinylated bases into DNA. The presence of biotins is subsequently revealed by immunocytochemistry. In our hands, the DNA label due to nick-translation is, in HeLa cells, observed in FCs as well as in DFCs (Fig. 3; Raska et al., 1992, Dundr et al., in preparation). Furthermore, we cannot reconcile the label in nick-translation experiments with sensitive DNase I sites only as we observe a uniform label on mitotic chromosomes under a mild DNase I treatment (Dundr et al., in preparation). We expect that a number of

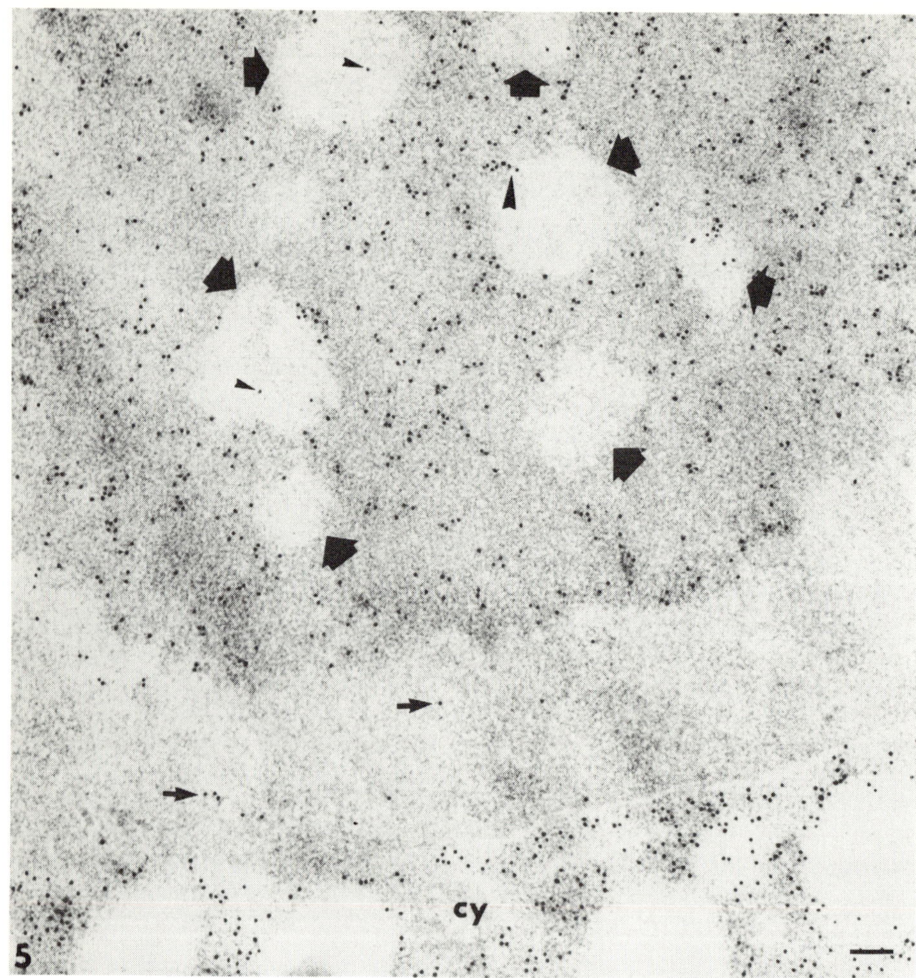

Fig. 5. Visualization of rRNAs by ISH using an *Arabidopsis* antisense
RNA probe encompassing 18S rRNA region. Meristematic onion root cells
have been aldehyde fixed and embedded in lowicryl. Most gold particles
are found over ribosomes and GCs and DFCs of the nucleolus. The area
corresponding to FCs (thick arrows) are in many cases not labelled and
in some cases where gold particles are present, the gold particles form
clusters (larger arrowheads) that also cover DFCs. The label closer to
the middle part of FCs usually corresponds to individual gold particles
(smaller arrowheads). A low label is also present in the nucleoplasm
(thin arrowheads) (Photo courtesy of A. Olmedilla and M.-C. Risueno).

additional sites for the nick-translation arises due to the cell processing for electron microscopy, particularly to the cutting of thin
sections (Dundr et al., in preparation).

In the TdT method, thin sections are incubated in a medium conta-

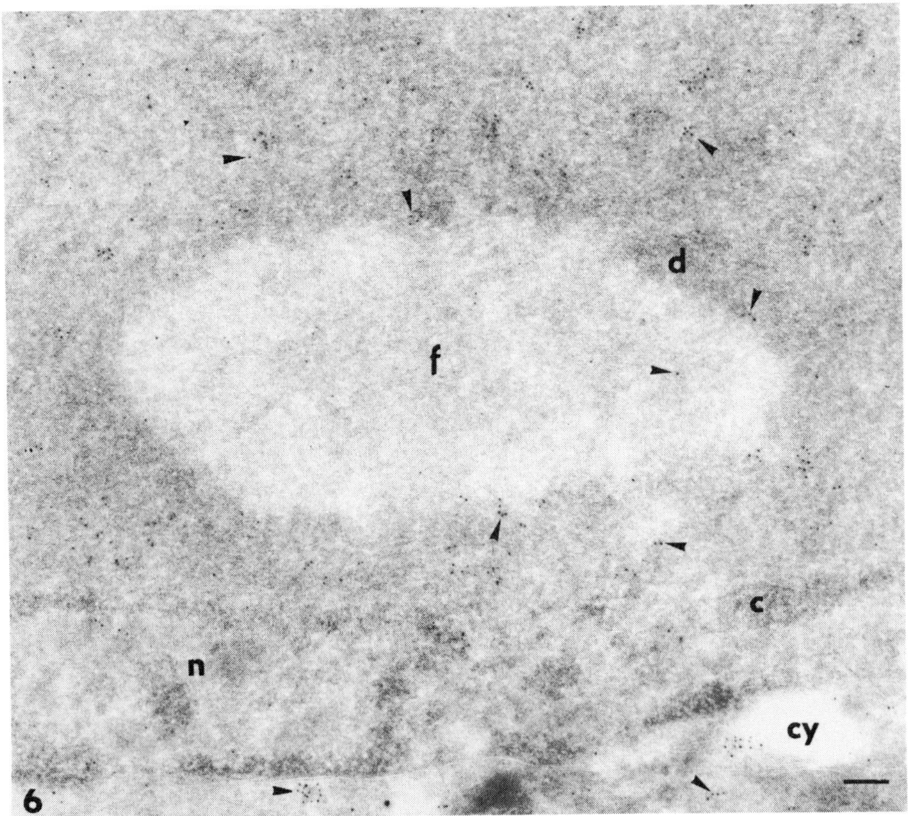

Fig. 6. Visualization of 28S rRNA by ISH employing a stable oligonucleotide probe in a HeLa cell cryosection. After aldehyde fixation, the
cells are infused with high molar sucrose, frozen in liquid nitrogen
and cut. After performing ISH, thin cryosections are postembedded in a
mixture of methyl cellulose with uranyl acetate (e.g. Tokuyasu et al.,
1986). High label (5 nm gold particles; arrowheads) is present over cytoplasm and over nucleolar GCs and DFCs. Only very few gold particles
are present within FCs.

ining bromodeoxyuridine triphosphate and TdT (Thiry, 1992). The free
3'-OH ends of DNA generated by the cutting procedure are labelled by
bromodeoxyuridine and the incorporated bases are detected immunocyto-
chemically. In a recent study, DNA has been demonstrated in FCs and not
in DFCs of nucleoli of several cell types (Thiry, 1992). We have repeated these experiments and reached a slightly different conclusion. In
our hands, the label is present both in FCs and in DFCs of HeLa cells
(Fig. 4; Dundr et al., in preparation).

3 Localization of nucleolar proteins, RNAs and RNPs

We shall limit the discussion to those nucleolar macromolecules and
macromolecular complexes the role of which in the metabolism of rRNA

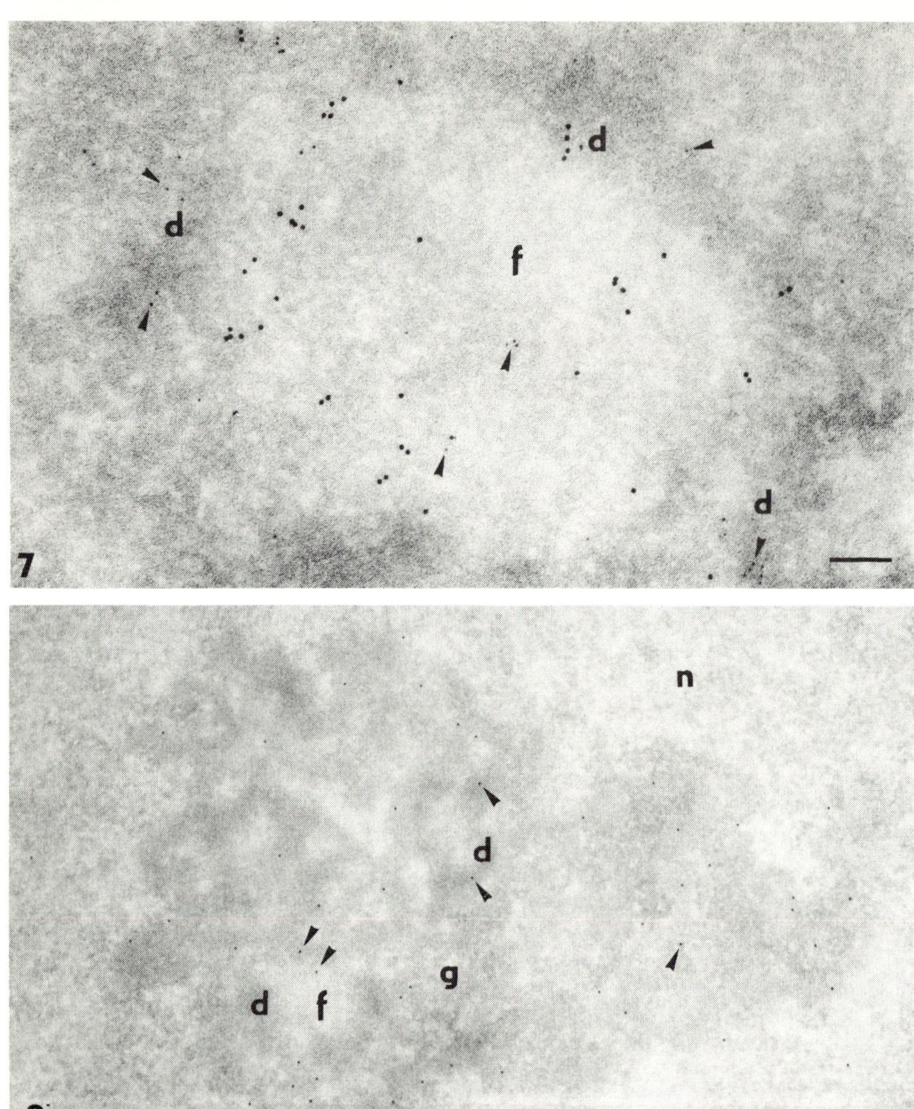

Fig. 7. Cryosection of a rapidly proliferating rat venous cell immuno-
labeled for RNA polymerase I (10 nm particles) and for DNA topoisomerase I (5 nm particles; arrowheads). Most 10 nm particles are present
within a distinct FC, whereas 5 nm particles (arrowheads) are located
more in DFCs.

Fig. 8. Cryosection of a HeLa cell nucleolus immunostained for DNA topoisomerase II. The label (10 nm particles; arrowheads) is enriched in
DFCs with respect to FCs.

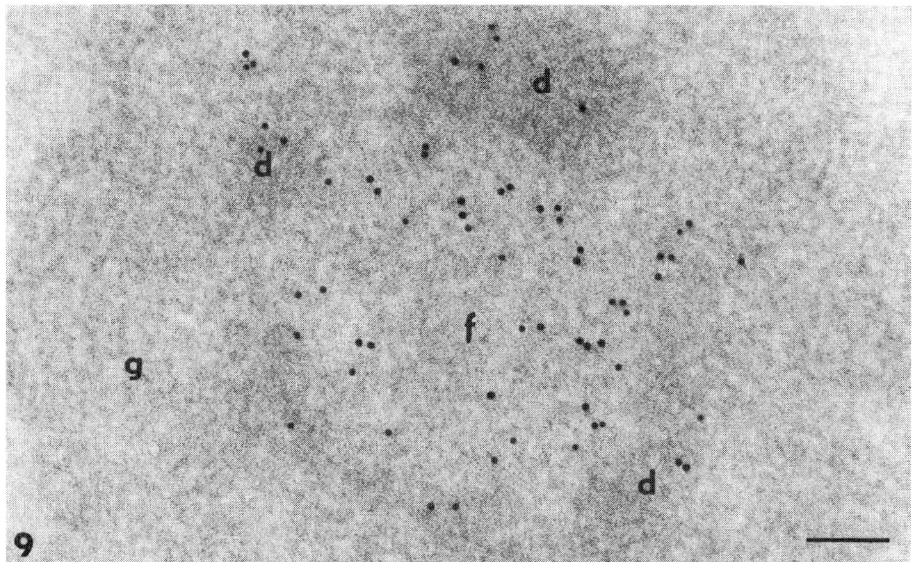

Fig. 9. Thin sectioned HeLa cell embedded in lowicryl is labelled with anti-NOR 90 antigen antibodies. The bound antibodies are visualized by protein A-gold complex. Gold partiles are present both in FC and DFCs (Photo courtesy of F.J. Moreno).

and the formation of preribosomes is more or less established. They are: rRNA, RNA polymerase I, DNA topoisomerases, upstream binding factor (UBF), snoRNAs and snoRNPs (fibrillarin), 5S rRNP, 5S rRNA, nucleolin, protein B23 and ribosomal proteins.

By means of ultrastructural ISH, various biotinylated probes to 18S and 28S rRNA have been mapped in both animal and plant cells (Fig. 5; Thiry and Thiry-Blaise, 1989, Escaig-Haye et al., 1989, - Puvion-Dutilleul et al., 1991b). With one exception (Escaig-Haye et al., 1989), most of the label is clearly localized over cytoplasmic ribosomes and nucleolar DFCs and GCs. A few particles are observed also over FCs (Fig. 5). In the paper of Escaig-Haye et al, (1989), the nucleolar label is enriched in DFCs, not in GCs. Interestingly, the labelling signal due to probes hybridizing the 5' external transcribed spacer of precursor rRNA, the cleavage of which is an early event in the maturation pathway of precursor rRNA, is confined mainly to DFCs (Puvion-Dutilleul et al., 1991b), or to FCs and DFCs (Fischer et al., 1991). We have used ISH approach employing a stable (2'-o-allyl RNA) antisense biotinylated oligonucleotide to 28S rRNA (Carmo-Fonseca et al., 1992). The results obtained are in harmony with those studies obtained with genomic probes in which most of the nucleolar label is associated both with DFCs and GCs (Fig. 6; Dundr et al., in preparation). An alternative EMI approach has been used to localize rRNA by means of a specific monoclonal antibody (Lerner et al., 1981). In this case, the label is enriched over cytoplasmic ribosomes and nucleoli. Within the nucleolus, the label present over DFCs and GCs is significantly higher than that in FCs (Raska et al., 1985).

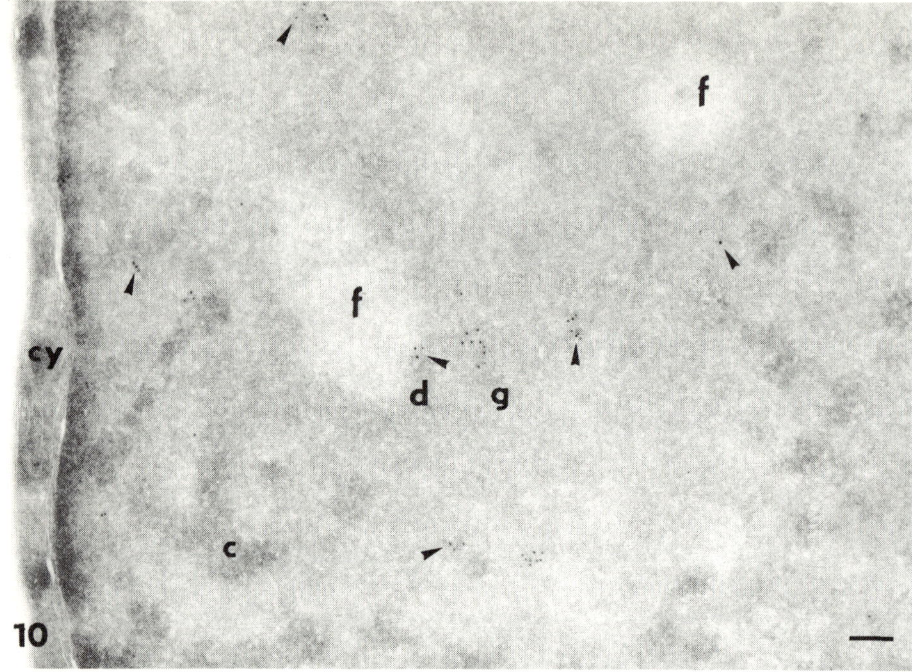

Fig. 10. Cryosection of a HeLa cell nucleolus. ISH with a stable antisense oligonucleotide to U3 snoRNA is perfomed. The label (5 nm particles; arrowheads) is present in GCs and DFCs. In this particular nucleolar section, FCs are not labelled. Interestingly, nuclear coiled bodies also exhibit a signal due to U3 snoRNA (Dundr et al., submitted). Coiled bodies should be then involved in the metabolism of both small nuclear RNAs (Raska et al., 1991) and snoRNAs.

RNA polymerase I is the enzyme transcribing rRNA genes. Immunolocalization of RNA polymerase I by means of specific antibodies (Reimer et al., 1987a) yields a prominent signal over FCs. This is particularly well seen in cells which exhibit clearly delineated compact nucleoli with distinct FCs, such as HeLa cells or rat PC12 cells (Reimer et al., 1987b, Raska et al., 1989, 1990). Nevertheless, a variable number of gold particles is also observed over DFCs. If expressed in area density i.e. the number of gold particles per unit area, the occurrence of gold particles in DFCs of HeLa cells is most frequently found to correspond to 10-30% of those encountered in FCs (Dundr et al., in preparation). In nucleoli with nucleolonemata (reticulated nucleoli) such as in rapidly proliferating rat venous cells or mouse P815 cells, the labelling of DFCs is increased relative to the findings with compact nucleoli (Fig. 7; Raska et al., 1989, 1990). Therefore, our conclusion is that the RNA polymerase I is mapped to FCs with some label being also present in DFCs.

DNA topoisomerase I is an enzyme activating rRNA genes (Wang, 1985, Rose et al., 1988). Müller et al. (1985) detected the increased inci-

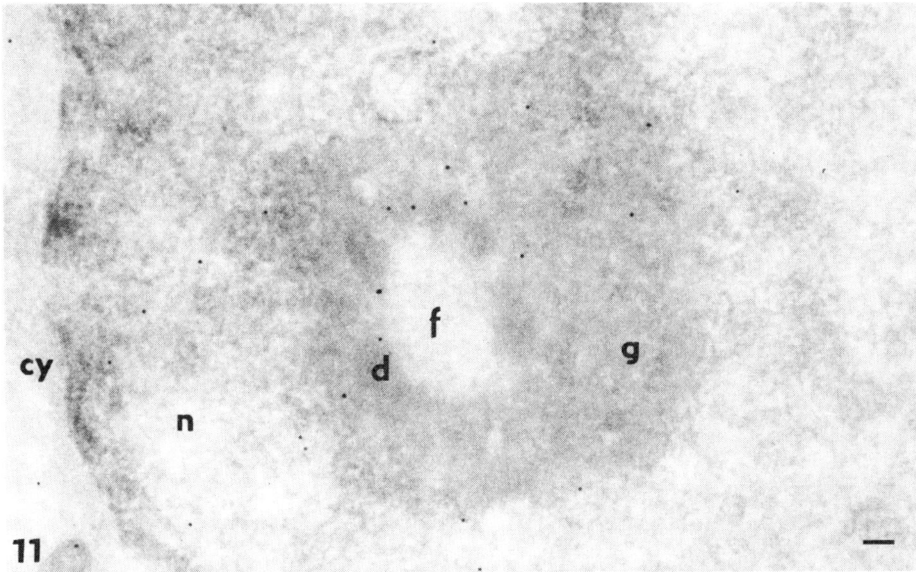

Fig. 11. Cryosectioned HeLa cell nucleolus. 5S rRNP complex is visualized by means of specific autoantibodies. 10 nm gold particles (arrowheads) are present in GCs and DFCs, not in FC.

dence of this enzyme in nucleoli of chicken cells, but not within FCs. In our hands, we observe in various cell lines an increased nucleolar signal, particularly within DFCs (Fig. 7; Raska et al., 1989, 1990). Furthermore, the 180-kD isoform of DNA topoisomerase II has been hypothesized to be involved in rRNA activation (Drake et al., 1989) and has been recently mapped to nucleoli of HeLa and K562 cells, mainly to DFCs (Zini et al. 1992). In our hands, the label due to this enzyme is compatible with the findings of Zini et al. (1992) and is higher in DFCs than in FCs (Fig. 8).

The UBF controlling the activity of RNA polymerase I (Sollner-Webb and Mougey, 1991) has been shown to be identical with the NOR 90 antigen (Rodriguez-Sanches et al., 1987, Chan et al., 1991). This antigen has been recently immunolocalized at the electron microscopical level both to FCs and DFCs of HeLa cells (Fig. 9; Rendon et al., 1992, Raska et al., 1992).

U3, U8, U13 and U14 snoRNAs contain a trimethylguanosine (m_3G) cap, and RNPs containing these snoRNAs are apparently involved in the processing of precursor rRNAs (Kass et al., 1990, Savino and Gerbi, 1990). ISH experiments by means of biotinylated probes to U3 snoRNA provide in various animal cell lines a nucleolar signal with gold particles being present over all nucleolar components. The highest incidence of gold particles is observed in DFCs, and perhaps, at the interface of FCs and DFCs (Fig. 10; Puvion-Dutilleul et al., 1991a, Fisher et al., 1991, Dundr et al., submitted). The EMI use of antibodies to the m_3G cap of snoRNAs (Lührmann et al., 1982) provides an information about the nucleolar distribution of all snoRNAs. Interestingly, the labelling pat-

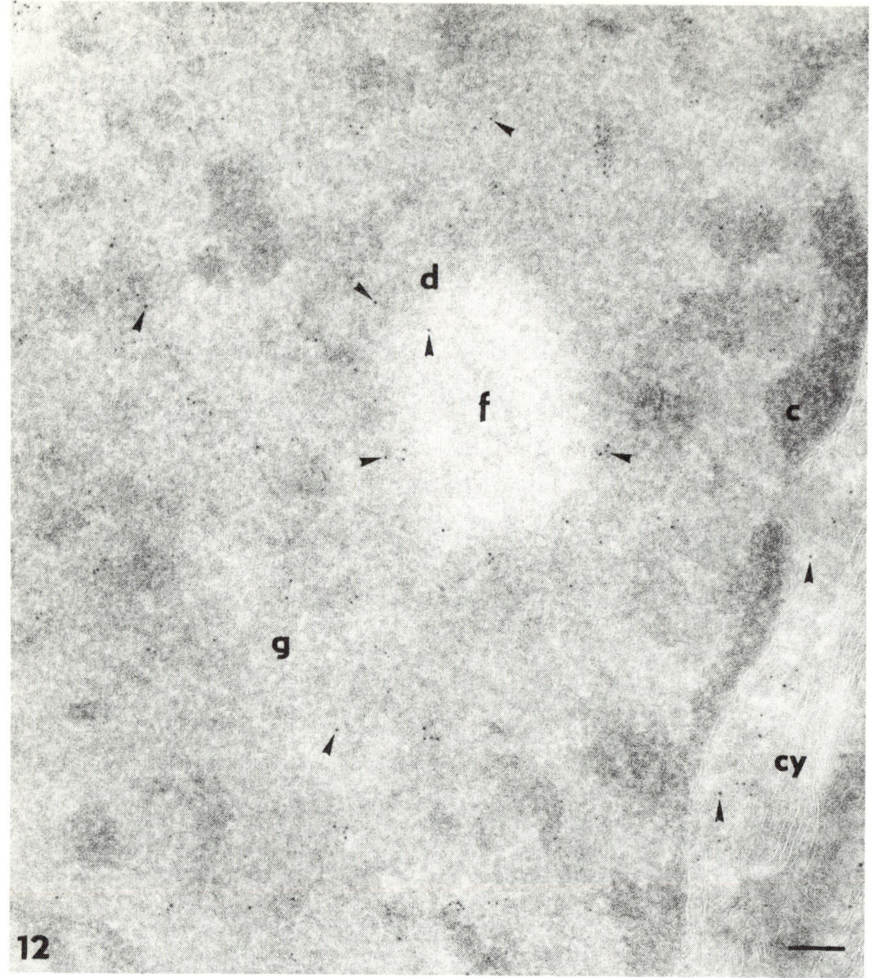

Fig. 12. Cryosectioned HeLa cell nucleolus. The label due to ISH of a stable oligonucleotide to 5S rRNA (5 nm particles; arrowheads) is increased over GCs and DFCs. The label is almost absent in FC.

tern is to some extent compatible with ISH of U3 snoRNA, confining most of the label to DFCs (Raska et al., 1989, 1990). The most studied protein of snoRNPs is fibrillarin (Ochs et al., 1985). EMI of fibrillarin provides the nucleolar labelling pattern which slightly differs from U3 snoRNA ISH and m₃G RNA immunocytochemistry. In this case, most gold particles are observed in DFCs, eventually at the interface of FCs or DFCs, of nucleoli of several mammalian cell types. Only a few particles are distributed over GCs and the interior of FCs (Reimer et al., 1987b, Raska et al., 1989, 1990, Puvion-Dutilleul et al., 1991a).

5S rRNA is synthesized in the nucleoplasm by RNA polymerase III and is imported to the nucleolus as RNP (Steitz et al., 1988). By EMI, the

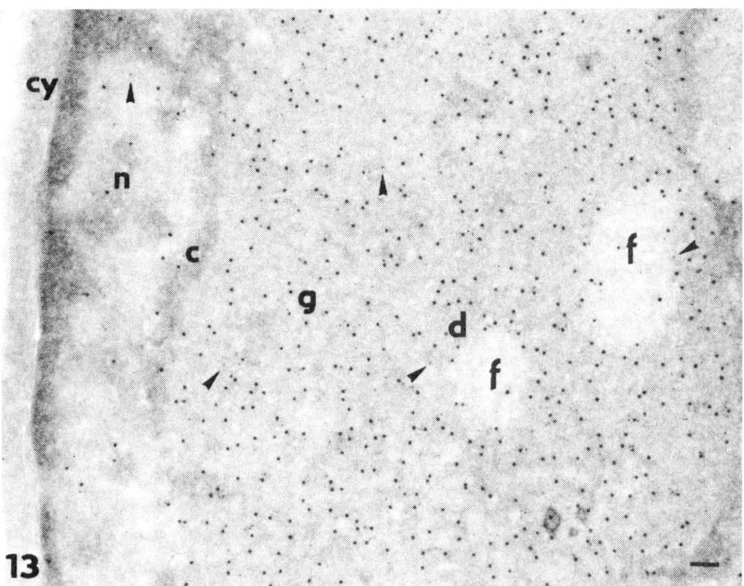

Fig. 13. Consecutive immunostaining of a cryosectioned mouse 3T3 cell nucleolus with rabbit anti-nucleolin (10 nm protein A-gold complex) and with rabbit anti-B23 antibodies (6 nm protein A-gold complex; arrowheads). Most particles are present over GCs and DFCs.

RNP complex containing 5S rRNA is localized in HeLa cell nucleoli to GCs and DFCs, but not to FCs (Fig. 11; Raska et al., 1990). Similar conclusion is reached concerning the nucleolar localization of 5S rRNA with a help of stable oligonucleotide probes tagged with biotins. Most of the label is found in GCs and DFCs, only a few particles are occasionally observed within FCs (Fig. 12; Dundr et al., in preparation).

Nucleolin is believed to be a multifunctional protein involved in the activation of ribosomal genes, the transcription and the processing of nucleolar precursoric rRNAs (Olson et al., 1990). Similarly to another protein, protein B23, nucleolin has been shown to shuttle between the cytoplasm and nucleolus, and to be a target of cdc2 kinase (Borer et al., 1989, Peter et al., 1990). Within nucleoli, both nucleolin and protein B23 are mainly mapped in DFCs and GCs, the occurrence of gold particles in FCs due to nucleolin and/or protein B23 is very low (Fig. 13; Escande et al., 1985, Biggiogera et al., 1989, Raska et al., 1990, 1992, Dundr et al., in preparation).

The available battery of antibodies to various ribosomal proteins represents in our opinion a unique tool to study the compartmentalization of the nucleolus. As far as we know, only a few ribosomal proteins have been mapped (e.g. Hügle et al., 1985). In our hands, the ribosomal protein S1 is mapped mainly to GCs, interestingly with some label being present also in DFCs (Raska et al., 1990, 1992). The presence of at least some ribosomal proteins in DFCs is substantiated as the localization of ribosomal proteins by means of characterized human autoantibodies yields the labelling picture with an increased signal observed over

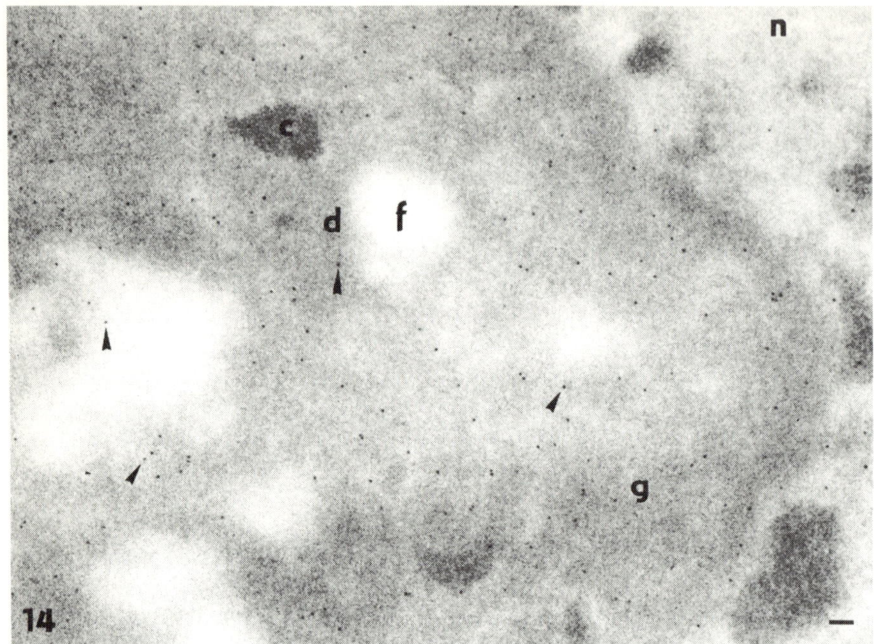

Fig. 14. Cryosectioned HeLa cell nucleolus is immunostained with auto-antibodies reacting specifically with several ribosomal proteins. The nucleolar label (5 nm particles; arrowheads) is present over GCs and DFCs, FCs are almost not labelled.

both GCs and DFCs (Fig. 14).

The documented results witness to the functional compartmentalization of the nucleolus. However, the majority of data do not provide clearcut information which would permit one to locate basic metabolic events of rRNA synthesis and maturation in just one or another nucleolar component. Most results of this review depict a "static" situation. We cannot differentiate between macromolecules, e.g. enzymes, participating directly in a metabolic process and those being recycled or being present in a storage pool.

Concerning the identification of sites of transcribing rRNA genes, the localization of involved proteins, i.e. RNA polymerase I, UBF, DNA topoisomerase I and possibly nucleolin and DNA topoisomerase II, are the most important. The majority of label due to RNA polymerase I and UBF is observed over FCs, while all other proteins as well as rRNA molecules exhibit a higher incidence in DFCs than in FCs. The first two protein factors are restricted to NOR regions of mitotic cells which are the only remaining nucleolar-like structures resembling FCs. One may expect rRNA genes being exclusively associated with FCs.

Could the presence of RNA polymerase I and UBF in DFCs of interphase cells represent a sort of storage pool or a site for assembly of transcription factor complexes? But the question may be reversed and we may consider their presence in FCs as reflecting such a hypothesized

storage pool of transcription factors, including perhaps those RNA polymerase I and UBF molecules sitting on DNA without, however, being involved in the transcription. Finally, a third speculation may be that
both FCs and DFCs accommodate transcribing rRNA genes. We are of opinion that FCs and DFCs form a single functional unit concerning rRNA
transcription, with nascent transcripts becoming "automatically" a part
of DFCs (Raska et al., 1989, 1990, 1992).

In summary, whatever the actual distribution of transcribing rRNA
genes is, the documented results of the present study do not allow us
to localise them either only to FCs or only to DFCs. We emphasize that
the only approach depicting the active process of rRNA transcription
has been documented by EMARG of tritiated uridine incorporation.
However, this is a tedious method the reproducibility and resolution of
which are far behind those provided by non-isotopic methods. Therefore,
the endeavour to standardize an experimental *in vivo* assay, in which
RNA polymerase I efficiently incorporates bases suitable for a subsequent immunolocalization (Wachtler, personal communication), is of primordial importance in this respect .

Whereas the question of transcribing rRNA genes is ambiguous from a
morphological point of view, there is no doubt that DFCs play a key role in the maturation of rRNA. This has been documented by the absence
of external transcribed spacer sequences in GCs (Puvion-Dutilleul et
al., 1991b) and by an increased incidence of several macromolecules involved in the processing of rRNA, i.e. U3 snoRNAs, m3G RNAs, fibrillarin and nucleolin, within DFCs. On the other hand, as a low incidence
of these macromolecules is also present in FCs, it cannot be completely
ruled out that some processing events or events related to processing
of rRNA take place in FCs. The bulk of nucleolin molecules is observed
all over DFCs and GCs. Different parts of this molecule are involved in
consecutive steps of rRNA metabolism and preribosome formation during
which structural changes of nucleolin take place (Olson et al., 1990).

Concerning the vectorial process of preribosome formation, GCs are
identified as being, at least in part, preribosomes. However, this claim is mostly derived from biochemical findings (e.g. Hadjiolov, 1985)
and two observations are to be mentioned. First of all, Warner and Soeiro (1967) and Kumar and Warner (1972) have demonstrated that early
precursor ribosomal particles, containing still 45S rRNA, contain most
ribosomal proteins as well as many other nucleolar proteins. Secondly,
Chooi and Leiby (1981) have clearly shown by EMI performed on Miller's
spreads the association of two ribosomal proteins, one from the small
subunit and one from the large ribosomal subunit, with nascent transcripts of transcribing rRNA genes. Therefore, the finding of several ribosomal proteins, of 5S rRNA and of 5S rRNP complex by EMAC in DFCs is
not that surprising. These findings should be taken into account in future discussions as the vectorial process of preribosome formation may
not be as straightforward as anticipated.

Finally, it is important to mention one matter which is beyond the
scope of this review, as it introduces a question of rDNA localization.
The active rRNA gene generates a high local concentration of rRNPs
which we expect to appear in more or less compacted form. The rRNP concentration is two orders of magnitude higher than that of rDNA. The time of synthesis of the whole mammalian precursor rRNA in cycling cells

is in the order of 5 minutes with a complete renewal of nucleolar pre-ribosomal particles occuring approximately within 30 minutes (Hadjiolov, 1985). EMARG of serial nucleolar sections has demonstrated that the label due to tritiated uridine incorporation during 10 minutes pulse is mostly present in GCs and DFCs, with gold particles being also confined to the periphery of FCs (Derenzini et al., 1987). This picture is compatible with the interpretation that rRNA transcription occurs only in the peripheral part of FCs, DFCs representing rather the site of accumulation of rRNA transcripts (Scheer and Benavente, 1990, Thiry and Goessens, 1992). Such a structure-function polarized view on FCs may be genuine. On the other hand, however, due to the nature of rDNA ISH employing genomic probes of non-negligible size, it is difficult to accept that absolutely no label is observed over DFCs (Thiry and Goessens, 1992).

Acknowledgments: The authors would like to thank Professor K. Smetana and Dr. R. Jelínek for a critical reading of the manuscript, K. Koberna, J. Otcovsky and Mrs. V. Rohlenova for a preparation of the manuscript and photographs, Dr. V. Mandys and Mrs. H. Ederova for a help with the cell culturing. They thank Professor U. Aebi, Dr. M. Carmo-Fonseca, Professor J. Dubochet, Professor W. W. Franke, Dr. G. Griffiths, Prof. U.K. Laemmli, Dr. A.I. Lamond, Professor R. Lührmann, Professor B. Mechler, Dr. R.L. Ochs, Professor M.O.J. Olson, Dr. G. Reimer, Dr. A. Richter, Dr. M.-C. Risueno, Professor U. Scheer, Dr. J.A. Steitz, Dr. E.M. Tan, Dr. M. Tosi and Dr. T. Yario for generous gifts of antibodies, antisense oligonucleotides and various (immuno)chemicals. Drs. F.J. Moreno, A. Olmedilla and M.-C. Risueno are greatly acknowledged for their contributions of photographs. This work was supported by a grant no. 53901 of the Czechoslovak Academy of Sciences.

4 References

Bachellerie, et al. (1977). Topographical distribution of rDNA sequences and isolation of ribosomal transcription complexes. **Eur. J. Biochem.**, 79, 23-32.

Biggiogera, M. et al. (1989). Simultaneous immunoelectron microscopical visualisation of protein B23 and C23 distribution in the HeLa cell nucleolus. **J. Histochem. Cytochem.**,37, 1371-1374.

Borer, R.A. et al. Major nucleolar proteins shuttle between nucleus and cytoplasm. **Cell**, 56, 379-390.

Busch, H. and Smetana, K. (1970). **The nucleolus.** New York, London: Academic Press.

Carmo-Fonseca, M. et al. (1992). Transcription-dependent colocalization of the U1, U2, U4/U6, and U5 snRNPs in coiled bodies. **J. Cell Biol.**, 117, 1-14.

Cartwright, I.L. and Elgin, S.C.R. (1988). Chromatin of active and inactive genes. **Architecture of eukaryotic genome.** Weinheim, 283-300.

Chan, E.K.L. et al. (1991). Human autoantibody to RNA polymerase I transcription factor hUBF. Molecular identity of nucleolus organizer region autoantigen NOR-90 and ribosomal RNA transcription upstream binding factor. **J. Exp. Med.**, 174, 1239-1244.

Chooi, W.Y. and Leiby, K.R. (1981). An electron microscopic method for localization of ribosomal proteins during transcription of ribosomal DNA: A method for studying protein assembly. **Proc. Natl. Acad. Sci. U.S.A.**, 78, 4823-4827.

Conconi, A. et al. (1989). Two different chromatin structures coexist in ribosomal RNA genes throughout the cell cycle. **Cell**, 57, 753-761.

Derenzini, M. et al. (1987). Structure of ribosomal genes of the mammalian cells in situ. **Chromosoma**, 95, 63-70.

Derenzini, M., Thiry, M. and Goessens, G. (1990). Ultrastructural cytochemistry of the mammalian cell nucleolus. **J. Histochem. Cytochem.**, 38, 1237-1256.

Drake, F.H. et al. (1989). Biochemical and pharmacological properties of p170 and p180 forms of topoisomerase II. **Biochemistry**, 28, 8154-8160.

Escaig-Haye, et al. (1989). Detection ultrastructurale de l' ARN ribosomal par hybridation in situ a l' aide d' une sonde biotinylée sur coupes ultrafines de cellule animale en culture. **C. R. Acad. Sci. [III]**, 309, 429-434.

Escande, M.L., Gas, N. and Stevens, B.J. (1985). Immunolocalization of the 100 kD nucleolar protein in CHO cells. **Biol. Cell**, 53, 99-110.

Fakan, S. and Puvion, E. (1980). The ultrastructural visualization of nucleolar and extranucleolar RNA synthesis and distribution. **Int. Rev. Cytol.**, 65, 255-299.

Fisher, D., et al. (1991). Review: Assigning functions to nucleolar structures. **Chromosoma**, 101, 133-140.

Goessens, G. (1984). Nucleolar structure. **Int. Rev. Cytol.**, 87, 107-158.

Graaf, A. de et al. (1990) Three-dimensional distribution of DNase I -sensitive chromatin regions in interphase nuclei of embryonal carcinoma cells. **Eur. J. Cell Biol.**, 52, 135-141.

Granboulan, N. and Granboulan, P. (1965). Cytochimie ultrastructurale du nucleole. II. Etude des sites de synthese du RNA dans le nucleole et le noyau. **Exp. Cell Res.**, 38, 604-619.

Hadjiolov, A.A. (1985). The nucleolus and ribosome biogenesis. **Cell Biology Monographs,** Vol. 12, Springer-Verlag, Wien, New York.

Hernandez-Verdun, D. (1991). The nucleolus today. **J. Cell Sci.**, 99, 465-471.

Hügle, B. et al. (1985). Localization of ribosomal protein S1 in the granular component in the interphase nucleolus and its distribution during mitosis. **J.Cell. Biol.**, 100, 873-886.

Jordan, E.G. (1991). Interpreting nucleolar structure: where are the transcribing genes?. **J. Cell Sci.**, 98, 437-442.

Kass, S. et al. (1990). The U3 small nucleolar ribonucleoprotein functions in the first step of preribosomal RNA processing. **Cell**, 60, 897-908.

Kumar, A. and Warner, J.R. (1972). Characterisation of ribosomal precursor particles from HeLa cell nucleoli. **J. Mol. Biol.**, 69, 233-246.

Labhart, P. and Koller, T. (1982). Structure of the active nucleolar chromatin of Xenopus laevis oocytes. **Cell**, 28, 279-292.

Lerner, E.A. et al. (1981). Monoclonal antibodies to nucleic acid -containing cellular constitutents: Probes for molecular biology and autoimmune disease. **Proc. Natl. Acad. Sci. U.S.A.**, 78, 2737-2741.

Lührmann, R. et al. (1982). Isolation and characterization of rabbit anti-m$_3$²²⁷G antibodies. **Nucl. Acid. Res.**, 10, 7103-7113.

Müller, M.T. et al. (1985). Eukaryotic type I topoisomerase is enriched in the nucleolus and catalytically active on ribosomal DNA. **EMBO J.**, 4, 1237-1243.

Ochs, R.L. et al. (1985). Fibrillarin: A new protein of the nucleolus identified by autoimmune sera. **Biol. Cell**, 54, 123-134.

Olson, M.O.J. (1990). Role of proteins in nucleolar structure and function, in **The Eucaryotic Nucleus: Molecular Structure and Macromolecular Assemblies** (eds. Wilson, S.H. and Strauss P.), Vol. 2. Caldwell, New Jersey: pp. 520-559.

Peter, M. et al. (1990). Identification of major nucleolar proteins as candidate mitotic substrates of cdc2 kinase. **Cell**, 60, 791-801.

Puvion, E. and Moyne, G. (1981). In situ localization of RNA structures. **The cell nucleus** (ed. H. Busch). Academic Press, Inc., New York. Vol. VIII, 59-115.

Puvion-Dutilleul, F. et al. (1991a). Localization of U3 RNA molecules in nucleoli of HeLa and mouse 3T3 cells by high resolution in situ hybridization. **Eur. J. Cell Biol.**, 56, 178-186.

Puvion-Dutilleul, et al. (1991b). Nucleolar organization of HeLa cells as studied by in situ hybridization. **Chromosoma**, 100, 395-409.

Raska, I. et al. (1985). Ultrastructural localization of rRNA in HeLa cells, rat liver cells and Xenopus laevis oocytes by means of the monoclonal antibody-protein A-gold technique. **Histochem. J.**, 17, 925-938.

Raska, I. et al. (1988). Human autoantibodies as convenient probes for the nucleolar structure and function, in **12th Electron Microscopical Conference of DDR**, (eds. J. Heidenreich and H. Luppa), Dresden, January 18-20, pp. 28-30.

Raska, I. et al. (1989). Does the synthesis of ribosomal RNA take place within nucleolar fibrillar centers or dense fibrillar components? **Biol. Cell**, 65, 79-82.

Raska, I., Ochs, R.L. and Salamin-Michel, L. (1990). Immunocytochemistry of the cell nucleus. **Electron Microsc. Rev.**, 3, 301-353.

Raska, I. et al. (1991). Immunological and ultrastructural studies of the nuclear coiled body with autoimmune antibodies. **Exp. Cell Res.**, 195, 27-37.

Raska, I., et al. (1992). Structure-function subcompartmentalization of the mammalian cell nucleus as revealed by the electron microscopic affinity cytochemistry. **Cell Biol. Int. Rep.**, 16, 771-789.

Reeder, R.H. (1990). rRNA synthesis in the nucleolus. **Trends Genet.**, 6, 390-395.

Reeves, R. (1988). Active chromatin structure. **Chromosomes and chromatin.** Boca Raton, CRC Press, 109-131.

Reimer, G. et al. (1987a). Autoantibody to RNA polymerase I in scleroderma sera. **J. Clin. Invest.**, 79, 65-72.

Reimer, G. et al. (1987b). Human autoantibodies: probes for nucleolus structure and function. **Virchows Arch. B**, 54, 131-143.

Rendón, M.C. et al. (1992). Characterization and immunolocalization of a nucleolar antigen with anti-NOR serum in HeLa cells. **Exp. Cell Res.**, 200, 393-403.

Rodriguez-Sanchez, J.L. et al. (1987). Anti-NOR 90. A new autoantibody in scleroderma that recognized a 90-kDa component of the nucleolus -organizing region of chromatin. **J. Immunol.**, 139, 2579-2584.

Rose, K.M. et al. (1988). Association of DNA topoisomerase I and RNA polymerase I: a possible role for topoisomerase I in ribosomal gene transcription. **Chromosoma**, 96, 411-416.

Savino, R. and Gerbi, S. (1990). In vivo disruption of Xenopus U3 snRNA affects ribosomal RNA processing. **EMBO J.**, 9, 2299-2308.

Scheer, U. and Zentgraf, H. (1982). Morphology of nucleolar chromatin in electron microscopic spread preparations. **The Cell Nucleus** (ed. H. Busch). Vol. 11, New York, Academic Press, 143-176.

Scheer, U. and Benavente, R. (1990). Functional and dynamic aspects of the mammalian nucleolus. **BioEssays**, 12, 14-21.

Smetana, K. and Busch, H. (1974). The nucleolus and nucleolar DNA. **The Cell Nucleus** (ed. H. Busch). Vol.1. New York: Academic Press.

Sollner-Webb, B. and Mougey, E.B. (1991). News from the nucleolus: rRNA gene expresion. **Trends Biochem. Sci.**, 16, 58-62

Steitz, J.A. et al. (1988). A 5S rRNA/L5 complex is a precursor to ribosome assembly in mammalian cells. **J. Cell Biol.**, 106, 545-556.

Thiry, M. (1991). DNase I-sensitive sites within the nuclear architecture visualized by immunoelectron microscopy. **DNA Cell Biol.**, 10, 169-180.

Thiry, M. (1992). Highly sensitive immunodetection of DNA on sections with exogenous terminal doexynucleotidyl transferase and non -isotopic nucleotide analogues. **J. Histochem. Cytochem.**, 40, 411-419.

Thiry and Muller (1989). Ultrastructural distribution of histones within Ehrlich tumor cell nucleoli: a cytochemical and immunocytochemical study. **J. Histochem. Cytochem.**, 37, 853-862.

Thiry, M. and Thiry-Blaise, L. (1989). In situ hybridization at the electron microscope level: An improved method for precise localization of ribosomal DNA and RNA. **Eur. J. Cell Biol.**, 50, 235-243.

Thiry, M. and Goessens, G. (1991). Distinguishing the sites of pre-rRNA synthesis and acumulation in Ehrlich tumor cell nucleoli. **J. Cell Sci.**, 99, 759-767.

Thiry, M. and Goessens, G. (1992). Where, within the nucleolus, are the rRNA genes located? **Exp. Cell Res.**, 200, 1-4.

Tokuyasu, K.T. (1986). Application of cryoultramicrotomy to immunocytochemistry. **J. Microsc.**, 143, 139-149.

Wachtler, F., Mosgöller, W. and Schwarzacher, H.G. (1990). Electron microscopic in situ hybridization and autoradiography: localization and transcription of rDNA in human lymphocyte nucleoli. **Exp. Cell Res.**, 187, 346-348.

Wang, J.C. (1985). DNA topoisomerases. **Annu. Rev. Biochem.**, 54, 665-697.

Warner, J.L. and Soeiro, R. (1967). Nascent ribosomes from HeLa cells. **Proc. Natl. Acad. Sci. U.S.A.**, 58, 1984-1990.

Warner, J.R. (1990). The nucleolus and ribosome formation. **Curr. Opin. Cell Biol.**, 2, 521-527.

Zini, N. et al. (1992). The 180-kDa isoform of topoisomerase II is localized in the nucleolus and belongs to the structural elements of the nucleolar remnant. **Exp. Cell Res.**, 200, 460-466.

Evolution

9 Chromosomal divergence and speciation in grasshoppers

G.M. HEWITT

University of East Anglia, UK

1 Introduction

Many related species can be distinguished by obvious chromosomal differences and Michael White (1978) argued that perhaps only 1% of species were homosequential. Furthermore, speciose groups exhibit many chromosome differences, and in vertebrate groups the rate of chromosome evolution is correlated with the rate of speciation (Bush et al. 1977). These indicate a significant role for chromosome changes in speciation, and the meiotic problems of rearrangement heterozygotes offer the barrier to gene flow required by the Biological Species Concept (Mayr, 1970). Consequently several models of speciation have been proposed with chromosome changes as their cause or central feature; there are many individual variants; White (1978), and Sites and Moritz (1987) provide a recent comprehensive review. They recognise seven broad modes of chromosomal speciation, viz: Stasipatry, Invasive, Primary Chromosomal Allopatry, Chain Process, Cascade, Monobrachial Centric Fusions and Recombinational Breakdown, considering their properties and consequences. They conclude that we cannot distinguish them and need multidisciplinary studies at the population level, where ecological, cytogenetic and population genetic factors are assessed.

It is clear that chromosomes are not the whole story, since species may be distinguished by many different characters including morphological, biochemical, developmental, behavioural and reproductive aspects; sometimes chromosomes are not involved at all - as is elegantly demonstrated by the homosequential species groups of Hawaiian *Drosophila* (Carson, 1981). Gene flow between two groups of organisms may be prevented by a number of biological attributes. Very strong barriers are produced by behavioural differences in the time or place of mating, in the mating process itself, or in the process of fertilization and compatibility of gametes. Differences in developmental pathways and the control of adult functions may cause hybrid inviability or sterility, and recombination of the two genomes can cause similar effects in the F_2 or Bx generations. Recent reviews of speciation have tended to concentrate on topics such as the importance of geographic isolation and sympatric speciation, changes in behaviour and mating preference, the role of ecology and demography and in particular the importance of small island populations (eg Otte and Endler, 1989). In *Drosophila* progress is being made in identifying the genes controlling some of these characters that are involved in species differences (Coyne, 1992).

In as much as speciation is a process of divergence of genomes to ultimate genetic isolation, then chromosomes are part of this. The question is what part? This process of

Chromosomes Today   Volume 11. Edited by A.T. Sumner and A.C. Chandley. Published in 1993 by Chapman & Hall, London. ISBN 0 412 47670 3

divergence may follow many paths and produce genomes that are isolated to varying degrees. One way of assessing the nature of this divergence, and hence characters involved in speciation, is by analysis of hybrid zones which form between different forms, races and species. There is currently much interest in such studies and hybrid zones have been aptly described as 'natural laboratories' and 'windows on the evolutionary process' (Hewitt, 1988; Harrison, 1990).

2 Hybrid zones

Many species and species complexes are subdivided genetically and geographically into clearly distinct forms. Where the ranges of such forms meet they may mate, hybridize and produce a hybrid zone. Such zones vary in structure from very narrow bands of a few metres to broad swathes of a hundred kilometres, and the changeover from one form to the other may be smooth and clinal or patchy and reticulate. Most hybrid zones have been recognised initially on morphological grounds. However, if examined further virtually every zone comprises differences in a range of characters, be they morphological, chromosomal, allozymic or behavioural. Over one third of reported zones show clear karyotypic differences between the two forms and in many others this has not been examined thoroughly (Hewitt 1985). The pioneering work of Michael White (1978), on Australian morabine grasshoppers involved karyotyping thousands of males from testis sections in wide geographic surveys. The now classic case of the *viatica* group in South Australia revealed at least 13 distinct karyotypic taxa in *Moraba viatica*, parapatrically distributed, and all within a few hundred kilometres of Adelaide. Grasshopper studies have yielded a number of such unseen karyotypic taxa, as has work on small mammals and plants. It is difficult to gauge the full extent of the phenomenon, but since we are only just beginning to look at cryptic molecular characters we are probably just scratching the surface.

3 Chromosome involvement

There are now several well studied hybrid zones in Orthoptera and they provide a microcosm with which to illustrate the nature of chromosomal involvement (Table 1).

(i) The *viatica* group includes a number of taxa which show a variety of chromosomal rearrangements. The formal taxonomic description of all these types is not yet complete, but morphological differences between some of the karyotyic forms have been found and some are of specific status (Key 1976). Some of the zones are very narrow, apparently just tens of metres where measured, and the fitness of the hybrids is also very low.

(ii) The *Podisma pedestris* hybrid zone was also first recognised by an X-A centric fusion and is fairly narrow, but extensive investigation has not yet revealed any consistent difference in morphology or allozymes between the two races. However the hybrid embryos are only half as fit, and recent work has revealed a difference in the ribosomal DNA (Dallas et al. 1988).

(iii) The Torresian and Moreton taxa of *Caledia captiva* were also first recognised by major karyotypic differences, where an essentially acrocentric karyotype has been converted to a largely metacentric one with distinct interstitial C-banding. Further study showed allozyme, behaviour, mitochondrial DNA and ribosomal DNA differences, and severe BX and F2 hybrid inviability. This zone is also fairly narrow.

(iv) The two subspecies of *Chorthippus parallelus* were recognised on the basis of their distinct morphology and song. Considerable effort has revealed only one or two differences in allozyme loci and an extra NOR on the X chromosome of *C.p. parallelus* (in addition to the two autosomal NORs in *C. p. erythropus*). F1 male hybrids are completely sterile but F1 females are normal. While the zone is narrow for the X-NOR and some other characters, it is variously wider for other characters.

Table 1. Well studied Orthopteran Hybrid Zones, (see text for explanations).

Species Range	Character	Width km	Hybrid Fitness	References
Vandiemenella viatica (S. Australia ~ 1000km)	19/24xy 17/19 K. Is 17/24xy 17/19 Keith Morphology	~0.01 ~0.02 ~0.20 ~0.40	v. low (1/10) low (3/38) hybrids (6/20) few hybrids (5/41)	White 1978 Mrongovius 1979 Hewitt 1979
Podisma pedestris (Alps-USSR 800km - 300 km)	xo/xy rDNA	0.1-0.8 varies	50-80% lab and field	Barton and Hewitt 1985, 1989 Nichols and Hewitt 1988
Caledia captiva (E. Australia > 300km)	Chrom invs Allozymes mtDNA rDNA (not coinc)	0.35-0.80	F_2 inviable Bx 50%	Shaw *et al* 1985,1988 Kohlmann and Shaw 1991
Chorthippus parallelus (Pyrenees- W. Europe ~ 3000km)	Allozymes Syll length Strid. pegs Echeme Int X-NOR mtDNA scDNA	20 ~15-20 2-9 1.4 ~0.6	F_1 ♂ sterile Bx ♂ partly sterile	Butlin and Hewitt 1985 Hewitt 1990 Butlin *et al* 1991
Gryllus pennsylvanicus firmus (NES. USA ~ 2000 km)	Morphology Allozyme Behaviour mtDNA	~40 mosaic	Gf ♀ x Gp ♂ Sterile	Rand and Harrison 1989 Harrison and Rand 1989
Allonemobius fasciatus/socius (E. USA ~2000 km)	Allozymes Chirp Int	5-200? mosaic?	As ♀ x Af ♂ Less fit F_2 poor	Howard 1986

(v) The zone in *Gryllus* was recognised morphologically and behaviourally, and the two species have since been shown to differ somewhat in their allozymes and mitochondrial DNA. Interestingly this zone is a mosaic, where each genotype is associated with one of two soil types that are intermingled in a long band running NE - SW in new England. The F1 cross is sterile one way only.

(vi) The *Allonemobius* species are best distinguished on their allozyme differences. There is little difference in song, and the morphological similarity raises questions about their specific status; although there is hybrid unfitness. There are apparently no reports of chromosomal differences in these two crickets - but given the size of Nemobiine chromosomes, that does not surprise me! It would be worth looking with the newer techniques that are now available.

What does this tell us about the relative differentiation of characters in divergent taxa? In this limited sample there is no obvious correlation between the degree of differentation; there is a range of chromosome change, a range of allozyme distinctness and a range of behavioural difference - all from very little to near specific level. The same appears to be the case in other groups where a number of hybrid zones are recorded. However, these are generally biased in the range of characters studied. Thus in amphibia, song, morphology and allozymes differ between hybridizing taxa; in birds, morphology and song are preferred, and in small mammals it is chromosomes, allozymes and morphology. Sites and Moritz (1987), rightly called for more multidisciplinary population level studies of organisms undergoing chromosomally mediated speciation in order to test various hypotheses. Work on hybrid zones is providing some of this, but it is clear that to put chromosomes in context also requires that all categories of differentiation are examined in each species complex, whether or not chromosomes are involved. This, naturally, takes time.

4 Hybrid unfitness

One thing that is found in all these orthopteran examples is some form of hybrid unfitness, varying from F1 inviability to F2 breakdown. This is quite general and strong selection seems to be involved in most hybrid zones (Barton and Hewitt, 1989). We may ask, how much of this hybrid unfitness is due to chromosomal differences? The width of such tension zones is determined by the dispersal of the organism countered by the strength of selection against hybrids. Consequently, narrow zones like those in *V. viatica* could be produced by low dispersal and/or strong selection, whilst the clines of allozymes in *C. parallelus* are so wide that selection must be very low and/or dispersal very high to explain them. However, in *C. parallelus* clines for various characters differ in width with some song components and the X-NOR being of the order of only 1km; since insect dispersal is the same for all characters this means that selection is operating in these narrower clines. Given a dispersal of 30m/gen (Virdee and Hewitt, 1990), the selection required against X-NOR heterozygotes to maintain the zone width is ~0.7% - which is small and difficult to measure. Similar calculations apply to the X-A centric fusion in *P. pedestris* where the cline is some 800 m wide across suitable habitat with dispersal at 20m/gen this requires selection of the order of 0.5% against the heterozygote. This agrees with the low level of meiotic nondisjunction in heterokaryotypes measured from embryos (Barton and Hewitt, 1981), and with the generally good behaviour of centric fusion/fission heterozygotes in Orthoptera (Hewitt, 1979). Both the *P. pedestris* and *C. parallelus* zones have strong hybrid unfitness ~50%, due to embryo inviability and testis dysfunction respectively; so the chromosomal differences are contributing only a small amount to this isolation which is polygenically controlled in both cases (Barton and Hewitt, 1981; Virdee and Hewitt, 1993a).

The situation in *Caledia captiva* is different. Here most chromosomes have pericentric inversions which cause no problems to the F1 hybrid itself, but recombination between the homologues of the two taxa produces inviable recombinant F2 embryos. Some of this inviability is due to chiasmata occuring in unusual regions in the hybrid because the bivalents are inversion heterozygotes, but much of it is due to genic divergence between the two genomes (Shaw and Coates, 1983). Since there is straight pairing over the inversion, the hybrid unfitness is not due to differences within the inverted region, but outside it. Such straight pairing is usual in grasshoppers (Fletcher and Hewitt, 1978), and probably accounts for the often high frequency of pericentric inversions in these insects as compared with *Drosophila.*. Unfortunately we do not have as much detail for the *V. viatica* zones, which involve a variety of rearrangements. Several forms differ for fusion/fission trivalents, and a few for an inversion or translocation (White, 1978). The fusion/fission trivalents usually behave well in meiosis, but the races that differ for a number of different rearrangements show considerable aneuploidy in 2° spermatocytes, and form the narrowest zones, for review see Hewitt (1979).

This brief survey shows that the effects of chromosome differences on hybrid fitness are often not great, particularly for centric fusions. Greater sterility or inviability is produced where a number of rearrangements are combined, but even here they do not account for all the reduction in hybrid fitness. None the less, fairly narrow hybrid zones can be produced when a rearrangement causes even a 1% reduction in fertility - as for example a fusion/fission trivalent.

This leads to two questions. How do chromosome races form and how are hybrid zones formed? They are closely interrelated.

5 Origins of chromosome races

Standard theory accepts that chromosome rearrangements cause meiotic nondisjunction when heterozygous and thus reduce fitness. Such underdominant mutants have no chance of surviving unless fixed by drift in a small population. Because of the reproductive isolation they produce they rarely occur as polymorphisms and frequently separate races and species. However, there are some problems with this tenet.

Firstly, as we have seen certain chromosomal rearrangements do not malorientate at any significant frequency and thus should not cause any great reduction in fitness of the heterozygote. This is true for centric fusions. Inversions can avoid effects on disjunction if straight pairing or gamete selection occurs and many are polymorphic in grasshoppers, Chironomids and Drosophilids. On the other hand translocations are rarely polymorphic and apparently lacking in the Orthoptera (Hewitt, 1979), but they cause severe meiotic upset between races and species (eg John and Lewis, 1965). Complex structural hybridity has developed in a few diverse organisms probably as a response to inbreeding, (eg James, 1992). Secondly, their survival will also be enchanced if they show meiotic drive. However, there is little evidence for the involvement of meiotic drive in *Drosophila* (Coyne, 1989), or other organisms, although these have not been examined as carefully. Similarly, if the rearrangement is associated with favourable alleles it may confer an advantage on the homozygote which will increase its chance of survival. However, this should not happen often and probably the homozygote fitness advantage would rarely be stronger than the heterozygote disadvantage.

Whilst these considerations suggest that the overall selection against the new mutant need not be so strong in certain cases, the extremely low probability of fixation remains for any rearrangement causing significant reproductive isolation. Theoretical approaches require population sizes of 10 or less for any realistic chance of survival, and immigrants from surrounding populations will greatly reduce this (Lande, 1979). This major theoretical problem applies also to the fixation of genes for other characters causing some reproductive isolation (Barton and Charlesworth, 1984). Recently some new calculations by Barton and Rouhani (1991), show that the probability of fixation is much greater in a widespread population consisting of a series of small demes with gene flow between them, than it is in a series of discrete demes. Their model involves appreciable heterozygote unfitness and slight selection for the mutant homozygote. Interestingly where neighbourhood size is small such underdominant rearrangements can become fixed even with free gene flow. Initially the frequency of new mutants must reach a certain level locally - forming a 'critical bubble' - after which it will most likely be fixed through all the array of demes.

This means that such underdominant rearrangements will most probably arise in regions of low density and dispersal and will form a simple hybrid zone around their perimeter. However, they will not be able to invade regions of higher population density. This is an interesting property of hybrid tension zones; they cannot advance up quite modest increases in density because the dispersal of greater numbers of the original form counters the increased numbers of the mutant form arising from its selective advantage. Indeed, density troughs trap hybrid zones - they can not move in either direction (Hewitt, 1975; Nagylaki, 1975; Barton, 1979; Barton and Hewitt, 1989). There is increasing evidence from the field for this concept. For example, the hybrid zone in *Podisma pedestris* has been shown to follow regions of low density and barriers to dispersal in several places (Barton and Hewitt, 1989; Nichols and Hewitt, 1986), and recently Kohlmann and Shaw (1991) demonstrated that the zone in *Caledia captiva* has been shifted some 200 metres into a reduction in density caused by regrowth of the forest. The distributions of most organisms vary greatly in density, with many fairly permanent density troughs and barriers to gene flow such as mountain ridges or hot valleys. Consequently, when a chromosome rearrangement, or any other underdominant difference gets established in a local low density region, it will be unable to move out across habitats causing even quite small reductions in neighbourhood size and will be trapped by more permanent barriers. How then do these incipient races spread to fill large areas and produce the patchwork we see in so many species?

6 Range changes

In recent years our understanding of the distrubution of species during the last ice age and their subsequent range changes in the present warm interglacial period has been advanced greatly by fossil studies, particularly in pollen and beetles (Huntley, 1990; Coope, 1990). Much of North America and Northern Europe was covered with ice and tundra so that most of the hybrid zones we study could not have been in their present positions (Hewitt, 1985). Conditions nearer the equator and in the southern hemisphere have also varied considerably over this period. In Europe many currently widespread species would have occupied more southerly refugia around the Mediterranean and Caspian Seas from which they spread north as the climate warmed and the ice retreated. Hybrid zones would form as different taxa met spreading from different refugia, and it seems likely therefore that most zones have been produced by secondary contact.

During these range changes the population dynamics could well have provided conditions suitable for the fixation, establishment and spread of underdominant mutants and the production of races. In the ice age the northern limit of a species range might be in small pockets in Southern Iberia, Italy, Balkans or Transcaucasus. These would be subject to periodic reductions in size and genetic drift. When conditions improved the populations at the edge would expand into new territory and colonise it. This seems to provide the low density and small neighbourhood size in which mutants could be established, and allows them then to be spread to cover large areas (Hewitt, 1989). There were probably several refugia for most species and the leading edge of colonization ultimately expanded over many thousands of miles, providing multiple opportunities for establishment. Furthermore the climatic amelioration was not a smooth process, it had some major reversals and many minor fluctuations which would repeat the process; several times in one place.

Obviously the adaptations, dispersal and reproductive capabilities of species differ greatly, and so they may respond to these processes in various ways and over different scales. This may explain some of the differences in raciation between groups.

7 Secondary contact

When two diverged forms expanding from refugia meet and form a secondary contact zone, a number of things may follow. If they hybridize and there is some fertility in the F1 and BXs, then gene flow is possible and there will be a progressive intermixing of the genomes over time. For chromosome rearrangements and gene mutants that cause some reduction in fitness in hybrids, the tension zones which are established will be relatively narrow - proportionate to dispersal and selection. However, whilst such tension zones are barriers to gene flow, chromosomes and genes which do not contribute to hybrid unfitness will ultimately pass through and continue to introgress (Barton and Hewitt, 1989). This should produce clines of different width but coincident at the same place. The clines of most zones are coincident, but sufficient information to assess this in detail is only available from a few organisms, such as the house mouse *Mus musculus* (Nance et al. 1990), the firebellied toad, *Bombina bombina* (Szymura and Barton, 1991), and the grasshoppers *Caledia captiva* (Shaw et al. 1988), and *Chorthippus parallelus* (Butlin et al. 1991: Hewitt, 1993a). This shows that some clines are not strictly coincident.

In *C. parallelus* the French and Spanish subspecies met along the ridge of the Pyrenees around 9000BP as the ice cap retreated and each ascended from its own side. There is evidence from palynology for such a date (Reille, 1990; Hewitt, 1993a). Dispersal has been measured at some 30m/generation (Virdee and Hewitt, 1990), which for neutral alleles should produce a cline about 5km wide in 9000 years. Some clines like the X-NOR are narrower than this and probably subject to selection, but others are some 20km wide which would require excessive dispersal rates of 120m/yr. The explanation we favour for this paradox derives not from exceptional dispersal but from the dynamics of contact and mixing (Hewitt, 1993a). At the leading edge of the species expansion occasional dispersers will establish small colonies in patches of suitable habitat ahead of the main body. These pioneers will come from both subspecies when they meet and produce a wide band of interdigitation. As these colonies grow they will coalesce to give a patchwork and then wide clines for the character differences. Hybrid unfitness will then narrow the clines for those characters which cause it. These processes can be quantified by computer modelling (Nichols and Hewitt, 1993), and produce very different dynamics from the meeting of two genotypes as phalanxes.

Such meeting and mixing would apply to most species and could account for many wide hybrid zones. They also offer an explanation for the noncoincidence of some clines. When the two genomes meet as pioneer patches their character differences may more readily be reassorted in small hybrid populations. These recombined genotypes may expand differentially by chance. One result could be that clines for some characters have their centres displaced, as in *C. parallelus* (Butlin et al. 1991). Another result is observed in *Caledia* where some allozyme and mitochondrial DNA differences are displaced hundreds of kilometres from the main chromosomal zone (Shaw et al. 1988). A reassorted genotype produced in a pioneering hybridization could spread into the occupied territory, and this would be made easier if the expansion contained occasional contractions. This is more a 'touch and go' scenario. A similar process could have occurred in the origin of the *V. viatica* chromosomal taxa (Hewitt, 1979) and for the mtDNA introgression in *Mus musculus* in Sweden (Gyllensten and Wilson, 1987).

Once a hybrid zone has formed with normal population densities on either side then the rate of introgression for most of the genome will not be very great. As in *Chorthippus parallelus* apparently neutral characters have only mixed over 20kms in 9000 years and most hybrid zones are narrow relative to the species range (Hewitt, 1989). Where one allele or chromosome has an advantage it should in principle penetrate the hybrid zone, although it may be hindered due to linkage with underdominant loci. For example in *C. parallelus* there is asymmetrical homogamy when a female of the *parallelus* subsepecies is mated with both types of male and produces mostly *parallelus* offspring (Bella et al. 1992). This will tend to drive forward those components of the *parallelus* genome causing this effect. However they will be mixed and recombined in the hybrid zone and possibly separated so that their synergistic effect is lost, and they may not be able to penetrate far into the *erythropus* genome. The cline for the X-NOR C-band of *parallelus* is found some 7-15 kms beyond the centre of the hybrid zone into *erythropus* over two separate cols in the Pyrenees, and could well provide an example of this process. Also, in *Caledia captiva* an rDNA variant has introgressed from the Moreton into the Torresian chromosomal taxon and it is proposed that this may be due to some form of selection (Arnold et al. 1987). However, whilst these cases of reassortment and introgression are interesting, for hybrid zones in general the great majority of characters are broadly coincident suggesting that for complex multigenic zones there is little gene flow between the taxa.

8   Reinforcement of mating isolation

Even though chromosome rearrangements may not cause complete hybrid sterility, it has been argued that postzygotic reduction in fitness will lead to the development of prezygotic barriers and full speciation. Such reinforcement has been an integral component of most theories of speciation, but has recently come under critical scrutiny (Butlin, 1989). Postglacial secondary contact zones, with their hybrid unfitness, would seem to be just the place for reinforcement to occur, and yet a comprehensive survey of hybrid zones provides no good evidence for this process (Barton and Hewitt, 1985; Butlin, 1989). The zone in *C. parallelus* has been used extensively to look for evidence of this process (Butlin and Ritchie, 1991), since the two subspecies have diverged significantly showing positive assortative mating and severe hybrid disfunction. However, there is a simple transition for female preference in the zone that does not support the development of reinforcing mechanisms. Indeed in the zone the F1 testis dysfunction appears to be reduced and it is possible that either reassortment or amelioration has occurred (Hewitt, 1993a; Ritchie and Hewitt, 1993; Virdee and Hewitt, 1993b). Evidence for amelioration of hybrid unfitness is also coming from shrew chromosomal hybrid zones where the bad effects of complex monobrachial rearrangements are alleviated by either the presence of the basic acrocentrics (Searle, 1986), or homozygous segregants (Fedyk et al.1991).

9   Repeated range changes and accumulation

It has been argued that postglacial expansion from refugia provided suitable conditions for the initial fixation and subsequent spread of underdominant mutants, including chromosomal rearrangements. During the Pleistocene (~2 Myrs) there have been a series of major ice ages with shorter warm interglacials every 100,000 yrs, brought about by the regular variation in orbital eccentricity of the earth round the sun. Shorter cycles of 41000 years and 23000 years in axial tilt and axial precession overlay this and cause further variations in climate (Bartlein and Prentice, 1989). Consequently these changing conditions would have been repeated many times, providing many opportunities for mutant establishment. Furthermore, when two such diverging genomes met they may well have formed a hybrid zone, which as we have seen can be a barrier to gene flow and can prevent the intermingling of the two genomes. Repeated expansions and contractions of these semipermeable bubbles would generate and accummulate underdominant differences producing more complex and stronger hybrid zones. In time the diverged genomes may reach species status (Hewitt,1989). Since the orbital cycles go back through geologic time, such a process could have operated over many millenia.

Such range changes will not only fix underdominant mutations but also produce selection for adaptations to divergent conditions in different parts of the range, (Hewitt, 1993b). For example in *C. parallelus* the Iberian refuge would have had a different climate from a Balkan or Asian one, and *C.p. erythropus* has a number of morphological and behavioural characters which indicate a different ecological adaptation from *C. p. parallelus*. In *Caledia captiva* the two forms appear adapted to different climate regimes as a result of their separate histories. As a result we would expect two emergent genomes to accumulate many kinds of character differences. Of course there will be occasions when diverging forms hybridise and intermix completely after contact is reestablished, particularly when the differences are not great. Furthermore, the importance of this process will vary with the attributes of the organism, eg its dispersal, host specificity, mating system and cytogenetic properties. For example, divergence and speciation in fruit flies or treehoppers which are very host specific may not require much, if any, geographic rearrangement (Diehl and Bush, 1989; Wood and Keese, 1990). None the less a large variety of organisms do have genomes subdivided by hybrid zones.

10   Relative contribution of chromosomes

From these considerations focussed on the Orthoptera, we may construct a likely evolutionary history to account for the prevalence of species divided and subdivided by hybrid zones which range in complexity from very simple to virtually complete barriers to gene flow. An underdominant chromosome mutant can become fixed in locations where the density and dispersal are low (small neighbourhood size) and they can then be spread over greater areas by changes in the organism's range caused by environmental changes. Because such differences cause some hybrid unfitness when two types meet thcy will form a hybrid zone, which will restrict gene flow to some extent. Repeated range changes could accummulate such underdominant differences until the two genomes form complex hybrid zones which largely prevents gene flow. When their hybrids are effectively unviable or sterile they may be considered two species, but to become sympatric they must have also developed differences in mate recognition and ecological adaptation.

Such a process would apply to all underdominant genes and characters together, so that we would expect the two diverging forms to differ for chromosomal, developmental, morphological and behavioural characters, any or all of which could contribute to the hybrid unfitness. The Orthopteran examples vary in the relative importance of different characters in reducing gene flow; in some, chromosomes are a major component while in others they are not. It may be chance which character differences are established together in local neighbourhoods, but maybe there could be a tendency for certain changes to be established in particular cases. Where two forms experience very different ecological conditions when separate, then morphological and behavioural differences might be expected to dominate. The karyotypes of different organisms may be more prone to certain types of mutation, and certain mutants may be more likely to survive. Thus in the Orthoptera interchanges are the most frequent spontaneous mutation recorded, but they rarely distinguish subspecies; while fusion/fission mutants occur less often and species are frequently polytypic for them (Hewitt, 1979). Interestingly interchange heterozygotes produce unbalanced gametes while fusion/fission trivalents nearly always orientate alternately. This would suggest that it is not the relative frequency of mutation which determines the type of karyotypic subdivision, but rather the chance of establishment in a small neighbourhood, which itself depends on only slight underdominance. Comparing groups within the Orthoptera, the Morabinae and some Acrididae contain a range of fusion/fissions distinguishing species; maybe their karyotypes or population dynamics favour this type of change. It is worth noting that the sex chromosome is more frequently involved in these rearrangements, which fits with the general case that is emerging for sex linked speciation genes (Coyne, 1992).

It has been argued that divergence and speciation involves changes in a complex of characters which are variably combined, and chromosome differences are just one of these. But we are still left with the question that prompts chromosome models of speciation: why should chromosomes nearly always be involved? One answer is that chromosome mutation rates are relatively high so that when a population reduction occurs providing good conditions for fixation there are often rearrangements around. Also certain chromosome rearrangements may have a low level of underdominance, which more readily allows fixation but is enough to produce hybrid zones between taxa. There is evidence that mutation frequencies may be somewhat higher in natural populations than is classically accepted (eg. Porter and Sites, 1987; Parker et al. 1988; Stevens and Bougourd, 1991), and the reductions in fitness caused by centric fusion/fissions and some inversions may be small (eg. Hewitt, 1979; Searle, 1990; Coyne et al. 1991). Perhaps it's not such a mystery after all.

11   Acknowledgements

I am very grateful to Drs. Colin Ferris and Diane Howland for their help with the manuscript. The research was supported by grants from the NERC and SERC.

12   References

Arnold, M.L., Shaw, D.D. and Contreras, N. (1987) Ribosomal RNA-encoding DNA introgression across a narrow hybrid zone between two subspecies of grasshopper. **Proc. Natl. Acad. Sci. USA**, 84, 3946-3950.

Bartlein, P.J. and Prentice, I.C. (1989) Orbital variations, climate and paleoecology. TREE, 4, 195-199.

Barton, N.H. (1979) The dynamics of hybrid zones. **Heredity**, 43, 341-359.

Barton, N.H. and Charlesworth, B. (1984) Genetic revolutions, founder effects and speciation. **Ann. Rev. Ecol. Syst.**, 15, 133-164.

Barton, N. H. and Hewitt, G.M. (1981) The genetic basis of hybrid inviability between two chromosomal races of the grasshopper *Podisma pedestris*. **Heredity**, 47, 367-383.

Barton , N.H. and Hewitt, G.M. (1985) Analysis of hybrid zones. **Ann. Rev. Ecol. Syst.**, 16, 113-148.

Barton, N.H. and Hewitt, G.M. (1989) Adaptation, speciation and hybrid zones. **Nature**, 341, 497-503.

Barton, N.H. and Rouhani, S. (1991) The probability of fixation of a new karyotype in a continuous population.**Evolution**, 45, 499-517.

Bella, J.L., Butlin, R.K. Ferris, C. and Hewitt, G.M. (1992) Asymmetrical homogamy and unequal sex ratio from reciprocal mating-order crosses between *Chorthippus parallelus* subspecies. **Heredity**, 68, 345-352.

Bush, G.L. et al. (1977) Rapid speciation and chromosomal evolution in mammals. **Proc. Nat. Acad. Sci.**, 74, 3942-3946.

Butlin, R.K. (1989) Reinforcement of premating isolation, in **Speciation and its Consequences** (eds D.Otte and J.A.Endler), Sinauer Assoc., Sunderland Mass, pp. 158-179.

Butlin, R.K. and Hewitt, G.M. (1985) A hybrid zone between *Chorthippus parallelus parallelus* and *C.p.erythropus* (Orthoptera: Acrididae): Morphological and electrophoretic characters. **Biol. J. Linn. Soc.**, 26, 269-285.

Butlin, R.K. and Ritchie, M.G. (1991) Variation in female mate preference across a grasshopper hybrid zone. **J. Evol. Biol.**, 4, 227-240.

Butlin, R.K., Ritchie, M.G. and Hewitt, G.M. (1991) Comparisons among morphological characters and between localities in the *Chorthippus parallelus* hybrid zone (Orthoptera: Acrididae). **Phil. Trans. R. Soc. B**, 344, 297-308.

Carson, H.L. (1981) Homosequential species of Hawaian *Drosophila*, in **Chromosomes Today**, vol. 7, (eds M.D. Bennett, M. Bobrow and G.M. Hewitt), Allen and Unwin, London, pp. 150-164.

Coope, G.R. (1990) The invasion of Northern Europe during the Pleistocene by Mediterranean species of Coleoptera, in **Biological Invasions in Europe and the Mediterranean Basin** (eds F.Di Castri, A.J.Hausen and M. De Bussche), Kluwer, Dordrecht, pp. 203-215.

Coyne, J.A. (1989) A test of the role of meiotic drive in chromosome evolution. **Genetics**, 123, 241-243.

Coyne, J.A. (1992) Genetics and speciation. **Nature**, 335, 511-515.

Coyne, J.A., Aulard, S. and Berry, A. (1991) Lack of underdominance in a naturally occurring pericentric inversion in *Drosophila melanogaster* and its implications for chromosome evolution. **Genetics**, 129, 791-802.

Dallas, J.F., Barton, N.H. and Dover, G.A. (1988) Interracial rDNA variation in the grasshopper, *Podisma pedestris*. **Mol. Biol. Evol.**, 5, 660-674.

Diehl, S.R. and Bush, G.L. (1989) The role of habitat preference in adaptation and speciation, in **Speciation and its Consequences** (eds. D. Otte and J.A. Endler), Sinauer Assoc., Sunderland Mass., pp. 345-365.

Fedyk, S., Chetnicki, W. and Banaszek, A. (1991) Genetic differentiation of Polish populations of *Sorex araneus* L. III. Interchromosomal recombination in a hybrid zone. **Evolution**, 45, 1384-1392.

Fletcher, H.L. and Hewitt, G.M. (1978) Non-homologous synaptonemal complex formation in a heteromorphic bivalent in *Keyacris scurra* (Morabinae, Orthoptera) **Chromosoma**, 65, 271-281.

Gyllensten, U. and Wilson, A.C. (1987) Interspecific mitochondrial DNA transfer and the colonization of Scandinavia by mice. **Genet. Res.**, 49, 25-29.

Harrison, R.G. (1990) Hybrid zones: windows on evolutionary process, in **Oxford Surveys in Evolutionary Biology Vol. 7** (eds D. Futuyma and J. Antonovics), Oxford University Press, New York, pp. 69-128.

Harrison, R.G. and Rand, D.M. (1989) Mosaic hybrid zones and the nature of species boundaries, in **Speciation and its Consequences** (eds D.Otte and J.A.Endler), Sinauer Assoc., Sunderland, Mass., pp. 111-113.

Hewitt, G.M. (1975) A sex chromosome hybrid zone in the grasshopper *Podisma pedestris* (Orthoptera: Acrididae). **Heredity**, 35, 375-387.

Hewitt, G.M. (1979) **Animal Cytogenetics, III, Orthoptera.** Gebruder Borntrager, Stuttgart.

Hewitt, G.M. (1985) The structure and maintenance of hybrid zones - with some lessons to be learned from alpine grasshoppers, in **Orthoptera Vol. I**, (eds J. Gosalvez, C. Lopez-Fernandez and C. Garcia de la Vega), Fundacion Ramon Areces, Madrid, pp 15-54.

Hewitt, G.M. (1988) Hybrid zones - natural laboratories for evolutionary studies. **TREE**, 3, 158-167.

Hewitt, G.M. (1989) The subdivision of species by hybrid zones, in **Speciation and its Consequences**, (eds D. Otte and J.A. Endler), Sinauer Associates, Sunderland Mass., pp. 85-110.

Hewitt, G.M. (1990) Divergence and speciation as viewed from a hybrid zone. **Can. J. Zool.**, 68, 1701-1715.

Hewitt, G.M. (1993a) After the Ice - *parallelus* meets *erythropus* in the Pyrenees, in **Hybrid Zones and the Evolutionary Process** (ed. R.G.Harrison), Oxford University Press, Oxford (in press).

Hewitt, G.M. (1993b) Postglacial distribution and species substructure: lessons from pollen, insects and hybrid zones. **Biol. J. Linn. Soc.**, (in press).

Howard, D.J. (1986) A zone of overlap and hybridization between two cricket species. **Evolution**, 40, 34-43.

Huntley, B. (1990) European vegetation history: paleovegetation maps from pollen data -13000 yr BP to present. **J. Quatern. Sci.**, 5, 103-122.

James, S.H. (1992) Inbreeding, self-fertilization, lethal genes and genomic coalescence. **Heredity**, 68, 449-456.

John, B. and Lewis, K.R. (1965) Genetic speciation in the grasshopper *Eyprepocnemis plorans*. **Chromosoma**, 16, 308-344.

Key, K.H.L. (1976) A generic and supragenic classification of the Morabinae (Orthoptera: Eumastacidae) with description of the type species and a bibliography of the subfamily. **Austral. J. Zool. Ser.**, No. 37, pp 185.

Kohlmann, B. and Shaw, D.D. (1991) The effect of a partial barrier on the movement of a hybrid zone. **Evolution**, 45, 1606-1617.

Lande, R. (1979) Effective deme sizes during long term evolution estimated from rates of chromosomal rearrangement. **Evolution**, 33, 234-251.

Mayr, E. (1970) **Populations, Species, and Evolution**. Harvard University Press, Cambridge, Mass.

Mrongovius, M.J. (1979) Cytogenetics of the hybrids of three members of the grasshopper genus *Vandiemenella* (Orthoptera: Eumastacidae: Morabinae). **Chromosoma**, 71, 81-107.

Nagylaki, T. (1975) Conditions for the existence of clines. **Genetics**, 80, 595-615.

Nance, V.et al. (1990) Chromosomal introgression in house mice from the hybrid zone between *M.m.domesticus* and *M.m.musculus* in Denmark. **Biol. J. Linn. Soc.**, 41, 215-227.

Nichols, R.A. and Hewitt, G.M. (1986) Population structure and the shape of a chromosome cline between two races of *Podisma pedestris* (Orthoptera: Acrididae). **Biol. J. Linn. Soc.**, 29, 301-316.

Nichols, R.A. and Hewitt, G.M. (1988) Genetical and ecological differentiation across a hybrid zone. **Ecol. Entomol.**, 13, 39-49.

Nichols, R.A. and Hewitt, G.M. (1993) The genetic consequences of long distance dispersal during colonisation. **Heredity** (in press).

Otte, D. and Endler, J. (1989) **Speciation and its Consequences** (eds), Sinauer Assoc., Sunderland, Mass.

Parker, J.S., Wilby, A.S. and Taylor, S. (1988) Chromosome stability and instability in plants, in **Kew Chromosome Conference III** (ed P.E. Brandham), HMSO, London, pp. 131-140.

Porter, C.A. and Sites, J.W. (1987) Evolution of *Sceloporus grammicus* complex (Sauria: Iguanidae) in central Mexico II. Studies on rates of nondisjunction and the occurence of spontaneous chromosomal mutations. **Genetica**, 75, 131-144.

Rand, D.M. and Harrison, R.G. (1989) Ecological genetics of a mosaic hybrid zone: mitochondrial, nuclear and reproductive differentiation of crickets by soil type. **Evolution**, 43, 432-449.

Reille, M. (1990) La Tourbiére de la Borde (Pyrénées orientales, France): un site clé pour l'etude du Tardiglaciaire Sud-Européan. **Compt. Rend. Acad. Sci. Paris ser. II,** 310, 823-829.

Ritchie, M.G. and Hewitt, G.M. (1993) Outcomes of negative heterosis, in **Essays in Honour of H.E.H.Paterson** (eds D. Lambert, H. Spencer and J. Masters), Johns Hopkins University Press, (in press).

Searle, J.B. (1986) Factors responsible for a karyotypic polymorphism in the common shrew, *Sorex araneus*. **Proc. R. Soc. Lond. (B)**, 229, 277-298.

Searle, J.B. (1990) A cytogenetical analysis of reproduction in common shrews (*Sorex araneus*) from a karyotypic hybrid zone. **Hereditas**, 113, 121-132.

Shaw, D.D. and Coates, D.J. (1983) Chromosomal variation and the concept of the coadapted genome- a direct cytological assessment, in **Kew Chromosome Conference II** (eds P.E. Brandham and M.D. Bennett), George, Allen and Unwin, London, pp. 207-216.

Shaw, D.D., Coates, D.J. and Wilkinson, P. (1985) Temporal variation in the chromosomal structure of a hybrid zone and its relationship to karyotypic repatterning. **Heredity**, 55, 293-306.

Shaw, D.D. et al. (1988) Chromosomal rearrangements, ribosomal genes and mitochondrial DNA: contrasting patterns of introgression across a narrow hybrid zone, in **Kew Chromosome Conference III** (ed P.E. Brandham), HMSO, London, pp. 121-129.

Sites, J.W.J. and Moritz, C. (1987) Chromosomal Evolution and speciation revisited. **Syst. Zool.,** 36, 153-174.

Stevens, J.P. and Bougourd, S.M. (1991) The frequency and meiotic behaviour of structural chromosome variants in natural populations of *Allium schoenoprasum* L. (wild chives) in Europe. **Heredity,** 66, 391-401.

Szymura, J.M. and Barton, N.H. (1991) The genetic structure of the hybrid zone between the fire-bellied toads *Bombina bombina* and *B. variegata* : comparisons between transects and between loci. **Evolution,** 45, 237-261.

Virdee, S.R. and Hewitt, G.M. (1990) Ecological components of a hybrid zone in the grasshopper *Chorthippus parallelus* (Zetterstedt) (Orthoptera: Acrididae). **Bol. San. Veg. Plagas (Fuera de Serie),** 20, 299-309.

Virdee, S.R. and Hewitt, G.M. (1993a) Postzygotic isolation and Haldane's Rule in a grasshopper. **Heredity,** (in press).

Virdee, S.R. and Hewitt, G.M. (1993b) Clines for hybrid dysfunction in a grasshopper hybrid zone. **Evolution,** (in press).

White, M.J.D. (1978) **Modes of Speciation.,** W.H.Freeman & Co., San Francisco.

Wood, T.K. and Keese, M.C. (1990) Host-plant-induced assortative mating in *Echenopa* treehoppers. **Evolution,** 44, 619-628.

10 Chromosomal variation in the *Sceloporus grammicus* complex (Sauria, Phrynosomatidae)

J.W. SITES, Jr
Brigham Young University, USA

1 Introduction

The Sceloporus grammicus complex offers a valuable resource for studies of interrelated questions about population structure, the origin and fixation of new chromosomal rearrangements, the fitness consequences of structural chromosomal heterozygosity in hybrid zones, and the possible role of such rearrangements in reducing gene flow between chromosomally pure types in regions of parapatric contact. This complex, which ranges from extreme southern Texas south across most of mainland Mexico (see map in Sites et al., 1987), is composed of a number of distinct chromosome races (= cytotypes) with different 2n's resulting largely from fixed simple Robertsonian rearrangements. Some of these cytotypes have diverged to the species level, as assessed by independent molecular markers, while others are less distinct. Similar chromosomal polytypy has been described in other vertebrates, notably eutherian mammals (Searle, 1992), but in contrast to most of these cases, many cytotypes in the S. grammicus complex exhibit extensive within-population polymorphism as well as between-population fixation for the same kinds of rearrangements. These observations, coupled with estimates of very low levels of isozyme divergence (see below), suggest that processes relevant to the origin, spread, and fixation (or maintenance as polymorphisms) of chromosomal rearrangements may still be actively occurring in this group. The apparent narrowness of a number of hybrid zones also suggests that even pairs of cytotypes capable of hybridizing and backcrossing are behaving as distinct gene pools, and that the between-race chromosomal differences may contribute significantly to speciation potential in this group (Hall, 1983). This paper reviews some aspects of recent work on a number of these issues by myself, students, and colleagues.

Out studies for the past several years have been concentrated on a small portion of the range in central Mexico, chosen because of earlier reports of extensive chromosomal polymorphism in a relatively small area, extreme

Chromosomes Today   Volume 11. Edited by A.T. Sumner and A.C. Chandley. Published in 1993 by Chapman & Hall, London. ISBN 0 412 47670 3

divergence of diploid numbers (32 to 46), and the documentation of zones of parapatric hybridization (Hall, 1980; Hall and Selander, 1973). Our own field work has confirmed and extended these findings (Porter and Sites, 1986; Arévalo et al., 1991), and has now revealed a total of 8 different cytotypes (two previously unknown), and at least seven distinct hybrid zones between different combinations of diploid numbers (including the three originally reported by Hall, 1980). Three other localities exhibit extreme chromosomal polytypy from single sampling points geographically positioned between different cytotypes, but transect sampling has not yet been carried out in these regions to determine if these also represent points of parapatric contact. The distributions of the 8 central Mexico cytotypes, hybrid zones, and polymorphic populations are summarized in Fig. 1, and provide the baseline data set from which we have sought to evaluate more explicit hypotheses of chromosomal evolution and speciation. Specifically, recent studies have centered on (or are still ongoing) population cytogenetics, molecular estimates of population structure, male meiosis in chromosomal homozygotes and heterozygotes in nonhybrid zone contexts, male meiosis and female fecundity in a hybrid zone, the gene flow dynamics of the same zone as assessed through nuclear and mitochondrial markers, and phylogenetic relationships of all 8 central Mexico cytotypes.

2 Population Cytogenetics

Population genetic structures have been estimated for all cytotypes on the basis of distributions and frequencies of chromosomal polymorphisms. Hierarchical F-statistics analyses show that 67% of the overall chromosomal polymorphism could be attributed to within-race variation relative to the between-race component. Within samples, genotype ratios generally conform to Hardy-Weinberg proportions, and where this is not the case, deviations always occur in the direction of heterozygote excess, although not at statistically significant numbers. Some cytotypes show significant between-locality heterogeneity in chromosomal polymorphism frequencies, and spatial autocorrelation analyses suggest an isolation-by-distance structure for some races, and a random pattern in others in which drift may be the overriding force determining population structure (details in Arévalo et al., 1991).

3 Independent Assessments of Population Structure

Many models of chromosomal evolution and speciation are predicated on the assumption of strong underdominance for a new chromosomal rearrangement in the heterozygous state, and a corollary of this assumption is the necessity of a strong "Wrightian" population structure to permit the occasional fixation of such

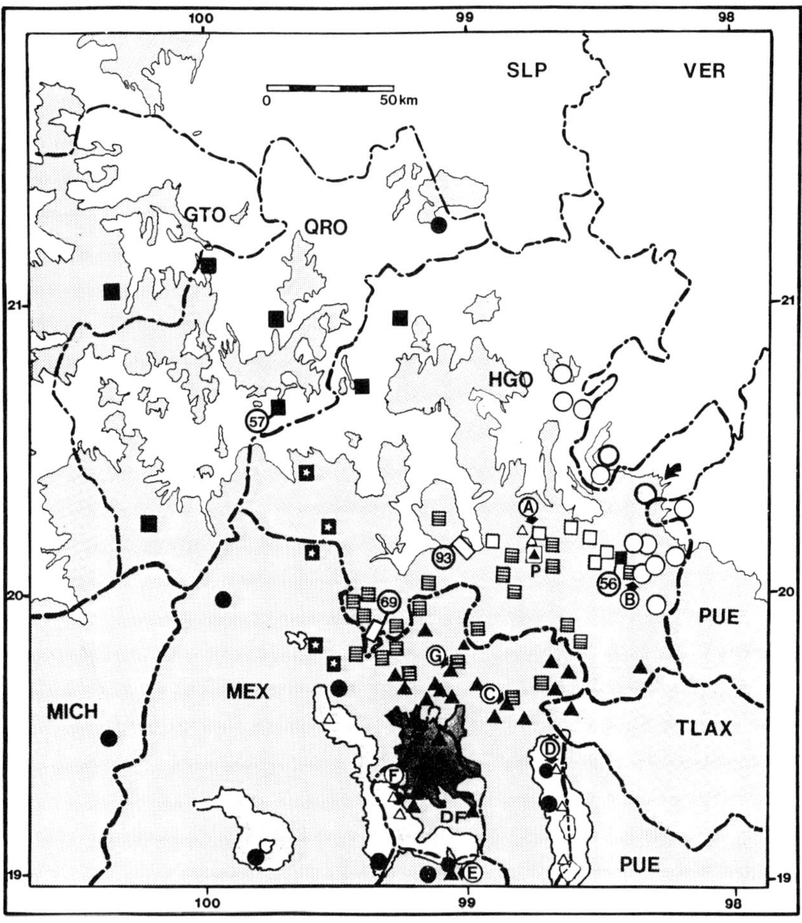

Fig. 1. Distribution of 8 cytotypes of <u>S</u>. <u>grammicus</u> in central Mexico. The 2000, 3000, and 4000 m contours are identified by extensive gray shading, limited fine-grained gray, and white volcanic peaks, respectively, while darkest shading identifies Mexico City. All other political units are abbreviated as in Arevalo et al. (1991). Cytotypes are identified as **low (LS)** and **high standard (HS,** both 2n = 32) - solid and open **triangles,** respectively; **fission 6 (F6)** and F5 (both 2n = 34) - solid and open **circles,** respectively; **F5 + 6,** (2n = 36) - solid squres; **FM3** (2n = 38) - open squares; **FM1** (2n = 42) - squares with stars; and **FM2** (2n = 46) - hatched squares. The arrow shows location of another cytogenetically distinct populations, but further sampling is needed to clarify its status. Numbers enclosed in circles (56, 57, and 69) identify extremely polymorphic populations, and capital letters in circles show hybrid zones (modified from Arevalo et al., 1993).

rearrangments by peak shifts to new homozygous states (Sites and Moritz, 1987; Searle, 1992). We have therefore carried out isozyme-based comparative studies of population structure in parallel with the cytogenetic studies, and to date these studies suggest that at least some cytotypes may not have the strongly subdivided population structure conducive to fixation of underdominant rearrangements by peak shifts (Thompson and Sites, 1986). Similar conclusions are taken from computer simulations of the range of electrophoretic "signatures" expected in breeding populations that occasionally fixed underdominant chromosomal mutations (Sites et al., 1988a). Not all central Mexico populations have yet been examined, but ongoing isozyme and mtDNA restriction mapping studies (Arévalo and Sites, Arévalo et al., in prep.) will include all cytotypes mapped in Fig. 1. The work completed to date collectively suggests that the rearrangements frequently seen as within-race polymorphisms in non-hybrid zone locations may be segregating as quasi-neutral markers, and therefore are being maintained in demes that approximate random mating and that are at least occasionally interconnected by gene flow.

4 Meiotic Studies

Direct evidence of the possible fitness consequences of structural chromosomal heterozygosity has been obtained from meiotic studies of reproductively active males representing both chromosomally homozygous and heterozygous individuals. Since genetic background may influence meiosis (reviewed in Searle, 1992), preliminary studies were centered on non-hybrid zone localities to establish controls for interpretations of meiotic behavior of heterozygotes in hybrid zones. In these early studies, Porter and Sites (1985, 1987) examined over 2,000 MII cells from 30 adult S. grammicus representing several cytotypes, and over half of these (n=16) segregated completely euploid MII cells (including both homozygous and heterozygous individuals). The remaining 14 lizards, also including homozygotes and heterozygotes, produced aneuploid MII cells at relatively low frequencies (0.6% - 7.1%). When all homozygotes were considered together as a sample, the mean frequency of nondisjunction/aneuploid MII formation was just above 1.0% (0.013, n=14), while lizards heterozygous for one or two simple fissions showed a mean aneuploid MII frequency of 0.024 (n=16). If the 1.0% aneuploid frequency of the homozygotes is used as a control to establish the "background" level of nondisjunction, then heterozygosity for simple fissions raises the nondisjunction frequency by just under 1.5%, at least when assessed on similar genetic backgrounds. These uniformly low rates of aneuploid gamete production are not likely to have serious fitness consequences (Nachman and Myers, 1989, for example), and corroborate our earlier conclusions about the nature of these polymorphisms from isozyme studies of population structure. These studies alone, however, provide no information to the meiotic behavior or possible fitness consequences of chromosomal heterozygotes in a hybrid zone.

The successful application of high resolution surface-spreading and electron microscopic studies of meiosis to <u>Sceloporus</u> (Reed et al., 1990) has permitted visualization of very early stages of synapsis and pairing behavior. Reed et al. (1992a) established criteria for substaging of zygonema and pachynema and characterization of the synaptonemal complex, which provided more detailed descriptions of orientation and segregation of both autosomes and sex chromosomes than those collected by Porter and Sites with light microscopy. An example of these data is presented in Fig. 2. Reed et al. (1992a) examined non-hybrid samples of the F5 cytotype (2n = 34, all acrocentric for pr 5) using SC methods, and determined two patterns of autosomal synapsis. In bivalents of biarmed chromosomes early pairing was biterminal, being initiated from both telomeric regions and progressing medially. In acrocentric bivalents pairing was uniterminal, involving only the noncentromeric distal ends of the homologs. The sex chromosomes (an X_1X_2Y system which pairs as a trivalent, Fig. 2) did not display conspicuous pairing and staining characteristics useful for substaging, but did pair synchronously with the autosomes. MI and MII data showed low levels of diakinetic irregularities and balanced segregation of autosomal bivalents and sex chromosome trivalents.

These studies were extended to examine meiotic behavior of a large autosomal (pr 4) pericentric inversion heteromorphism segregating at low frequencies in the F5 race (Reed et al., 1992b), and a simple fission heteromorphism occurring at high frequencies in this same pair in the FM2 race (Reed et al., 1992c). Analysis of the SCs in the inversion showed that homologously paired inversion loops were not formed, and synapsis of the inverted regions proceeded directly to nonhomologous straight pairing. Examination of Giemsa- and silver-stained diakinetic (MI) nuclei indicated that crossing-over was limited to the noninverted portion of the heteromorphic bivalents, and counts of MII nuclei revealed normal disjunction and balanced segregation of these elements (Reed et al., 1992b).

SC studies of surface-spread zygotene/pachytene nuclei of the pr 4 fission heteromorphism in FM2 showed that these elements initiated pairing at distal telomeric regions, but with delayed completion of synapsis relative to homomorphic autosomal bivalents. Associations between fission trivalents and other autosomal and sex-chromosome elements occurred in about one-third of the pachytene nuclei examined. However, distal synaptic initiation sites and unidirectional synapsis proceeding toward the centromere probably facilitates

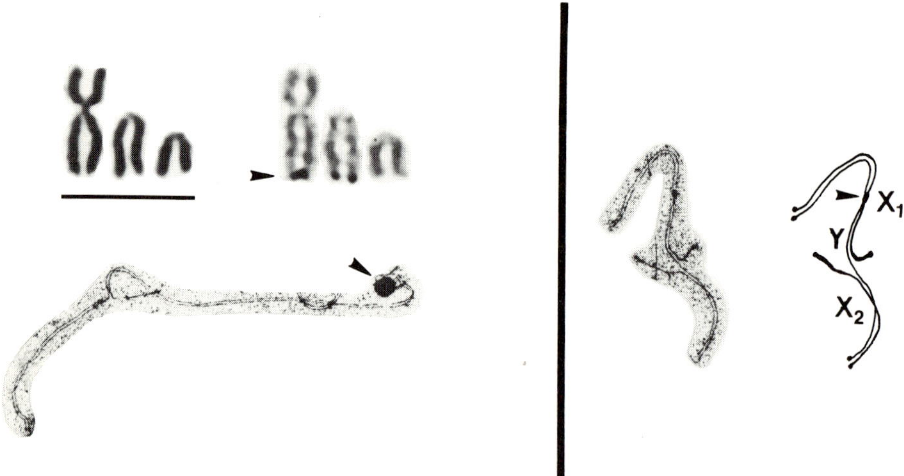

Fig. 2. Chromosomal preparations from a hybrid male S. grammicus (BYU 39988) taken from the Tulancingo transect (see below). Left panel shows Giemsa- and silver-stained chromosomes (bar = 10 um) and SC trivalent (reproduced at 3000x) from an individual heterozygous for a recombinant morphology of chromosome 2 (details in Reed et al., 1992c). Arrowheads denote position of the NOR. Right panel is an electron micrograph and line drawing of the X_1X_2Y sex chromosome trivalent (reproduced at 2500x); centromeres of the Y and X_1 are aligned.

proper orientation (White, 1973), and analysis of MII cells revealed very low levels of nondisjunction in fission heterozygotes and no meiotic deficits (Reed et al., 1992c).

5 Phylogenetic Studies

Recently renewed efforts to synthesize systematics and ecology (Brooks and McLennan, 1991; Harvey and Pagel, 1991) have emphasized the importance of well-corroborated phylogenetic hypotheses in comparative studies. In the case of the chromosomal polytypy in the S. grammicus complex, it is important to determine the geographic boundaries of phylogenetically "diagnosable" units (Nixon and Wheeler, 1990) with molecular markers, and then determine the extent to which these are geographically concordant with the distribution of the chromosome races. Results showing all cytotypes to be recognizable on the basis of other classes of genetic markers would indicate that the chromosomal races had diverged to the point of independently evolving lineages, whereas a finding that two or more races were contained within a single phylogenetic unit would suggest that karyotypic rearrangements may be manifestions of peak shifts within a single species. Establishing the phylogenetic relationships of the cytotypes will also provide an independent test of the direction of chromosomal change within this complex, inform interpretations about the origins of the hybrid zones (primary vs. secondary), and identify apomorphic molecular markers between races involved in parapatric contacts for studies of gene flow dynamics of hybridization.

Of the 8 cytotypes depicted in Fig. 1, the $2n = 32$ karyotype has been postulated to be the ancestral karyotype, while the races with higher diploid numbers are presumably derived by successive fission events in the macrochromosome pairs, which generates a linear series of $2n$'s ranging almost continuously from 34 to 46 (Hall, 1980, 1983). Allozyme studies (Sites et al., 1988b) and restriction-site mapping studies of nuclear (ribosomal) and mitochondrial DNAs have revealed phylogenetically informative markers in most of the central Mexico cytotypes (Sites and Davis, 1989). The study by Sites and Davis was preliminary in that one cytotype was not included, and several others were inadequately sampled, but results were encouraging because the molecular-based cladograms were robust and generally confirmed the topology of the chromosomally-based hypothesis. Ongoing studies are extending the earlier work by including mtDNA sequence data, and additional populations of all 8 central Mexico cytotypes in more extensive isozyme and RFLP data sets (Arévalo and Sites, in prep.; Arévalo et al., in prep.).

6. Hybrid Zone Structure

Studies of the dynamics of hybrid zones in the S. grammicus complex will fill crucial gaps in our understanding of the consequences of chromosomal heterozygosity, and ultimately provide deeper insight into general aspects of adaptation and speciation (Barton and Hewitt, 1989; Harrison, 1991). We have established single transects across two of the contacts illustrated in Fig. 1 (B and F) to provide comparative perspectives on the genetic structure and patterns of gene flow in this group. These zones differ greatly in the degree of chromosomal divergence (32 x 34 in F; 34 x 46 in B), and their overall structures are described in Arévalo et al. (1993, zone F) and Sites et al. (1993, zone B) on the basis of morphological, chromosomal, isozyme, mtDNA and rDNA restriction-site data sets. In zone B, three unambiguous chromosome markers showed sharp, concordant transitions from one race to the other over a distince of less than 2 km, and virtually no introgression of diagnostic chromosome markers occurred outside of the area immediately adjacent to the point of contact. This transition was also reflected in diagnostic mtDNA haplotypes, albeit with slight asymmetry of flow in the direction of F5, while rDNA repeats introgressed more extensively in both directions (Sites et al., 1993). Sample sizes were small, however, and sufficient only to characterize the general patterns of this transect. Because of ease of access and the high divergence in karyotypes involved in this contact, it was selected for more intensive sampling during the summer of 1989 (to maximize meiotic data collected from males) and the winter of 1991 (to collect clutch size data from females). All collecting was carried out with surveying equipment to map all capture points and major structural features of the transect, and maps were prepared on a scale of 1:1000. This effort netted karyotypes from a total of 558 lizards (including some from the earlier study), meiotic data on almost 180 males, fecundity data from 110 females, and molecular data in the form of diagnostic mtDNA and rDNA restriction sites scored for a large subset (ca. 525 animals) of the total. Analyses of these data sets are still in progress, but when completed will provide estimates of selection and dispersal derived from cline shapes for the diagnostic chromosome markers, meiotic malsegregation potential in males and reproductive success in females of hybrid and backcross genotypes from the central region of the transect, and gene flow patterns for mtDNA and rDNA markers. These data sets collectively will offer deeper insight into factors likely responsible for maintenance of this zone in the S. grammicus complex.

ACKNOWLEDGMENTS

I thank Elisabeth Arévalo, Calvin Porter, Kent Reed, and Pam Thompson for their participation and input over the past several years on different aspects of the research effort on the S. grammicus complex. These investigations have been supported by grants from the American Museum of Natural History (Theodore Roosevelt Fund), Brigham Young University, the National Geographic Society (nos. 2803-84 and 3088-85), and the National Science Foundation (BSR 85-09092 and 88-22751).

REFERENCES

Arévalo, E., et al. 1991. Population cytogenetics and evolution of the Sceloporus grammicus complex (Iguanidae) in central Mexico. **Herp. Monogr.** 5:79-115.

Arévalo, E., et al. 1993. Parapatric hybridization between chromosome races of the Sceloporus grammicus complex (Phrynosomatidae): structure of the Ajusco transect. **Copeia** 1993:(in press).

Barton, N.H., and G.M. Hewitt. 1989. Adaptation, speciation, and hybrid zones. **Nature** 341:497-503.

Brooks, D.R., and D. McLennan. 1991. **Phylogeny, Ecology, and Behavior.** Univ. Chicago Press, Chicago.

Hall, W.P. 1980. Chromosomes, speciation, and the evolution of Mexican iguanid lizards. **Nat. Geog. Res.** 12:309-329.

Hall, W.P. 1983. Modes of speciation and evolution in the sceloporine iguanid lizards. I. Epistemology of the comparative approach and introduction to the problem, in **Advances in Herpetology and Evolutionary Biology** (eds A.G.J. Rhodin and K. Miyata), Mus. Comp. Zool., Harvard Univ., Cambridge, pp. 643-679.

Hall, W.P., and R.K. Selander. 1973. Hybridization of karyotypically differentiated populations in the Sceloporus grammicus complex (Iguanidae). **Evolution** 27:226-242.

Harvey, P., and M. Pagel. 1991. **The Comparative Method in Evolutionary Biology.** Oxford Univ. Press, New York.

Harrison, R.G. 1991. Hybrid zones: windows on evolutionary process, in **Oxford Surveys in Evolutionary Biology** (eds. D. Futuyma and J. Antonovics), Oxford Univ. Press, London, pp. 69-128.

Nachman, M.W., and P. Myers. 1989. Exceptional chromosomal mutations in a rodent population are not strongly underdominant. **Proc. Nat. Acad. Sci. USA** 86:6666-6670.

Nixon, K.C., and Q.D. Wheeler. 1990. An amplification of the phylogenetic species concept. **Cladistics** 6:211-223.

Porter, C.A., and J.W. Sites, Jr. 1985. Normal disjunction in Robertsonian heterozygotes from a highly polymorphic lizard population. **Cytogenet. Cell Genet.** 39:250-257.

Porter, C.A., and J.W. Sites, Jr. 1986. Evolution of the Sceloporus grammicus complex (Sauria, Iguanidae) in central Mexico: population cytogenetics. **Syst. Zool.** 35:334-358.

Porter, C.A., and J.W. Sites, Jr. 1987. Evolution of Sceloporus grammicus complex (Sauria: Iguanidae) in central Mexico. II. Studies on rates of nondisjunction and the occurrence of spontaneous chromosomal mutations. **Genetica** 75:131-144.

Reed, K.M., et al. 1990. Synaptonemal complex analysis of sex chromosomes in two species of Sceloporus. **Copeia** 1990:1092-1099.

Reed, K.M., J.W. Sites, Jr., and I.F. Greenbaum. 1992a. Chromosomal synapsis and the meiotic process in male mesquite lizards, Sceloporus grammicus complex. **Genome** 35:398-408.

Reed, K.M., J.W. Sites, Jr., and I.F. Greenbaum. 1992b. Synapsis, recombination, and meiotic segregation in the mesquite lizard, Sceloporus grammicus complex. I. The pericentric inversion heteromorphism in the F5 cytotype. **Cytogenet. Cell Genet.** 61:40-45.

Reed, K.M., J.W. Sites, Jr., and I.F. Greenbaum. 1992c. Synapsis, recombination, and meiotic segregation in the mesquite lizard, Sceloporus grammicus complex. II. The fission heteromorphism of the FM2 cytotype and the evolution of chromosome 2. **Cytogenet. Cell Genet.** 61:46-54.

Searle, J. 1992. Chromosomal hybrid zones in eutherian mammals, in **Hybrid Zones and the Evolutionary Process** (ed R.G. Harrison), Oxford Univ. Press, New York, pp.xx-xx.

Sites, J.W., Jr., and C. Moritz. 1987. Chromosomal evolution and speciation revisited. **Syst. Zool.** 36:153-174.

Sites, J.W., Jr., and S.K. Davis. 1989. Phylogenetic relationships and molecular variability within and among six chromosome races of Sceloporus grammicus (Sauria, Iguanidae) based on nuclear and mitochondrial markers. **Evolution** 43:296-317.

Sites, J.W., Jr., C.A. Porter, and P. Thompson. 1987. Population genetic structure and chromosomal evolution in the Sceloporus grammicus complex (Sauria, Iguanidae). **Nat. Geog. Res.** 3:343-362.

Sites, J.W., Jr., R.K. Chesser, and R.J. Baker. 1988a. Population genetic structure and the fixation of chromosomal rearrangements in Sceloporus grammicus (Sauria, Iguanidae): a computer simulation study. **Copeia** 1988:1045-1055.

Sites, J.W., Jr., et al. 1988b. Allozyme variation and genetic divergence within and between three cytotypes of the Sceloporus grammicus complex (Sauria, Iguanidae) in central Mexico. **Herpetologica** 44:297-307.

Sites, J.W., Jr., et al. 1993. Parapatric hybridization between chromosome races of the Sceloporus grammicus complex (Phrynosomatidae): structure of the Tulancingo transect. **Copeia** 1993:(in press).

Thompson, P., and J.W. Sites, Jr. 1986. Population structure in chromosomally polytypic versus monotypic lizards of the genus Sceloporus (Sauria, Iguanidae) in relation to chromosomally-mediated speciation. **Evolution** 40:303-314.

White, M.J.D. 1973. **Animal Cytology and Evolution.** Cambridge Univ. Press, London.

Cancer

11 Deletions of chromosome 5 in malignant myeloid disorders

M.M. LE BEAU
University of Chicago, USA

1 Introduction

1.1 Role of chromosomal abnormalities in the pathogenesis of human leukemias

Recurring chromosomal abnormalities are characteristic of human malignant diseases, particularly the leukemias and lymphomas (Mitelman, 1991). Molecular analysis has revealed that alterations in the function of the genes that are located at the breakpoints of the recurring structural rearrangements play an integral role in the process of malignant transformation. The transforming genes that are involved in chromosomal translocations fall into several functional classes, including tyrosine protein kinases (ABL), serine protein kinases (BCR), cell surface receptors (TAN1), growth factors (IL3), and inner mitochondrial membrane proteins (BCL2). However, the largest class encode transcriptional regulating factors (PBX1, E2A); these genes have been implicated in the pathogenesis of T cell and B cell neoplasms as well as some myeloid leukemias.

1.2 Tumor Suppressor Genes

In addition to the recurring translocations which result in the activation of an oncogene in a dominant fashion, the loss of genetic material has been identified in many tumor types. Such a loss may result from chromosomal loss or deletion as well as by other genetic mechanisms. The consequence of these abnormalities is the development of hemizygosity resulting in a gene dosage effect or in the unmasking of a recessive allele on the structurally "normal" homologue (Klein, 1988). Retinoblastoma is the prototypic model for the study of tumor suppressor genes; however, they have been implicated in the pathogenesis of a number of other tumors for which allele loss has been demonstrated. The identification of recurring chromosomal deletions in acute myeloid leukemia (AML) suggests that tumor suppressor genes may be involved in the pathogenesis of some malignant myeloid disorders.

Chromosomes Today   Volume 11. Edited by A.T. Sumner and A.C. Chandley. Published in 1993 by Chapman & Hall, London. ISBN 0 412 47670 3

2   Therapy-related acute myeloid leukemia (t-AML)

2.1 Clinical features

The occurrence of a myelodysplastic syndrome (MDS) or AML
has been recognized as a late complication of cytotoxic
therapy used in the treatment of malignant diseases
(Koeffler and Rowley, 1985). Therapy-related MDS or AML
(t-MDS/t-AML) typically presents with a latency period of
approximately five years after treatment. Two-thirds of
these patients are first recognized by evidence of myelo-
dysplasia, bone marrow failure, and pancytopenia. Frequently, all three hematopoietic cell lines (erythroid,
myeloid, and megakaryocytic) are involved in the myelodysplastic process. Survival times of patients with
t-AML are usually short (median 8 mos.).

2.2 Chromosomal abnormalities

The association of abnormalities of chromosomes 5 and/or
7 with these diseases was first noted by Rowley et al.
(1981) who observed loss of an entire chromosome 5 or 7
or a deletion of the long arm of these chromosomes
[del(5q)/del(7q)] in cells from 23 of 26 patients (88%)
examined. The high incidence of these specific
chromosomal abnormalities has been confirmed by other
investigators. In our series of 129 consecutive patients
with t-MDS/t-AML, 120 (93%) had a clonal chromosomal
abnormality (Le Beau et al., 1986 and unpublished work),
and 97 (75%) had a clonal abnormality leading to the loss
or deletion of chromosomes 5 and/or 7. Overall, 55
patients (43%) had abnormalities of chromosome 5. A
del(5q) was the most common structural aberration in our
series.

Other recurring chromosomal abnormalities identified in
this series include rearrangements of band 11q23,
particularly the t(9;11), and the t(3;21)(q26;q22).
These rearrangements define newly-recognized cytogenetic
subsets of t-MDS/t-AML and are associated with different
clinical and morphological features than are t-MDS or
t-AML with abnormalities of chromosome 5 or 7.

3   Abnormalities of chromosome 5 in other myeloid disorders

In addition to t-AML, a -5/del(5q) has also been observed
in the malignant cells of 10% of patients with AML de
novo (Nimer and Golde, 1987). These abnormalities have
been observed in all subtypes of AML; however, the
frequency of a -5 or del(5q) is particularly high in AML-
M6 (erythroleukemia, 70% of M6 patients in our series,
Olopade et al., unpublished work). As in t-AML, AML de
novo with abnormalities of chromosome 5 is characterized

by prominent trilineage dysplasia, and many of these patients have had a pre-existing MDS. Clinically, they tend to be older individuals and they respond poorly to cytotoxic therapy (Samuels et al., 1988). These patients frequently have had significant exposure to environmental carcinogens, leading to the suggestion that abnormalities of chromosome 5 or 7 may be markers of mutagen-induced leukemia.

Abnormalities of chromosome 5 are observed in ~15% of patients who have MDS arising de novo. These disorders have a wide variety of clinical manifestations which range from minimal symptoms to an aggressive pre-leukemia. The frequency of a -5/del(5q) is highest in refractory anemia (RA, 29%, see the 5q- syndrome described below) and the more aggressive subtypes of MDS, refractory anemia with excess blasts (RAEB) and RAEB in transformation to acute leukemia (RAEB-T) (27%) and lowest in RA with ringed sideroblasts (6%) and chronic myelomonocytic leukemia (2%).

A distinct clinical syndrome associated with a del(5q) is also seen in older patients, especially in females (Van den Berghe et al., 1985). Clinically, this disorder, termed the "5q- syndrome" is characterized by RA. A consistent finding in the bone marrow is the presence of abnormal megakaryocytes with mono- or bilobulated nuclei. These patients have a del(5q) as their sole abnormality, and they tend to have a relatively mild course that usually does not progress to acute leukemia, whereas in patients with AML, the del(5q) is usually accompanied by additional abnormalities (Van den Berghe et al., 1985).

4   Molecular hypothesis for abnormalities of chromosome 5 in MDS/AML

A deletion may result in a reduction in the level of a gene product (gene dosage effect), or in the loss of a wild type allele. In the latter case, loss of function of both alleles may occur, in one instance through a detectable chromosomal deletion and, in the other, as a result of a mutation. In the case of retinoblastoma and Wilms' tumor, genetic evidence supported the hypothesis of a two-step mechanism; individuals with familial retinoblastoma may inherit one mutated allele, whereas in sporadic cases, two mutations affecting both alleles of the retinoblastoma gene (RB1) occur in a somatic cell. Subsequent cloning of the RB1 gene confirmed this theory, and revealed that the mutations identified in tumor cells resulted in inactivation of the RB1 alleles or in the production of a non-functional protein.

We and other investigators have proposed that 5q contains a myeloid tumor suppressor gene, and that this gene is likely to be located within a commonly deleted segment in patients who have a del(5q). In t-MDS/t-AML, the latency period between the time of exposure to bone marrow dysfunction is 4-5 years; this long latent period is compatible with a two-step mechanism, in which both mutations must occur in a myeloid progenitor cell.

A molecular hallmark of tumor suppressor genes is allele loss. The majority of tumors for which allele loss has been demonstrated represent neoplasms that occur primarily in adults, such as colon, lung, or breast carcinoma. MDS and AML, particularly those subtypes with abnormalities of chromosomes 5 and/or 7, are primarily diseases of the elderly. An interesting difference between myeloid neoplasms and other adult solid tumors associated with tumor suppressor genes is the observation that allele loss for loci on chromosomes 5 or 7 in MDS/AML usually results from major cytogenetic deletions or simple chromosome loss, rather than from mitotic recombination, or chromosome loss with duplication of the remaining homologue (Neuman et al., 1992a).

Familial forms of these adult tumors are rare, suggesting that, in most cases, both mutations are likely to occur in somatic cells. The exposure history of patients with t-MDS/t-AML is compatible with this hypothesis. These patients may have normal alleles at this locus initially, one of which becomes a mutant "leukemogenic" allele as the result of therapy. Subsequent loss of the other allele in a bone marrow myeloid stem cell would result in leukemia. Alternatively, these individuals may carry a predisposing mutated allele; subsequent exposure to mutagenic agents may induce the second mutation, i.e., the chromosomal abnormality giving rise to leukemia. If this is the case, characterization of the predisposing mutation will be important in identifying individuals who are at risk of developing t-AML, and in the selection of the appropriate therapy for the primary malignancy.

Although we cannot formally rule out the possibility that t-AML results from a gene dosage effect (one-step process), this is a much less satisfactory hypothesis. The long latency period argues in favor of a two-step process.

5 Identification of the critical region of chromosome 5

5.1 Cytogenetic delineation of the critical region

To determine the location of genes on 5q that may be involved in myeloid leukemogenesis, we previously examined the breakpoints and the extent of the deletions in 17 patients with t-MDS/t-AML. Our analysis revealed

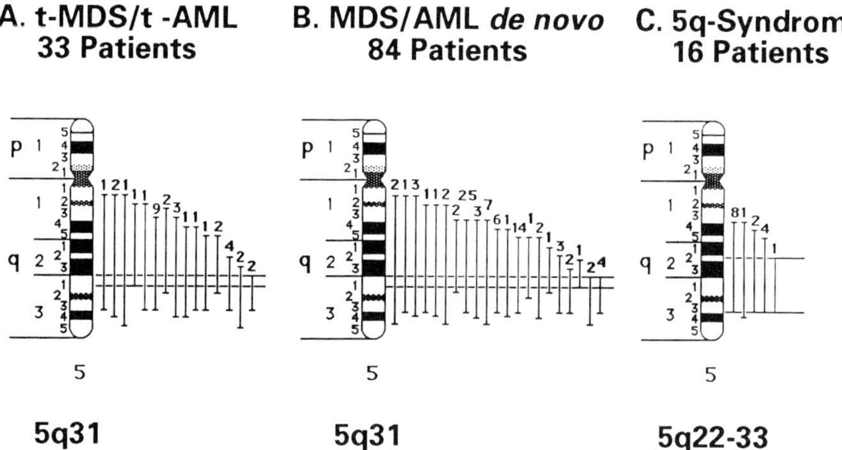

Fig. 1. Diagram of the banding pattern of chromosome 5 illustrating the breakpoints and deletions in 133 patients with myeloid disorders. The numbers above each vertical bar indicate the number of patients with this deletion. The dashed horizontal lines indicate the commonly deleted segment.

that these deletions were interstitial, with a proximal breakpoint commonly in q13-15, and a distal breakpoint in q33-35. We identified a segment (critical region) consisting of bands q23 and q31 that was deleted in each patient (Le Beau et al., 1986).

To refine the critical region of 5q, we have recently examined the breakpoints of the deletions in an expanded series of 133 patients (Le Beau et al., 1989, and unpublished data). This analysis included 33 patients who had t-MDS/t-AML, 84 patients who had de novo MDS or AML, and 16 patients who had the RA 5q- syndrome (Fig. 1). Ninety patients had a distal breakpoint in 5q33.3; less often, distal breakpoints were observed in 5q31 (9 patients), 5q32 (1 patient), 5q34 (17 patients), or 5q35 (16 patients). Eight patients had a <u>proximal</u> breakpoint in q31 and 9 patients had a <u>distal</u> breakpoint in this band allowing us to refine the critical region of 5q to q31. We have also identified four patients who have balanced translocations involving 5q31, providing further support that the critical region of 5q may actually be limited to q31. Thus, it is likely that loss of a gene(s) located within this band is involved in the pathogenesis of myeloid disorders characterized by a del(5q).

5.2 Identification of genes in the critical region

A striking number of genes encoding growth factors and growth factor receptors have been mapped to distal 5q. Five of these genes encode hematopoietic growth factors. By using in situ hybridization and the analysis of somatic cell hybrids, we and others previously localized the interleukin-3 (IL3), interleukin 4 (IL4), interleukin 5 (IL5), interleukin 9 (IL9), granulocyte-macrophage CSF (CSF2) (5q23-31), and CSF1R (5q33) genes to chromosome 5 (see Wasmuth et al., 1991 for references). Other genes that have been mapped to the critical region of 5q include the genes encoding the $\alpha 1$ and β_2-adrenergic receptors (ADRA1, ADRB2, 5q31-32), endothelial cell growth factor (FGFA, 5q31-32), the CD14 antigen, a myeloid-specific differentiation-associated antigen (5q23-31), the early growth response 1 protein (EGR1, 5q23-31), a T-cell specific transcription factor (TCF7), osteonectin (SPARC, 5q31-33), and the glucocorticoid receptor (GRL, 5q31-32) (Wasmuth et al., 1991).

Based on their biological activity, several of these genes are good candidates for a tumor suppressor gene, or they may contribute to gene dosage effects. Thus, to determine the order of these genes, and their relationship to the critical region of 5q, we used dual-color fluorescence in situ hybridization (FISH) analysis (Trask et al., 1991) (Fig. 2). The localization of some of the genes was refined and their order was determined to be cen-[IL3/CSF2-IL4/IL5]-IL9-TCF7-EGR1-CD14-FGFA-GRL-SPARC-ADRA1-tel. The ADRA1 gene is within q33, a band that is slightly distal to its previous localization. The order of the genes relative to the cosmid clones mapped to 5q31 and the critical region is also illustrated in Fig. 2. This order is consistent with that obtained for the IL4/IL5, IL3/CSF2, and FGFA genes by Huebner et al. (1990).

5.3 Molecular delineation of the critical region

Several experimental approaches can be used to identify a putative tumor suppressor gene. These include candidate gene and physical mapping approaches. In the first approach, one seeks to identify mutations of genes that are known to map to the region of interest in tumor cells, whereas the second approach involves the translation of the cytogenetic map of the critical region into a physical map, in which the deleted regions are identified by molecular probes with defined physical and genetic locations. Such a map could be used then to screen DNA from leukemia cells with abnormalities of chromosome 5 to identify alterations of genomic sequences within the critical region. These two approaches are not mutually exclusive.

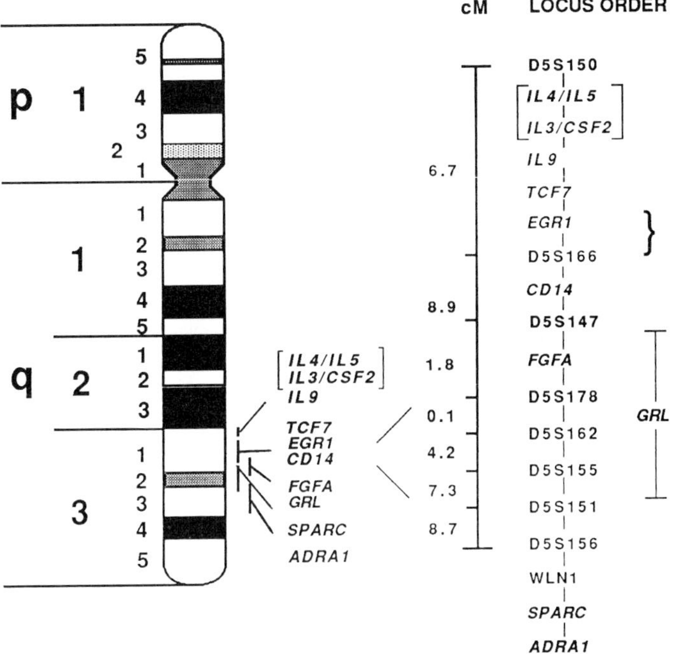

Fig. 2. Schematic diagram of the banding pattern of
chromosome 5 ilustrating chromosomal localization and
order of the IL3/CSF2, IL4/IL5, IL9, TCF7, EGR1, CD14,
FGFA, GRL, SPARC, and ADRA1 genes determined by fluorescence in situ hybridization, the physical order of
cosmid clones, and relationship of genes on 5q to the
critical region of 5q31. The brackets identify probes
for which the order is unknown.

As an initial step in preparing a physical map of 5q31,
we performed FISH of 70 cosmid and phage clones that were
mapped by genetic linkage analysis (Neuman et al.,
1992b). The results of these studies allowed us to
identify 11 probes that were located within 5q31. To
identify probes that were located within the commonly
deleted segment of 5q31, and to delineate the critical
region at the molecular level, we performed FISH of 9 of
these probes to metaphase cells from 17 patients who had
either a proximal breakpoint (10 patients) or distal
breakpoint (7 patients) within 5q31. Two yeast
artificial chromosome (YAC) clones containing the
CSF2/IL3 and IL4/IL5 genes, and probes for the IL9, TCF7,
and EGR1 genes were also included in this analysis. The
results of these hybridizations revealed that the
CSF2/IL3, IL4/IL5, IL9, and TCF7 genes are proximal to

the critical region and are not deleted in all patients, whereas the EGR1 gene lies within the critical region (Fig. 2). These results identified proximal and distal markers that flank the critical region as well as a marker within this region. We are currently examining patients who have translocation breakpoints within 5q31 to determine the relationship of the probes in 5q31 to the breakpoint.

To determine the order of the genes and cosmid probes in 5q31, we used the technique of dual-color FISH analysis of interphase and metaphase cells. We also used FISH interactively with genetic-linkage analysis to determine the order of these probes, and to estimate the genetic distances between them (Westbrook et al., 1991). The order of the cosmids within 5q31 was determined to be cen-D5S150-D5S166-D5S147-D5S178-D5S162-D5S155-D5S151-D5S156-WLN1-tel (Fig. 2). The estimated genetic distance between D5S150 and D5S166 is 6.7 cM. Chromosome 5 corresponds to 250-300 cM, or 160 Mb (5q31 represents ~30 cM or 17 Mb); thus, the critical region is likely to represent less than 3 Mb.

6 Analysis of genes on 5q in myeloid leukemia

To determine whether rearrangements of genes in 5q31 had occurred, we analyzed DNA from leukemia cells with a del(5q) (3 patients), or other rearrangement involving 5q31 (1 patient). Southern blot analysis using pulsed field gel electrophoresis and probes for the CSF2, IL4, IL5, EGR1, CD14, and FGFA genes did not reveal the presence of rearrangements (Le Beau et al., 1989). More recently, Gilliland et al. (1991) found no mutations of the EGR1 gene by PCR analysis of clonally-derived mono-cytes in ten 5q- syndrome patients. At present, the status of the EGR1 gene in patients with t-MDS/t-AML or AML de novo is unknown. These analyses are by no means extensive; in many cases, only a few patients were examined and, of those genes examined, only the EGR1 gene lies within the critical region.

With respect to mutations of other genes on 5q in leukemia cells with deletions of chromosome 5, Boultwood et al. (1991) used restriction fragment length poly-morphism analysis to examine the CSF1R (FMS, 5q33) gene in 10 patients with primary MDS and a del(5q) charac-terized by a distal breakpoint in 5q33. The results were compatible with the loss of one CSF1R allele in each patient as a result of the chromosomal deletion. Of note, however, was the fact that the results of densi-tometric analysis were compatible with homozygous loss of the CSF1R alleles in 4 patients. In view of our cyto-genetic and molecular delineation of the critical region as consisting of sequences in 5q31, these results were unexpected. Nonetheless, they are intriguing, and suggest that the <u>CSF1R</u> gene should be examined in additional patients, including patients with AML or t-AML. Alternatively, these results raise the possibility that there are multiple genes on 5q that are involved in the pathogenesis or progression of these myeloid disorders.

7 Etiology of the 5q- syndrome vs. AML with a del(5q)

The relationship of the 5q- syndrome to other subtypes of MDS, AML de novo, or t-MDS/t-AML with a del(5q) is poorly understood. On the one hand, the clinical course of the 5q- syndrome is distinct from that of the other dis-orders, suggesting that the pathogenesis of this disease differs from that of the others. On other levels, there is considerable overlap, e.g., the distinctive changes in the megakaryocyte lineage, which argues for a common underlying defect at the molecular level. Most patients with RA and a del(5q) have a stable disease, which usually does not progress to acute leukemia. Is this because the mutations involve a different gene in patients with RA as compared to those who have leukemia, or that a different mutation of the same gene has occurred? Or is it that other genes in the multi-step process giving rise to leukemia have not been altered, e.g., mutations of a gene on chromosome 7, and that the deleted chromosome is sufficient to cause a hematologic problem, in this case anemia, but not to cause leukemia? It is unlikely that we will be able to resolve these issues until the gene(s) involved in the etiology of these subtypes of MDS and AML are identified.

8 Discussion

The results of cytogenetic and molecular mapping of the deletions of chromosome 5 in myeloid disorders suggest that the gene(s) that is involved in the pathogenesis of these disorders is likely to be located in 5q31. The identification of a cluster of genes encoding hemato-poietic growth factors in this region of 5q suggests a role for these genes in the autocrine growth of leukemia cells characterized by a del(5q). However, our results of FISH analysis indicate that the cluster of hemato-poietic growth factor genes is proximal to the critical region.

The properties of the <u>EGR1</u> gene and encoded protein suggest that this gene may be a more suitable candidate than those encoding growth factors for playing a role in the malignant transformation of myeloid cells with a del(5q). The EGR1 protein is a DNA-binding protein with transcriptional regulatory activity; expression of EGR1 has been found to be upregulated during terminal myeloid differentiation (Sukhatme et al., unpublished work). It is notable that the EGR1 protein binds to the same target DNA sequence as does the product of the WT1 gene, a candidate tumor suppressor gene in Wilms' tumor. To determine the role of the EGR1 gene in the pathogenesis of MDS and AML in patients with del(5q), a detailed molecular characterization of this locus in myeloid leukemia cells is necessary.

9   Acknowledgements

I thank Susan Jarman for secretarial assistance.

10 References

Boultwood, J. et al. (1991) Loss of both CSFIR (FMS) alleles in patients with myelo-dysplasia and a chromosome 5 deletion. **Proc. Natl. Acad. Sci. U.S.A.**, 88, 6176-6180.

Gilliland, D.G., Perrin, S., and Bunn, H.F. (1991) Analysis of clonality and EGR-1 structure/expression in patients with 5q- syndrome. **Leuk. Res.**, 15, (Suppl).

Huebner, K. et al. (1990) Order of genes on human chromosome 5q with respect to 5q interstitial deletions. **Am. J. Hum. Genet.**, 46, 26-36.

Klein, G. (1988) The approaching era of the tumor suppressor genes. **Science**, 238, 1539-31545.

Koeffler, H.P. and Rowley, J.D. (1985) Therapy-related acute nonlymphocytic leukemia, in (eds. P.H. Wiernik, G.P. Canellos, K.A. Kyle, and C.A. Schiffer). **Neoplastic Diseases of the Blood**, Churchill Livingstone, New York, vol. 1, pp. 357-381.

Le Beau, M.M. et al. (1986) Clinical and cytogenetic correlations in 63 patients with therapy-related myelodysplastic syndromes and acute nonlymphocytic leukemia: Further evidence for characteristic abnormalities of chromosomes 5 and 7. **J. Clin Oncol.**, 4, 325-345.

Le Beau, M.M. et al. (1989) Molecular and cytogenetic analysis of chromosome 5 abnormalities in myeloid disorders: Chromosomal localization and physical mapping of IL-4 and IL5. **Cancer Cells**, 7, 53-58.

Mitelman, F. (1991) **Catalog of chromosome abnormalities in cancer**, 4th edition, Wiley-Liss, New York.

Neuman, W.L. et al. (1992a) Chromosomal loss and deletion are the most common mechanisms for loss of heterozygosity from chromosomes 5 and 7 in malignant myeloid disorders. **Blood**, 76, 1501-1510.

Neuman, W.L. et al. (1992b) Physical localization of 70 genetically mapped loci to human chromosome 5 by fluorescence in situ hybridization. **Cytogenet. Cell Genet.**, in press.

Nimer, S.D. and Golde, D.W. (1987) The 5q- abnormality. **Blood**, 70, 1705-1712.

Rowley, J.D., Golomb, H.M., Vardiman, J.W. (1981) Nonrandom chromosome abnormalities in acute leukemia and dysmyelopoietic syndromes in patients with previously treated malignant disease. **Blood**, 58, 759-767.

Samuels, B.L. et al. (1988) Specific chromosomal abnormalities in acute non-lymphocytic leukemia correlate with drug susceptibility in vivo. **Leukemia**, 2, 79-83.

Trask, B.J. et al. (1991) Mapping of human chromosome Xq28 by two-color fluorescence in situ hybridization of DNA sequences to interphase nuclei. **Am. J. Hum. Genet.**, 48, 1-15.

Van den Berghe, H. et al. (1985) The 5q- anomaly. **Cancer Genet. Cytogenet.**, 17, 189-255.

Wasmuth, J.J., Bishop, D.T., and Westbrook, C. (1991) Report of the committee on the genetic constitution of chromosome 5. Human Gene Mapping 11. **Cytogenet. Cell Genet.**, 58, 261-294.

Westbrook, C.A. et al. (1992) High resolution physical and genetic map of 5q31 by interactive use of fluorescent in situ hybridization and genetic linkage. **Cytogenet. Cell Genet.**, in press.

12 Chromosomal abnormalities in solid tumours

S. HEIM

Odense University, Denmark, and University Hospital, Lund, Sweden

1 Acquired clonal chromosome abnormalities in neoplastic cells

The observation that tumours are the result of clonal expansions of neoplastic cells carrying karyotypic aberrations (Heim and Mitelman, 1987; Sandberg, 1990) lends convincing support to the somatic mutation theory of carcinogenesis. Such chromosomal changes were in the past often dismissed as epiphenomena of no pathogenetic importance, but latterly the paradigm has changed, and they are now generally accepted as crucial events in multistage tumourigenesis.

A wide array of consistent chromosomal abnormalities has been detected in haematological neoplasias (Heim, 1990). Because of the technical difficulties involved, progress has been much slower in solid tumour cytogenetics. Several characteristic karyotypic patterns have, nevertheless, in later years been associated with some of the diagnostic categories. Among them are anomalies no less pathognomonic for the neoplasms in question than are the best known leukaemia-associated changes. The following summary of what is known about the chromosomal aberrations of solid tumours is by no means exhaustive and, in particular, the referencing had to be severely restricted. Overviews providing a more extensive coverage of the field include Sandberg et al. (1988), Mitelman et al. (1990) and Heim and Mitelman (1992). A complete listing of all published karyotypic information on neoplastic cells was given by Mitelman (1991).

2 Epithelial tumours

2.1 Respiratory tract
Both cytogenetically unrelated clones, clones with simple reciprocal rearrangements and clones with complex changes (often including 11q13 rearrangements) have been detected in head and neck squamous cell carcinomas.

In lung cancer, deletions of the short arm of chromosome 3 were initially thought to be characteristic of small-cell carcinomas but are now known to occur frequently also in non-small-cell cancers.

Chromosomes Today   Volume 11. Edited by A.T. Sumner and A.C. Chandley. Published in 1993 by Chapman & Hall, London. ISBN 0 412 47670 3

2.2 Digestive tract

Adenomas of the salivary gland frequently have rearrangements of the chromosomal bands and regions 3p21,
8q12 and 12q13-15. Often the first two are recombined
as t(3;8). The most common structural aberrations in
pancreatic carcinomas have affected chromosome arms 6q,
1p, 17p, 17q, 3p and 8p. Colorectal adenomas have
mostly had numerical aberrations only; it is possible
that this is a feature also of adenomas of other sites.
Colorectal adenocarcinomas have chromosome numbers
ranging from the near-diploid to the massively aneuploid. Consistent karyotypic imbalances include partial
loss of the short arms of chromosomes 17, 1 and 8 and
loss of the entire chromosomes 18, 22 and Y. Trisomy 7
is the most common gain.

2.3 Urinary tract

The most characteristic cytogenetic abnormalities in
renal cell carcinomas are deletions of the short arm of
chromosome 3. In transitional cell carcinomas of the
bladder, the most frequent aberrations have been of
chromosomes 1 (variable changes), 3 (especially affecting the short arm), 5 (often isochromosome for the
short arm), 7 (trisomy), 9 (mostly loss of one copy)
and 10 (deletions of the long arm).

2.4 Breast

Among the characteristic karyotypic patterns that have
been recorded in breast cancers are gains of 1q - often
as i(1q) or combined with simultaneous loss of 16q
because of an unbalanced t(1;16) - and interstitial
deletions of 3p. In a fairly high number of cases,
cytogenetically unrelated clones are seen, perhaps
indicating polyclonal carcinogenesis.

2.5 Ovary

The most common aberrations in ovarian carcinomas - as
indeed in cancers of many organs - are of chromosome 1,
but with a wide range of breakpoints and imbalances as
the result. More characteristic than the chromosome 1
changes have been addition of unknown material to 19p
to form 19p+ markers and structural changes leading to
loss of material from 6q and 11p. In adenomas and other
benign tumours of the ovary, a solitary trisomy 12 is a
frequent finding.

2.6 Testis

The most common chromosomal aberration in testicular
germ cell tumours is i(12p).

2.7 Prostate

In prostatic adenocarcinomas, breakpoint clusters have been found in 7q, 8p and 10q. The rearrangements often lead to loss of genetic material from these regions.

2.8 Thyroid

Rearrangements of 10q seem to characterize nonmedullary carcinomas, especially tumours that are papillary.

3 Neurogenic tumours

Most meningiomas are cytogenetically characterized by monosomy 22. The malignant gliomas are not equally uniform in their karyotypic appearance: the most common changes seem to be trisomy 7, loss of a sex chromosome, loss of one chromosome 10 or 22, loss of material from chromosome arms 9p and 17p, and double minute chromosomes (dmin). The formation of an isochromosome for the long arm of chromosome 17 may define a subgroup of primitive neuroectodermal tumours of the central nervous system in children. The characteristic karyotypic features of neuroblastomas are deletions of the short arm of chromosome 1 and cytogenetic evidence of gene amplification, i.e., homogeneously staining regions (hsr) and dmin. In retinoblastoma, the most common cytogenetic change is not, as might have been expected, any visible change of the Rb locus in 13q14, but the formation of an i(6p).

4 Mesenchymal tumours

4.1 Ewing´s sarcoma

The reciprocal translocation t(11;22)(q24;q12) is found in all or nearly all Ewing´s sarcomas and may be the only change. The most common secondary aberrations are trisomy 8 and der(16)t(1;16)(q21;q13). Peripheral neuroepithelioma, Askin tumour, and esthesioneuroblastoma have the same t(11;22) as Ewing´s sarcoma, indicating that all these diagnostic categories represent variations on the same pathogenetic theme.

4.2 Rhabdomyosarcoma

A reciprocal t(2;13)(q35;q14) has been reported in rhabdomyosarcomas of all histologic subtypes.

4.3 Chondromatous tumours

No specific chromosomal abnormality has been found in chondrosarcomas of bone, but in extraskeletal myxoid chondrosarcoma, t(9;22)(q22;q12) seems to be a recurring abnormality. Chondromas often have rearrangements of 12q13-15.

4.4 Synovial sarcoma

These tumours carry the only cancer-associated chromosomal rearrangement known to affect a sex chromosome,
namely a t(X;18)(p11;q11). Half of the tumours have the
t(X;18) as the only aberration. The secondary changes
often involve chromosomes 1 and 12.

4.5 Malignant fibrous histiocytoma

Telomeric associations, rings, dicentric chromosomes,
hsr and dmin are frequent, albeit not entirely specific, karyotypic phenomena in this group of sarcomas.
Structural rearrangements often affect 19p (giving rise
to 19p+ markers), 11p (mostly deletions) and 1q and 3p
(mostly deletions).

4.6 Infantile fibrosarcoma

These tumours seem to be characterized by numerical
aberrations, mostly gains of whole chromosomes.

4.7 Dermatofibrosarcoma protuberans

Supernumerary ring chromosomes have been a feature of
these tumours.

4.8 Clear-cell sarcoma of tendons and aponeuroses

The balanced, reciprocal translocation t(12;22)
(q13;q13) characterizes these tumours.

4.9 Hemangiopericytoma

Various rearrangements of 12q13-15 have been seen in
the few hemangiopericytomas hitherto analysed.

4.10 Leiomyoma

Only uterine leiomyomas have been examined. Several
cytogenetic subgroups have been delineated: 1. Tumours
with t(12;14)(q14-15;q23-24), probably t(12;14)
(q15;q24.1), or other 12q14-15 changes. 2. Tumours with
7q-, mostly del(7)(q21.2q31.2). 3. Tumours with trisomy
12. 4. Tumours with rearrangements of 6p. 5. Tumours
with other cytogenetic changes. Clonal evolution has
often involved telomeric associations and ring clos-
ures, especially of chromosome 1, and loss of one
chromosome 22.

4.11 Lipoma

The multiple lipomas of lipomatosis patients usually
have normal karyotypes. Typical, sporadic lipomas often
exhibit rearrangements of 12q13-15, either as t(3;12)
(q27-28;q13-14) or as variants recombining 12q with
other chromosome arms than 3q. Atypical lipomas, on the
other hand, do not have 12q changes. They are cytogenetically characterized by supernumerary ring chromosomes

of unknown origin.

4.12 Liposarcoma
Well differentiated liposarcomas often have supernumerary ring markers. It is possible that atypical
lipomas and well differentiated liposarcomas are but
two different terms for the same biological entity; at
least cytogenetically there is no difference between
them. All or almost all myxoid and mixed liposarcomas
share a t(12;16)(q13;p11), at subband level interpreted
as t(12;16)(q13.3;p11.2). It is at present uncertain
whether the exact breakpoint in 12q in liposarcomas is
different from that in the same chromosome arm in
lipomas or, for that matter, in the other solid tumours
with 12q aberrations.

5 Cancer cytogenetics in research and clinical medicine

The scientific aim of solid tumour cytogenetics is to
identify the karyotypic changes that characterize all
tumour types and all stages of multistep tumourigenesis. The aberrations can be subdivided into primary
changes, which are important in establishing a neo-
plasm, and secondary changes, which accumulate during
clonal evolution and, at least in part, are responsible
for the phenotypic transformations that characterize
the later, more malignant tumour stages. Some changes
occur in many different tumours and probably reflect
nonspecific, "final-common-path" mechanisms in neoplastic transformation. Other aberrations are pathognomonic; they are seen in only one tumour type, indicating an intimate and probably causal relationship to
the histogenesis of those particular tumours. The
detection of consistent abnormalities at the cytogenetic level helps molecular geneticists to search in the
right direction for the molecular equivalents of the
chromosomal changes. This research strategy has in the
haematological malignancies repeatedly proved its
validity - the molecular dissection of the Philadelphia
chromosome in chronic myelogenous leukaemia and the
t(8;14) in Burkitt´s lymphoma are but two examples -
and similar investigations are now being undertaken on
solid tumours.

On the other hand, cancer cytogenetics also has
intimate links with clinical medicine. The karyotypic
findings can be utilized as disease markers that, even
without knowledge of the pathogenetic mechanisms they
reflect, may have a profound diagnostic and prognostic
impact. The diagnostic role is obvious: pathognomonic,
or at least characteristic, rearrangements are found in

many tumour types. Cytogenetics is therefore in a
position to supplement the classical histopathological
methods in several differential diagnostic situations,
for example in the diagnosis of the small-cell, round-
cell tumours of childhood. That the chromosomal constitution of solid tumours is a prognostic indicator has
been shown for several diagnostic categories, including
bladder tumours (Sandberg, 1986), neuroblastomas (Kaneko et al., 1987; Christiansen and Lampert, 1988),
malignant melanomas (Trent et al., 1990), malignant
fibrous histiocytomas (Rydholm et al., 1990) and prostatic (Lundgren et al., 1992) and ovarian cancers
(Pejovic et al., 1992). However, only when the cytogenetic data bases are significantly increased and the
findings have been correlated with pathological and
clinical parameters in a prospective manner, can the
true prognostic impact of the various aberration patterns be determined. Then one will also be able to
determine to what extent the chromosomal constitution
of tumours is an independent prognostic indicator.

6 References

Christiansen, H. and Lampert, F. (1988) Tumour karyotype discriminates between good and bad prognostic
outcome in neuroblastoma. **Br. J. Cancer,** 57, 121-
126.
Heim, S. (1990) Cytogenetics in the investigation of
haematological disorders, in **Baillière's Clinics in
Haematology, vol. 3, number 4** (ed I. Cavill), Baillière Tindall, London, pp. 921-948.
Heim, S. and Mitelman, F. (1987) **Cancer Cytogenetics.**
Alan R. Liss Inc., New York.
Heim, S. and Mitelman, F. (1992) Cytogenetics of solid
tumours. **Recent Adv. Histopathol.,** 15, 37-66.
Kaneko, Y. et al. (1987) Different karyotypic patterns
in early and advanced stage neuroblastomas. **Cancer
Res.,** 47, 311-318.
Lundgren, R. et al. (1992) Chromosome abnormalities are
associated with unfavorable outcome in prostatic
cancer patients. **J. Urol.,** 147, 784-788.
Mitelman, F. (1991) **Catalog of Chromosome Aberrations
in Cancer. 4th Ed.** Wiley-Liss, New York.
Mitelman, F., Kaneko, Y. and Trent, J. (1990) Report of
the committee on chromosome changes in neoplasia.
Cytogenet. Cell Genet., 55, 358-386.
Pejovic, T. et al. (1992) Prognostic impact of chromo
some aberrations in ovarian cancer. **Br. J. Cancer,**
65, 282-286.
Rydholm, A. et al. (1990) Malignant fibrous histiocytomas with a 19p+ marker chromosome have an increased

relapse rate. **Genes Chrom. Cancer,** 2, 296–299.
Sandberg, A.A. (1986) Chromosome changes in bladder cancer: Clincal and other correlations. **Cancer Genet. Cytogenet.,** 19, 163–175.
Sandberg, A.A. (1990) **Chromosomes in Human Cancer and Leukemia.** Elsevier, New York.
Sandberg, A.A., Turc-Carel, C. and Gemmill, R.M. (1988) Chromosomes in solid tumors and beyond. **Cancer Res.,** 48, 1049–1059.
Trent, J.M. et al. (1990) Relation of cytogenetic abnormalities and clinical outcome in metastatic melanoma. **N. Engl. J. Med.,** 322, 1508–1511.

13 Non-radioactive *in situ* hybridization to metaphase and interphase nuclei of malignant cells

R. BERGER

Institut de Génétique Moléculaire, Paris, France

Introduction

Recent advances in fluorescence *in situ* hybridization (FISH) techniques with non radioactive probes have provided new possibilities for the investigation of malignant cells in three important ways: cytogenetic analysis of metaphases allowing the identification of rearrangements difficult to analyse with usual banding techniques; delineation of chromosomal breakpoints of acquired rearrangements by using molecular probes previously localized on rearranged chromosome bands; cytogenetic analysis of non dividing cells that makes it unnecessary to obtain of metaphases and allows analysis of more cells than classical cytogenetics does. Consequently, applications of FISH now appear wider, allowing, for instance, studies on gene amplification and follow-up of patients in medical practice.

Techniques

Various techniques of FISH have recently been developed depending on the probes and the fluorochromes used. The probes may be unique DNA sequences used with or without their vector (phage, plasmid). The limitation of their use is the size of the insert since a DNA probe less than 1 kb long is usually difficult to analyse because of the insufficient resolution of the techniques without possible new technical improvements in image analysis techniques. Chromosome-specific repeated sequences can also be used as probes either on metaphase chromosomes or on interphase nuclei. The most commonly used are alphoid sequence probes that are particularly important for hybridization to interphase nuclei.

Chromosomes Today   Volume 11. Edited by A.T. Sumner and A.C. Chandley. Published in 1993 by Chapman & Hall, London. ISBN 0 412 47670 3

More recently, larger DNA sequences have been inserted within cosmids or yeast artificial chromosomes (YAC) and used as probes. For this purpose, competitive in situ hybridization (CISS) with an excess of DNA is necessary to lower the background due to hybridization of repeated sequences invariably present in DNA probes of large size. These probes may also be used to hybridize to metaphase chromosomes or nuclei. More recently whole chromosome specific probes have been prepared from somatic cell hybrids with only one human chromosome or from sorted chromosomes by flow cytometry to specifically colour one chromosome. This "chromosome painting" is particularly interesting in the study of chromosome rearrangements.

Several techniques of FISH have been developed. The most widely used, initiated by Ward's group (Langer et al., 1981) and improved by Pinkel et al. (1986), is based upon the preparation of biotinylated probes revealed immunologically (biotin-avidin system). Simultaneous hybridization with several probes each detectable as a different colour has been proposed to allow their simultaneous analysis. By combination of several fluorochromes, up to seven colours are available, i.e. seven different probes can be analysed on the same preparation (Ried et al., 1992a).

Cytogenetic study of gene amplification

Because of its rapidity and its resolution, FISH has been used to visualize gene amplification and show the identity of the genes implicated. The diversity of mechanisms implicated may be illustrated by two examples. In one study on CHO cells (Trask and Hamlin, 1989), it was suggested that early dihydrofolate reductase gene amplification events usually occur on the same chromosome arm as the original locus. In another (Cherif et al., 1989), on MYC gene amplification in the breast carcinoma cell line SW613, several localizations of amplified MYC sequences were demonstrated, but only one was shown to be selected after passage in nude mice. All the localizations were outside chromosome 8 where the MYC gene is normally located. In another study FISH provided evidence of periodic structure of HSR in neuroblastoma cells (Garson et al., 1987).

Analysis of acquired chromosomal rearrangements

Repetitive DNA probes such as a satellite DNA are useful to define translocations involving pericentromeric regions. These probes as well as simple satellite specific probes have been used in five cases of hematological disorders with whole arm translocations (Speleman et al., 1991a). In the so-called t(1;7)(p11;p11) of four patients with hematopoietic disorders (Kibbelaar et al., 1992a), the question was to identify the origin of the centromeres of the rearranged chromosomes. Colocalization of alphoid DNA sequences specific for chromosomes 1 and 7 demonstrated that the translocation was t(1;7)(cen;cen). Satellite and alphoid probes have also been used to analyse complex translocations of cancer cells with several chromosome rearrangements (Smit et al.,1990).

Use of whole chromosome specific probes (chromosome painting)

Whole chromosome specific probes are now available that allow specific fluorescent staining of the chromosome studied after in situ hybridization to metaphase chromosomes. This can be used to better define chromosome rearrangements insufficiently identified with usual banding techniques. These types of probes have been used to study constitutional translocations (Rosenberg et al, 1992; Carter et al., 1992), and they can obviously be used for analysis of cancer cells. We have used a chromosome 11-specific whole painting probe to study acute monocytic leukemias with rearrangements of band 11q23. In one case, a previously recognized complex translocation t(1;6;11) could be demonstrated, and in another, FISH allowed the definition of a t(9;11;17) unrecognized with banding techniques which defined the rearrangement as a possible del(11)(q23) (Cherif et al., 1992b). Chromosome markers may be identified with FISH, e.g. for prostatic cancer (Brothman and Patel, 1992).

Delineation of chromosome breakpoints with FISH

DNA probes corresponding to fragments already localized on chromosome bands may serve to refine the chromosome breakpoints in rearrangements of cancer cells. This was done to characterize marker chromosomes in the Burkitt lymphoma cell line Namalwa (Ruppersberger et al., 1991), to identify a marker chromosome in an acute myeloid leukemia (Wullich et al., 1991), to define the breakpoints in Ewing sarcoma (Selleri et al., 1991) and an Ewing sarcoma with complex t(10;22;11) translocation (Speleman et al., 1991a), and to ascertain a t(6;11) in an acute monocytic leukemia (Derré et al, 1990). Interestingly, use of cosmid clones demonstrated that the breakpoints on chromosome 16 short arm were different in leukemia with inv(16)(p13q12) and t(8;16)(p11;p13) (Wessels et al., 1991). It is also noteworthy that cosmid probes corresponding to 16p13 were used to detect rapidly chromosome 16 inversion, specific to acute myelomonocytic leukemia with bone marrow eosinophilia (Dauwerse et al., 1990), a rearrangement that may be difficult to ascertain with banding techniques. The breakpoints of t(8;21), characteristic of a subtype of acute myeloblastic leukemia (AML), have been delineated by YAC probes encompassing these breakpoints (Kearney et al., 1991; Popp et al., 1991; Gao et al., 1991). These YACs may be used to better define them as well as to detect the chromosome rearrangement in AMLs. Similarly, it has been shown that the breakpoints of four different translocations of acute leukemias involving band 11q23 could be detected by a single YAC probe, suggesting that the breakpoints were adjacent (Rowley et al., 1990). However, using cosmid probes in FISH experiments, others found that the breakpoint on 11q23 of t(11;19) AML was distal to those of three other translocations of acute leukemia, t(4;11), t(6;11), and t(9;11) (Cherif et al., 1992a). Recently, FISH was also used to delineate the breakpoints in t(2;8) of the Burkitt lymphoma cell line JI (Ried et al., 1992b).

FISH on interphase nuclei
It has been shown that a given chromosome at interphase occupies a domain of the nucleus that can be visualized by in situ hybridization using chromosome-specific DNA probes corresponding to DNA alphoid sequences. The main interest of using FISH on interphase nuclei is that no cell culture is necessary, and the number of cells analysable is higher than that studied with cytogenetic methods on metaphase chromosomes. The technique can also be applied when few nuclei are analysable and, therefore, the probability to yield a sufficient number of metaphases very low. Another important application is the possibility to combine FISH and surface

marker studies, a technique particularly important in hematopoietic disorders (Tiainen et al, 1992, Kibbelaar et al, 1992b, Perez Losada et al., 1991,Weber-Matthiesen et al., 1992). This has been used, for instance, to show that trisomy 12 in chronic lymphocytic leukemia is only found in the malignant B-cells (Perez Losada et al., 1991) and to correlate trisomy 17 and cell morphology in acute leukemia (Anastasi et al., 1991b). In tumor cells, FISH may be of prognostic value on interphase nuclei, e.g. for trisomy 7 in bladder cancer (Waldman et al., 1991). Its usefulness in the analysis of the chromosome content of cancer cells has been confirmed on various types of malignancies: brain tumors (Arnoldus et al., 1991; Cremer et al., 1988b), neuroectodermal tumors (Cremer et al., 1988a), breast cancer (Balazs et al., 1991, Devilee et al., 1988, Viegas-Péquignot et al., 1989), gastric tumors (van Dekken et al., 1989), colon cancer (Nederlof et al., 1989, Viegas-Péquignot et al., 1989), bladder cancer (Hopman et al., 1989, 1991, Waldman et al., 1991), germ cell tumors (Mukherjee et al., 1991, Suijkerbuijk et al., 1991), ovarian cancer (Smit et al., 1990), hematopoietic disorders (Anastasi et al., 1992, Kibbelaar et al.,1991,1992b; Kolluri et al., 1990, Nederlof et al., 1989; Poddighe et al., 1991, Ried et al., 1992b, Ruppersberger et al., 1991). Another possibility is to use FISH on nuclei to detect recurrent translocations, e.g. t(9;22) with probes containing BCR and ABL gene sequences (Arnoldus et al., 1990; Tkachuk et al., 1990).

FISH and follow-up of patients with malignant disorders

FISH has been proposed to detect residual disease mainly in patients treated for blood disorders (Anastasi et al., 1991a,b). In practice, however, the method is not sensitive enough to be used without restriction. This is particularly the case for monosomy detection since in normal cell populations, the frequency of nuclei with one spot instead of two is about 5% with alphoid probes (Baurmann et al., unpublished). On the contrary, detection of a low proportion of trisomic nuclei is theoretically possible. FISH has also been used in patients who had had sex-mismatched bone marrow transplantation (van Dekken et al., 1989) and the method may be complemented by other molecular biology methods.

R. Berger

References

Anastasi, J. et al.(1991a) Interphase cytogenetic analysis detects minimal residual disease in a case of acute lymphoblastic leukemia and resolves the question of origin of relapse after allogeneic bone marrow transplantation. Blood, 77, 1087-1091.

Anastasi, J. et al. (1991b) Direct correlation of cytogenetic findings with cell morphology using in situ hybridization : An analysis of suspicious cells in bone marrow specimens of two patients completing therapy for acute lymphoblastic leukemia. Blood, 77, 2456-2462.

Anastasi, J. et al. (1992) Detection of trisomy 12 in chronic lymphocytic leukemia by fluorescence in situ hybridization to interphase cells: A simple and sensitive method. Blood, 79, 1796-1801.

Arnoldus, E.P.J. et al. (1990) Detection of the Philadelphia chromosome in interphase nuclei. Cytogenet. Cell Genet., 54, 108-111.

Arnoldus, E.P.J. et al. (1991) Interphase cytogenetics of brain tumors. Genes Chrom. Cancer, 3, 101-107.

Balazs, M. Mayall, B.H. and Waldman, FM. (1991) Interphase cytogenetics of male breast cancer. Cancer Genet. Cytogenet., 55, 243-247.

Brothman, A.R. and Patel,.M. (1992) Characterization of 10 marker chromosomes in a prostatic cancer cell line by in situ hybridization. Cytogenet. Cell Genet., 60, 8-11.

Carter, N.P. et al. (1992) Reverse chromosome painting: a method for the rapid analysis of aberrant chromosomes in clinical cytogenetics. J. Med. Genet., 29, 299-307.

Cherif, D. et al. (1989) Selection of cells with different chromosomal localizations of the amplified c-myc gene during in vivo and in vitro growth of the breast carcinoma cell line SW613-S. Chromosoma, 97,327-333.

Cherif, D. et al. (1992a) The 11q23 breakpoint in acute leukemia with t(11;19)(q23;p13) is distal to those of t(4;11), t(6;11) and t(9;11). Genes Chrom. Cancer, 4,107-112.

Cherif D. et al. (1992b) Chromosome painting in acute monocytic leukemia. Genes Chrom Cancer (in press).

Cremer, T. et al. (1988a) Rapid interphase and metaphase assessment of specific chromosomal changes in neuroectodermal tumor cells by in situ hybridization with chemically modified DNA probes. Exp. Cell Res., 176,199-220.

Cremer, T. et al. (1988b) Detection of chromosome aberrations in metaphase and interphase tumor cells by in situ hybridization using chromosome-specific library probes. Hum. Genet., 80,235-246.

Dauwerse, J.G. et al. (1990) Rapid detection of chromosome 16 inversion in acute nonlymphocytic leukemia, subtype M4: regional localization of the breakpoint in 16p. Cytogenet. Cell Genet., 53, 126-128.

Derré, J. et al. (1990) In situ hybridization ascertains the presence of a translocation t(6;11) in an acute monocytic leukemia. Genes Chrom. Cancer, 2, 341-344.

Devilee, P. et al. (1988) Detection of chromosome aneuploidy in interphase nuclei from human primary breast tumors using chromosome-specific repetitive DNA probes. Cancer Res., 48, 5825-5830.

Gao, J. et al. (1991) Isolation of a yeast artificial chromosome spanning the 8;21 translocation breakpoint t(8;21)(q22;q22.3) in acute myelogenous leukemia. Proc. Natl. Acad. Sci. USA, 88, 4882-4886.

Garson, J.A. van den Berghe, J.A. and Kemshead, J.T. (1987) High-resolution in situ hybridization technique using biotinylated NMYC oncogene probe reveals periodic structure of HRSs in human neuroblastoma. Cytogenet. Cell Genet., 45,10-15.

Griffin, D.K. Leigh, S.E.A. and Delhanty, D.A. (1990) Use of fluorescent in situ hybridisation to confirm trisomy of chromosome region 1q32-qter as the sole karyotypic defect in a colon cancer cell line. Genes Chrom. Cancer, 1, 281-283.

Hopman, A.H.N. et al. (1989) Detection of numerical chromosome aberrations in bladder cancer by in situ hybridization. Am. J. Pathol., 135, 105-117.

Hopman, A.J.N. et al. (1991) Numerical chromosome 1, 7, 9, and 11 aberrations in bladder cancer detected by in situ hybridization. Cancer Res., 51, 644-651.

Kearney, L. et al. (1991) DNA sequences of chromosome 21-specific YAC detect the t(8;21) breakpoint of acute myelogenous leukemia. Cancer Genet. Cytogenet., 57, 109-119.

Kibbelaar, R.E. et al. (1991) Detection of trisomy 8 in hematological disorders by in situ hybridization. Cytogenet. Cell Genet., 56, 132-136.

Kibbelaar, R.E. et al. (1992a) Non radioactive in situ hybridisation of the translocation t(1;7) in myeloid malignancies. Genes Chrom. Cancer, 4,128-134.

Kibbelaar, R.E. et al. (1992b) Combined immunophenotyping and DNA in situ hybridization to study involvement in patients with myelodysplastic syndromes. Blood, 79, 1823-1824.

Kolluri, R.V. et al. (1990) Detection of monosomy 7 in interphase cells of patients with myeloid disorders. Am. J. Hematol., 33, 117-122.

Langer, P.R. Waldrop, A.A. and Ward, D.C. (1981) Enzymatic synthesis of biotin-labeled polynucleotides: Novel nucleic acid affinity probes. Proc. Natl. Acad. Sci. USA 78, 6633-6637.

Mukherjee, A.B. et al. (1991) Detection and analysis of origin of i(12p), a diagnostic marker of human male germ cell tumors, by fluorescence in situ hybridization. Genes Chrom. Cancer, 3, 300-307.

Nederlof, P.M. et al. (1989) Detection of chromosome aberrations in interphase tumor nuclei by nonradioactive in situ hybridization. Cancer Genet. Cytogenet., 42, 87-98.

Perez Losada, A. et al. (1991) Trisomy 12 in chronic lymphocytic leukemia: an interphase cytogenetic study. Blood, 78, 775-779.

Pinkel, D. Straume, T. and Gray, J.W. (1986) Cytogenetic analysis using quantitative, high sensitivity, fluorescence hybridization; Proc. Natl. Acad. Sci. USA 83, 2934-2938.

Poddighe, P.J. et al. (1991) Interphase cytogenetics of hematological cancer: comparison of classical karyotyping and in situ hybridization using a panel of eleven chromosome specific DNA probes. Cancer Res., 51, 1959-1961.

Popp, S. et al. (1991) Translocation (8;21) in acute nonlymphocytic leukemia delineated by chromosomal in situ suppression hybridization. Cancer Genet. Cytogenet., 57, 103-107.

Ried, T. et al. (1992a) Simultaneous visualization of seven different DNA probes by in situ hybridization uisng combinatorial fluorescence and digital imaging microscopy. Proc. Natl. Acad. Sci. USA, 89, 1388-1392.

Ried, T. et al. (1992b) Specific metaphase and interphase detection of the breakpoint region in 8q24 of Burkitt lymphoma cells by triple color fluoresence in situ hybridization. Genes Chrom. Cancer, 4, 69-74.

Rosenberg, C. et al. (1992) Analysis of reciprocal translocations by chromosome painting: Applications and limitations of the technique. Am. J. Hum. Genet., 50, 700-705.

Rowley, J.D. et al. (1990) Mapping band 11q23 in human acyute leukemia with biotinylated probes: Identification of 11q23 translocation breakpoints with a yeast artificial chromosome. Proc. Natl. Acad. Sci. USA 87, 9358-9362.

Ruppersberger, P. et al. (1991) Characterization of marker chromosomes in Namalwa cells by chromosomal in situ suppression (CISS) hybridization and R-banding. Genes Chrom. Cancer, 3, 394-399.

Selleri, L. et al. (1991) Molecular localization of the t(11;22)(q24;q12) translocation of Ewing sarcoma by chromosomal in situ suppression hybridization. Proc. Natl. Acad. Sci. USA, 88, 887-891.

Smit, V.T.H.B.M. et al. (1990) Combined GTG-banding and nonradioactive in situ hybridization improves characterization of complex karyotypes. Cytogenet. Cell Genet., 54, 20-23.

Speleman, F. et al. (1991a) Analysis of whole-arm translocations in malignant blood cells by nonisotopic in situ hybridization. Cytogenet. Cell Genet., 56, 14-17.

Speleman, F. et al. (1991b) Molecular cytogenetic analysis of a complex t(10;22;11) translocation in Ewing's sarcoma. Genes Chrom. Cancer, 4, 188-191.

Suijkerbuijk, R.F. et al. (1991) Demonstration of the genuine iso-12p character of the standard marker chromosome of testicular germ cell tumors and identification of further chromosome 12 aberrations by competitive in situ hybridization. Am. J. Hum. Genet., 48, 269-273.

Tiainen, M. et al. (1992) Chromosomal in situ suppression hybridization of immunologiocally classified miotic cells in hematologic malignancies. Genes Chrom. Cancer, 4, 135-140.

Tkachuk, D.C. et al. (1990) Detection of bcr-abl fusion in chronic myelogenous leukemia by in situ hybridization. Science,250, 559-562.

Trask, B.J. and Hamlin, J.L. (1989) Early dihydrofolate reductase gene amplification events in CHO cells usually occur on the same chromosome arm as the original locus. Gene Develop., 3, 1913-1925.

van Dekken, H. Hagenbeck, A. and Bauman, J.G.J. (1989) Detection of host cells following sex-mismatched bone marrow transplantation by fluorescent in situ hybridization with a Y-chromosome specific probe. Leukemia 3, 724-728.

Viegas-Péquignot, E. et al. (1989) Detection of 1q polysomy in interphase nuclei of human solid tumors with a biotinylated probe. Hum. Genet., 81, 311-314.

Waldman, F.M. et al. (1991) Centromeric copy number of chromosome 7 is strongly correlated with tumor grade and labeling index in human bladder cancer. Cancer Res., 51, 3807-3813.

Weber-Matthiesen, K. et al. (1992) Simultaneous fluorescence immunophenotyping and interphase cytogenetics: a contribution to the characterization of tumor cells. J. Histochem. Cytochem., 40, 171-175.

Wessels, J.W. et al. (1991) Two distinct loci on the short arm of chromosome 16 are involved in myeloid leukemia. Blood, 77, 1555-1559.

Wullich, B. et al. (1991) Nonradioactive in situ hybridization. A rapid approach for the identification of marker chromosomes: Study of a case of acute leukemia with a Yq specific DNA probe. Cancer Genet. Cytogenet., 52, 165-172.

Genome analysis

14 Microdissection of GTG-banded chromosomes and PCR-mediated cloning

U. CLAUSSEN
Institut für Humangenetik, Erlangen, Germany
H.-J. LÜDECKE
Institut für Humangenetik, Essen, Germany
H. SPIELVOGEL
Institut für Humangenetik, Erlangen, Germany
H. TELENIUS
Cambridge University, UK and Karolinska Institute, Sweden

H.-C. HENNIES
Institut für Humangenetik, Berlin, Germany
G. SENGER
Institut für Humangenetik, Erlangen, Germany
A. REIS
Institut für Humangenetik, Berlin, Germany
A. MAZUR
Institut für Humangenetik, Erlangen, Germany
and B. HORSTHEMKE
Institut für Humangenetik, Essen, Germany

1 Introduction

With the use of DNA polymorphisms as genetic linkage markers, an increasing number of disease loci have been mapped to specific chromosome regions. The way from linkage to the gene, however, is tedious and has not yet been accomplished for a greater number of diseases. This is mainly due to the lack of narrowly spaced DNA markers for defined regions of the human genome. Physical dissection of metaphase chromosomes is the most direct approach to overcome this limitation.

The first scientific paper in this field was published by Scalanghe and coworkers in 1981. This group dissected and cloned DNA from <u>Drosophila</u> polytene chromosomes. In the time to come, microdissection and microcloning was applied to various chromosome regions in mouse and man. The few reports and their results showed, however, that the dissections were too coarse and inefficient to saturate a specific chromosome region with DNA markers.

One limitation of this conventional technique was due to the problem of the optical condition. Conventional microdissection was performed under phase contrast microscopy, in oil, on unbanded chromosomes. Most mammalian chromosomes, however, cannot be identified in unstained preparations.

A further limitation of the conventional microdissection and microcloning technique was the low efficiency of introducing recombinant DNA into bacteria. Even if phage vectors like lambda gt10 or gt11 and in vitro packaging were used, transformation efficiencies of more than 10^8 transformants/μg recombinant DNA could not be achieved.

To overcome these limitations, we modified and improved the microdissection technique and introduced the polymerase chain reaction into the cloning procedure (Lüdecke et al., 1989). In this review, we summarize the efforts that have been made in our laboratories to further expand the range of applications for microdissection and microcloning. Future aspects are discussed with respect to new techniques of micromanipulation of chromosomes.

2 Material and Methods

2.1 Microdissection of GTG-banded chromosomes

Microdissection is performed on an inverted microscope (IM Zeiss) with the help of a micromanipulator (MR Mot; Zeiss or micromanipulator 5171, Eppendorf) as desribed by Lüdecke et al. (1989) and Senger et al. (1990). Briefly, mitotic cells

Chromosomes Today   Volume 11. Edited by A.T. Sumner and A.C. Chandley. Published in 1993 by Chapman & Hall, London. ISBN 0 412 47670 3

are harvested for chromosome preparations from amniotic fluid cell cultures by the use of siliconized micropipettes (Pipette method; Claussen, 1980; Claussen and Hansmann 1984; Claussen et al., 1986). Fixed mitotic cells are dropped onto clean, wet coverslips. Thus, as an advantage, the coverslips contain only metaphases and are free of DNA from ruptured interphase cells. Furthermore, the time of fixation can be reduced to a few seconds which is important to avoid depurination caused by the acetic acid (Mezzanotte et al. 1988). After evaporation of the fixative, coverslips are washed and stored over night in 70% ethanol at -20°C. Prior to microdissection, the chromosomes are GTG-banded.

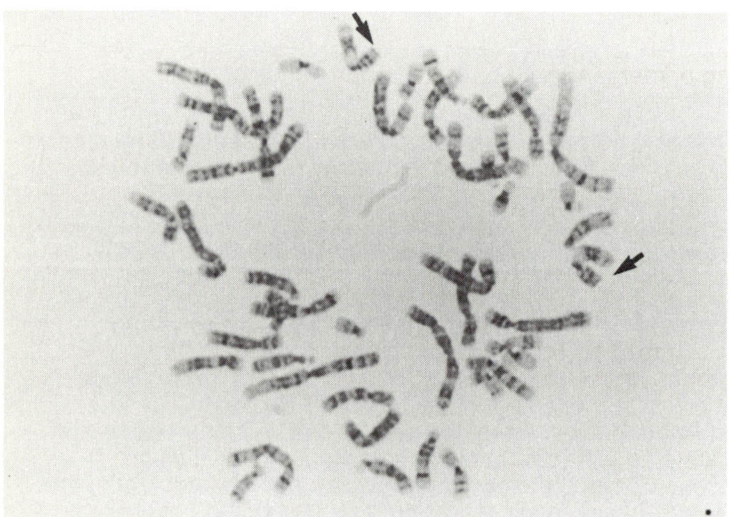

Fig. 1: Chromosome microdissection of 11p15.5. Arrows
indicate the excised bands of chromosome 11.

Chromosomes are dissected at a magnification of 1250 times (see Fig. 1). The fragment of interest is pushed out of the chromosome with the tip of an extended, siliconized glass needle. Thereafter, the fragments sticking to the tip of the needle are transferred to a 1-nl drop containing proteinase K and sodium dodecyl sulfate (SDS). To avoid evaporation, this droplet is placed inside a small moist chamber constructed from a cap of a freezing tube with a filling of wet viscose wadding inside. The volume of the collection drop inside the moist chamber can readily be controlled by the heat of the microscope lamp. For the construction of band-specific libraries twenty to forty excised chromosome fragments are collected.

2.2 Microcloning

2.2.1 Cloning of RsaI or HpaII fragments
Microcloning of RsaI or HpaII digested chromosomal fragments is performed as described by Lüdecke et al. (1989, 1990) and Horsthemke et al. (1992). Briefly, the collection drop is overlaid with liquid paraffin and transferred to a small

siliconized microdish within a new Petri dish also filled with liquid paraffin. This measure is taken, because subsequent microreactions require constant and precisely controlled volumes. Microreactions under oil are performed with the help of extended, siliconized Pasteur pipettes at a magnification of 30 times. All reagents are added by fusing drops in a volume ratio of 1:1. Proteins are again digested with proteinase K, and the chromosomal DNA is purified by three phenol extractions. Residual phenol is removed by diffusion into oil over night (Edström et al., 1987). Then, the extracted chromosomal DNA is digested with RsaI or HpaII. Again three phenol extractions are performed. The RsaI or HpaII fragments are then ligated to a SmaI or ClaI cut pUC plasmid, respectively. Since linear, rather than circular, molecules are more efficiently primed, polyethylene glycol is included in the ligation step to obtain long concatemeres. Non recombinant polylinker sequences are recut with SmaI or ClaI, respectively. Ligated inserts are PCR-amplified with the universal M13/pUC forward and reverse seqencing primers. After removal of the primer binding sites by EcoRI digestion and gel filtration, the amplified DNA fragments are ligated into pUC13. Ligation products are then used to transform competent DH5⍺ cells. Recombinant clones comprising white and white/blue colonies are isolated and analyzed in detail.

2.2.2 Cloning by DOP-PCR
After microdissection, the 1 nl collection drop containing the excised fragments is transferred directly into a PCR tube containing all reagents for DOP-PCR as described by Telenius et al. (1992b). The reaction is interrupted after the 10 initial low annealing temperature cycles. An aliquot is then diluted serially, and fractions empirically containing one to ten unique DNA fragments re-amplified by high temperature DOP-PCR. The PCR clones are isolated by excision of single fragments from Low Melting Point (LMP) agarose gel.

2.3 Band-specific chromosomal localization of very small DNA probes
Microdissection is performed as described above. The dissected chromosome fragments (n=6-7) are subsequently transferred to collection drops inside the moist chamber. Next, the collection drops are transferred directly into the PCR mixture containing primers flanking the DNA probe of interest. The presence of a sequence tagged site (STS) in a specific chromosome band can then easily be detected after electrophoresis of an aliquot of the amplification reaction on agarose gel (Spielvogel et al., 1992).

2.4 Chromosome stretching
Mitotic cells in the metaphase stage are harvested using the pipette method for microdissection as decribed above. The short time of fixation is crucial for the proper consistency of the chromosomes. Thereafter, chromosomes are soft and elastic under wet conditions. Chromosome stretching using extended glass needles is performed at a magnification of 1250 times.

3 Results and Discussion

Using the cloning procedure based on RsaI or HpaII digestion and ligation in pUC vector, we have successfully constructed libraries for the following regions of human chromosomes: 3p14, 5q13, 5q22, 8q13, 8q2305-24.1, 8q24.1, 11p13, 11p15, 11p15.5, 11q23, 12q14, 15q11-q13, 22q11.2, 22q12-13.1, Xp22.3, and Xq27-2805. Furthermore we have constructed libraries for mouse and plant chromosomes. These libraries are also useful for FISH as described by Trautmann et al. (1991) and Lengauer et al. (1991).

On average, the insert size is about 170 bp. The amount of single copy sequences varies from band to band. Most libraries contain a high number of single copy clones (39-80%), and some of them appear to be enriched for gene sequences (Lüdecke et al., 1990; Buiting et al., 1990; MacKinnon et al., 1990; Newsham et al., 1991; Puech et al., 1992). This holds true especially for libraries based on cloning of HpaII tiny fragments (Horsthemke et al., 1992).

In order to obtain contiguous DNA from large chromosome regions, DNA probes of these libraries can be translated into STSs and used for the screening of YAC libraries (Lüdecke et al., 1992).

Recently, it has been shown that generation of region-specific probes can also be achieved by random PCR on microdissected material, without the need for the microcloning procedure (Hadano et al., 1991; Meltzer et al., 1992). By introducing a few initial cycles at a low annealing temperature, random PCR enables direct enzymatic amplification of DNA from excised chromosome fragments. We have constructed region-specific libraries based on general DNA amplification using DOP-PCR first described by Telenius et al. (1992a,b). As a new and very rapid way to obtain single clones from these libraries, we included a dilution step after the first ten low temperature cycles. The high temperature cycles are then performed on aliquots containing 1 to 10 unique PCR-fragments. After excision from gel, these `PCR clones` can immediately be used as hybridization probes without any further treatment (Telenius et al., manuscript in preparation). This procedure hence omits all steps involved in classical cloning, such as restriction enzyme digestion, ligation to vectors, and transformation of bacterial cells. At present, four different microdissection libraries have been constructed in this manner. Preliminary results indicate that contamination has occured in at least one of these, and this problem will prompt further attention, as we aim to generate libraries from as few as one excised chromosome fragment.

Microdissection and subsequent PCR can also be used for band- and subband-specific mapping of small DNA probes even smaller than 80 bp. Mapping the microsatellite at D13S71 to the distal part of band 13q32 was accomplished. The amplification products of the microsatellite at D13S71 have a size range from 67 to 79 bp. The microsatellite is highly polymorphic and can easily be used for linkage analysis. This approach allows a size independent band- or subband-specific mapping of single copy or repetitive DNA probes, if an overlapping microdissection strategy is used (Spielvogel et al., 1992).

4 Future aspects

Within the next few years, most of the human genome will be covered with YAC contigs. Gaps between these contigs may be closed by specific microdissection and microcloning if these regions. Therefore, we aim at dissecting regions as small as 1 to 3 Mbp, for the generation of non-overlapping subband-specific libraries constructed from one excised chromosome fragment each. For excision of subbands, however, the use of high resolution chromosomes prepared by standard protocols is not recommendable. These chromosomes show many overlaps with other chromosomes, are often distorted, and therefore are not easy to identify. Another method thus needs to be devised for the preparation of high-resolution chromosomes. Stretching of metaphase chromosomes seems to be the solution to this problem. We have developed a chromosome stretching procedure based on the observation that moist chromosomes are highly elastic and can easily be stretched with the help of glass needles. Fig. 2 shows a softened chromosome 7 at the 300 to 400 band level (Fig. 2a,b) with its proximal q-arm stretched to the 850 band level (Fig. 2c). We have found that chromosome stretching can even be performed on chromosomes after fluorescence in situ hybridisation. We are now investigating as to what extent this finding could contribute to the fine physical mapping of narrowly spaced DNA probes.

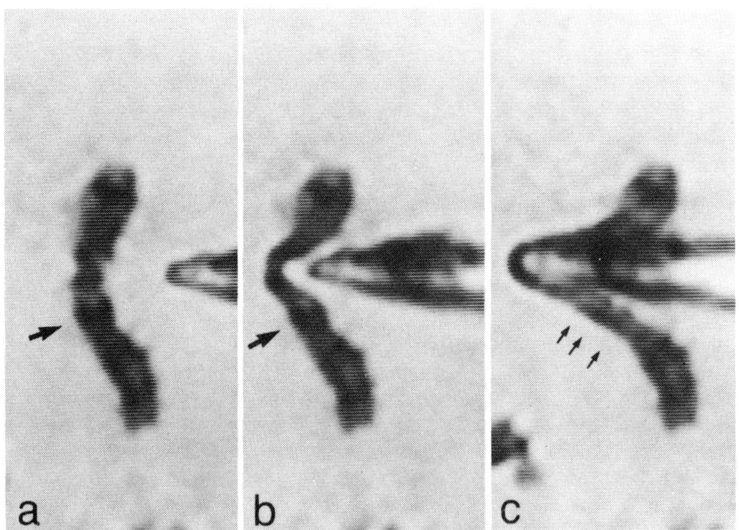

Fig. 2: Stretching of the proximal part of human chromosome 7q. Arrows indicate 7q21 (a,b) and its dark subbands (c).

Besides analyzing the human genome, the construction of band-specific DNA libraries for other species is of interest. Here, chromosome- and band-specific, narrowly spaced DNA probes and YAC contigs will not be readily available within the next few years.

From an evolutionary point of view, the construction of band-specific DNA-libraries of the whole human karyotype at the 400 band level for example is desirable. This holds true also for other mammals.

Dark and light bands can be analyzed in greater detail, considering the differences between their distribution of coding and noncoding DNA sequences. Centromere- and telomere-sites can also be investigated on the DNA level to give a better understanding of the mechanisms that link genome organisation and chromosome structure.

From the cytogenetic point of view, nearly all limitations that used to prevent the excision of chromosome fragments containing less than 3 Mbp, are solved. In genome analysis, YAC libraries with very large average inserts (1 Mbp) are starting to appear. Therefore, chromosome stretching for microdissection, coupled with rapid PCR amplification and -cloning, may provide the bridge between these advances in cytogenetics and molecular genome analysis.

U. Claussen, H.-J. Lüdecke, H. Spielvogel et al.

5 Acknowledgements

We thank Gabriele Schmitt and Sonja Hesse for excellent techniqual
assistance. Part of this work was supported by Wilhelm Sander-Stiftung and
Deutsche Forschungsgemeinschaft.

6 References

Buiting, K. et al. (1990) Microdissection of the Prader-Willi syndrome
chromosome region and identification of potential gene sequences.
Genomics, 6, 521-527.
Claussen, U. (1980) The pipette method: A new rapid technique for
chromosome analysis in prenatal diagnosis. **Hum. Genet.**, 54, 277-278.
Claussen, U. und Hansmann, M. (1984) Die "Pipettenmethode" zur schnellen
Karyotypisierung bei sonographischen Verdachtskriterien für eine
Chromosomenanomalie. **Gynäkologe**, 17, 33-40.
Claussen, U., Klein, R. and Schmidt, M. (1986) A pipette method for
rapid karyotyping in prenatal diagnosis. **Prenatal Diagnosis**, 6, 401-408.
Claussen, U. et al. (1992) Cloning defined regions of the human genome
by microdissection of banded chromosomes and enzymatic DNA
amplification. in **Advances in Genome Biology**, 1. (ed R.S. Verma),
Jai Press, Greenwich, London, pp 9-36.
Edström, J.E., Kaiser, R. and Röhme, D. (1987) Microcloning of mammalian
metaphase chromosomes. **Methods in Enzymology**, 151, 503-516.
Hadano, S. et al. (1991) Laser microdissection and single unique primer
PCR allow generation of regional chromosome DNA clones from a single
human chromosome. **Genomics**, 11, 364-373.
Horsthemke, B. et al. (1992) PCR-mediated cloning of HpaII tiny fragments
from microdissected human chromosomes. **PCR Methods and Application**,
1, 229-237.
Lengauer, C. et al. (1991) Comparative chromosome mapping in primates by in
situ suppression hybridization of band specific DNA microlibraries.
J. Human Evol., 6, 67-71.
Lüdecke, H.-J. et al. (1989) Cloning defined regions of the human genome
by microdissection of banded chromosomes and enzymatic amplification.
Nature, 338, 348-350.
Lüdecke, H.-J. et al. (1990) Construction and characterisation of
band-specific DNA libraries. **Hum. Genet.**, 84, 512-516.
Lüdecke, H.-J. et al. (1992) Generation of region-specific probes by
microdissection and universal enzymatic DNA amplification. in **Techniques
for the Analysis of Complex Genomes** (ed R. Anand), Academic Press,
London, San Diego, New York, Boston,Sydney, Tokyo, pp 215-229.
MacKinnon, R. N. et al. (1990) Microdissection of the fragile X region.
Am. J. Hum. Genet., 47, 181-187.
Meltzer, P.S. et al. (1992) Rapid generation of region specific probes
by chromosome microdissection and their application. **Nature Genet.**,
1, 24-28.
Mezzanotte, R. et al. (1988) Ageing of fixed cytological preparations
produces degradation of chromosomal DNA. **Cytogenet. Cell Genet.**,
48, 60-62.
Newsham, I. et al. (1991) Microdissection of chromosome band 11p15.5:
Characterization of probes mapping distal to the HBBC locus. **Genes**

Chrom. Cancer, 3, 108-116.
Puech, A. et al. (1992) 11p15.5 specific libraries for identification
  of potential gene sequences involved in Beckwith-Wiedemann syndrome and
  tumorgenesis. **Genomics**, 13, 1274-1280.
Scalanghe, F. et al. (1981) Microdissection and cloning of DNA from a
  specific region of Drosophila melanogaster polytene chromosomes.
  Chromosoma, 82, 205-216.
Senger, G. et al. (1990) Microdissection of banded human chromosomes.
  Hum. Genet., 84, 507-511.
Spielvogel, H. et al (1992) Band-specific localization of the
  microsatellite at D13S71 by microdissection and enzymatic
  amplification. **Am. J. Hum. Gent.**, 50, 1031-1037.
Telenius, H. et al. (1992a) Cytogenetic analysis by chromosome
  painting using DOP-PCR amplified flow-sorted chromosomes.
  Genes Chrom. Cancer, 4, 257-263.
Telenius, H. et al. (1992b) Degenerate oligonucleotide-primed PCR:
  General amplification of target DNA by a single degenerate primer.
  Genomics, 13, 718-725.
Trautmann, U. et al. (1991) Detection of APC region-specific signals by
  nonisotopic chromosomal in situ suppression (CISS)-hybridisation
  using a microdissection library as a probe. **Hum. Genet.**, 87, 495-497.

15 Molecular cytogenetics – biology and applications in plant breeding

J.S. HESLOP-HARRISON and T. SCHWARZACHER
John Innes Centre, UK

Karyobiology Group, John Innes Centre, Norwich NR4 7UJ, UK

1. Introduction

Cytogenetic manipulation of the chromosome complements of crop plants continues to be one of the most important methods available to plant breeders for introducing new variation into varieties of crops (see e.g., Law 1981, Gale and Miller 1987). Molecular cytogenetic analysis of the chromosomes of crops is a vital part of understanding their evolution, genetics, genetic recombination and karyotypic stability.

There are several important aspects of our research program that relate to plant breeding. These include detecting and characterizing alien chromosomes or chromosome segments in plant breeding lines with chromosome segments of various origins, developing markers for characters of interest to plant breeders, understanding physical chromosome behaviour, chromosome elimination, evolution and gene mapping: finding where genes and DNA sequences lie along chromosomes and within the three dimensions of the interphase nucleus. Here, we will review our work applying molecular cytogenetic analyses to the small grained cereals such as wheat, rye and barley. Like many groups of plants, the cereals include diploid, polyploid and hybrid species, and cultivated crops, their ancestors and wild relatives (see Table 1). The principles discussed here are applicable to other groups of crop plants, such as the Brassicas (oil seed rape, mustards, cabbages, turnips etc.).

Other groups of species where genomic *in situ* hybridization has been used to analyze genomic relationships (Table 1) include allopolyploid tobacco (*Nicotiana tabacum* L. 2n=4x=48; A. Kenton, A.S. Parokonny and M.D. Bennett, pers. comm.) and the Brassica species (G.E. Harrison, J. Maluszynska and J.S. Heslop-Harrison, pers. comm.).

Bread wheat (*Triticum aestivum* L.), the world's most important crop in terms of both economic and food value, is an allohexaploid (2n=6x=42) which arose from a cross between a cultivated tetraploid, Durum wheat (*T. durum* Desf.; 2n=4x=28) and a wild goat grass (similar to *Aegilops squarrosa* L.; 2n=2x=14) in the Middle East about 7000 B.C. Triticale (X *Triticosecale* Wittmark; 2n=6x=42) is a hybrid crop of more recent origin: it was first made by crossing a tetraploid wheat and diploid rye (*Secale cereale* L.; 2n=2x=14) in the 1930s with the aim of combining the hardiness of rye with the high yield of wheat. Progeny from more recent crosses, after inbreeding and selection of varieties, are widely grown on sandy soils in Poland and Canada.

Chromosomes Today   Volume 11. Edited by A.T. Sumner and A.C. Chandley. Published in 1993 by Chapman & Hall, London. ISBN 0 412 47670 3

2. Identification of chromosomes using genomic *in situ* hybridization

Figure 1 shows fluorescent micrographs of root tip metaphase chromosomes from the Polish triticale variety Lasko after genomic *in situ* hybridization. To distinguish the chromosomes originating from the two parents, wheat and rye, total genomic DNA from rye was labelled with digoxigenin, used as a probe, together with excess amounts of unlabelled DNA from wheat (Heslop-Harrison et al. 1988, 1990; Anamthawat-Jónsson et al. 1990). The 14 chromosomes of rye origin show strong

Table 1. Chromosome number, DNA content and genome size for some crop species and their wild relatives, in comparison to human, mouse and *A. thaliana*. DNA contents taken from Bennett and Smith (1976) for plants and Greilhuber et al. (1983) for mammals.

Species	Common name	Chromo-some number (2n)	Genome design-ation	1C DNA content (pg)	Genome size (Mbp)	Average chromo-some (Mbp)
Gramineae						
T. monococcum L. [1]	Einkorn	2x = 14	AA	6.2	6230	890
Ae. speltoides Tausch	Goat grass	2x = 14	SS	5.8	5 800	830
Ae. squarrosa L. [2]	Goat grass	2x = 14	DD	5.0	5 010	715
T. durum Desf.	Macaroni wheat	4x = 28	AABB	12.0	12 030	860
T. aestivum L.	Bread wheat	6x = 42	AABBDD	17.3	17 330	825
H. vulgare L.	Cultivated barley	2x = 14	II	5.5	5 550	790
H. chilense R.& S.		2x = 14	$H^{ch}H^{ch}$	5.4	5 450	780
H. bulbosum Nevski		2x = 14	II	5.4	5 400	770
S. cereale L.	Cultivated rye	2x = 14	RR	8.3	8 280	1185
S. africanum Stapf		2x = 14	RR	7.4	7 400	1060
Zea mays L.	Maize	2x = 20		2.4	2 350	235
Oryza sativa L.	Rice	2x = 24	"A"	0.6	600	50
Brassicaceae						
B. campestris L.	Chinese cabbage	2x = 20	aa	0.8	830	85
B. nigra (L.)Koch	Black mustard	2x = 16	bb	0.8	780	100
B. oleracea L.	Cabbage	2x = 18	cc	0.9	900	100
B. juncea (L.)Czern.		4x = 36	aabb	1.5	1 530	85
B. napus L.	Oil seed rape	4x = 38	aacc	1.6	1 600	85
B. carinata A.Braun		4x = 34	ccbb	1.6	1 580	95
A. thaliana (L.)Heynh.		2x = 10		0.15[3]	150	30
Solanaceae						
L. esculentum Miller	Tomato	2x = 24		0.9[4]	900	75
So. tuberosum L.	Potato	4x = 48		2.1	2 100	90
N. tomentosiformis G.		2x = 24	TT	1.8	1 850	155
N. sylvestris S.& C.		2x = 24	SS	2.1	2130	180
N. tabacum L.	Tobacco	4x = 48	TTSS	3.9	3 880	160
Mammalia						
Homo sapiens	Human	2x = 46		3.0	3 000	130
Mus musculus	Mouse	2x = 40		3.0	3 000	150

T. = *Triticum*, *Ae.* = *Aegilops*, *H.* = *Hordeum*, *S.* = *Secale*, *B.* = *Brassica*, *A.* = *Arabidopsis*, *L.* = *Lycopersicon*, *So.* = *Solanum*, *N.* = *Niocotiana*. R.& S. = Roem. & Schult., G. = Goodspeed, S.& C. = Speg. & Comes,

1) also *T. uratu* Tum., wild Einkorn, 2) also *T. dicoccum* Schübl., Emmer, 3) J.S. Heslop-Harrison pers. comm. 4) J.D.G. Jones pers. comm.

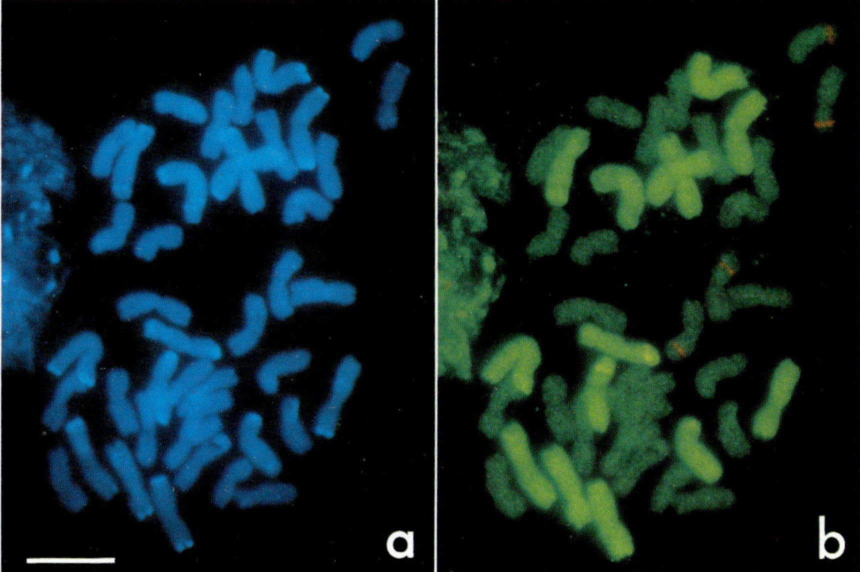

Figure 1. Genomic *in situ* hybridization using digoxigenin labelled rye DNA and biotinylated rDNA on a root tip metaphase chromosomes of hexaploid triticale cv. Lasko. **a.** The 42 chromosomes show uniform DAPI staining with some brighter fluorescing heterochromatin bands. **b.** Sites of probe hybridization were detected by fluorescein conjugated anti- digoxigenin (green fluorescence) and Texas Red conjugated avidin (red fluorescence). The 14 chromosomes of rye origin appear brightly yellow-green, while the 28 wheat chromosomes are dark green. The two pairs of major wheat rDNA sites are fluoresceing orange-red, while the pair of rye rDNA sites are yellow. We thank Nuno Neves for the preparation shown. Bar 10μm.

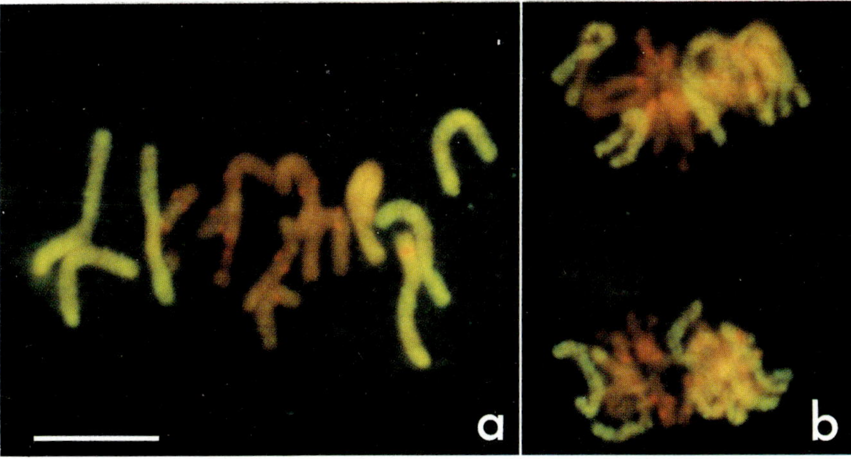

Figure 2. Genomic *in situ* hybridization enables the parental origin of the chromosomes to be determined in a sexual hybrid between barley x *S. africanum* (wild rye). The seven chromosomes of *S. africanum* origin are labelled and fluoresce yellow while the seven chromosomes from barley fluoresce orange with the propidium iodide counterstain. The two parental genomes lie in spatially separated domains at metaphase (**a**) and other stages of the cell cycle. At anaphase (**b**) all chromosomes have moved to the spindle poles. (Figure taken from Leitch et al. 1991). Bar 10μm.

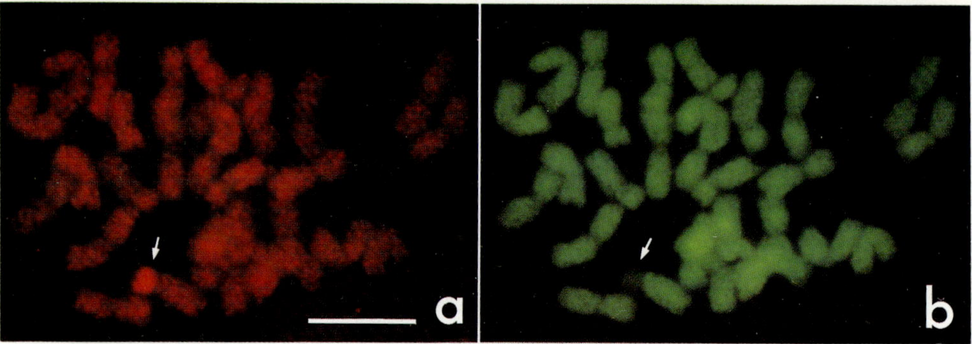

Figure 4. Genomic *in situ* hybridization to a wheat/*Hordeum chilense* translocation line. **a.** Using biotinylated *H. chilense* DNA as a probe and Texas Red detection labels the *H. chilense* chromosome arm (arrow) by bright red fluorescence. **b.** Simultaneous probing with digoxigenin labelled wheat DNA and fluorescein detection shows the wheat chromosomes green and the *H. chilense* chromosome arm (arrow) remains unlabelled. Bar 10μm.

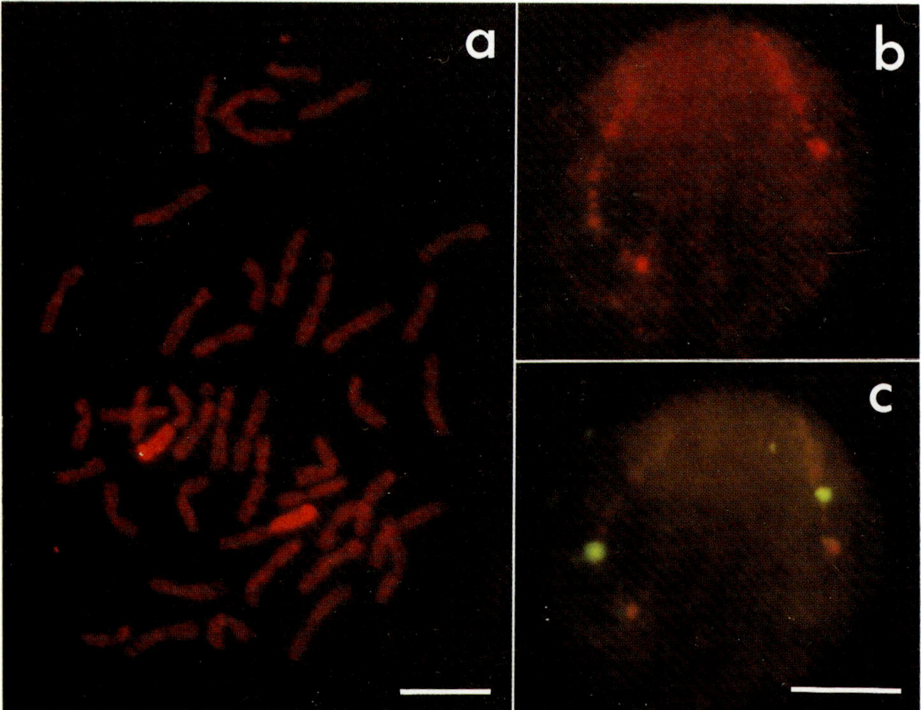

Figure 5. Genomic *in situ* hybridization to wheat lines which include a translocation with a rye chromosome (1BL/1RS). **a.** At metaphase the two rye chromosome arms are brightly fluorescing with direct rhodamine labelled rye DNA. **b.** At interphase, the two rye arms form distinct domains which ar not closely associated. **c.** Double exposure of the rhodamine labelled rye DNA (as in b) and a fluorescein labelled rDNA probe. The telomeric heterochromatin (red) and the rDNA locus (yellow) remain condensed, while the short segment between the NOR and the C-band is greatly extended. Bars 10 μm.

yellow-green fluorescence along their entire length. The 28 wheat-origin chromosomes show only weak fluorescence (Fig. 1b). The four major wheat nucleolus origin regions (NORs) show an orange-red signal of the simultaneously hybridized biotinylated rDNA probe. Genomic probing also was able to differentiate between chromosomes originating from two very closely related *Hordeum* species, *H. vulgare* and *H. bulbosum* in a F_1 (first generation) hybrid (Anamthawat-Jónsson et al. 1990, 1993).

3. Organization of genomes in hybrids

Formation of sexual hybrids between different species of the same or different genera, followed by polyploidization, is a common mechanism in plant speciation and evolution. We have looked at chromosome behaviour in several intra- and inter-generic hybrids between barley (*Hordeum*) and rye (*Secale*) species, and have found that their chromosomes were often non-randomly distributed. The chromosomes originating from the two parental genomes did not mix shortly after fertilization, but remained spatially separated (for review see Heslop-Harrison and Bennett 1984, 1990). Reports from other sexual and cell fusion hybrids in both plants and animals also show separation of the parental chromosome sets at metaphase (Finch et al. 1981, Gleba et al. 1987, Linde-Laursen and Jensen 1991, Ordartschenko and Keneklis 1973, Zelesco and Marshall Graves 1988).

Development of *in situ* hybridization techniques using total genomic DNA, noted above, enabled us to determine the parental origin of all chromosomes not only at metaphase, but at all stages of the cell cycle (Schwarzacher et al. 1989; Leitch et al. 1990, 1991 and Fig. 2). The studies showed that the parental sets of chromosomes in the hybrids *H. chilense* x *S. africanum* and barley x *S. africanum* lay in separate domains within the interphase nucleus and tended to maintain their positions throughout the cell cycle. They were organized in the Rabl orientation with the centromeres clustered at one end of the nucleus and the telomeres at the other end. Leitch et al. (1991) extended the two-dimensional analysis of spread preparations to three-dimensional reconstructions of serial sections of interphases of barley x *S. africanum*. Each 0.25 μm thick section was labelled by genomic *in situ* hybridization, photographed in the fluorescent light microscope, the photographs arranged in their consecutive order and individual nuclei reconstructed. Analysis of the reconstructions showed that the DNA was distributed in the nuclei to form a gradient from a region with high proportion of DNA to a region with a lower proportion of DNA as also described in other cereals (Anamthawat-Jónsson and Heslop-Harrison 1990). The distribution of the *in situ* hybridization signal, labelling the chromatin originating from *S. africanum*, did not follow the gradient, but labelled chromatin tended to occupy one, two or three domains to one side or surrounding the single barley domain (Leitch et al. 1991).

Another hybrid between the same female parent, *H. vulgare* (barley) and the closely related wild barley, *H. bulbosum*, is economically important. When crossed with certain genotypes of *H. bulbosum*, the hybrid eliminates the *H. bulbosum* chromosomes, giving a haploid barley plant. The haploid can be doubled in chromosome number to produce a homozygous plant which can be immediately selected in a breeding programme without problems of heterozygosity and gene segregation (Kasha and Kao 1970). Some barley x *H. bulbosum* hybrids tend to be more stable and retain most chromosomes from both parents. In this hybrid combination we again found that the barley genome was central, while the *H. bulbosum* genome lay in a more peripheral domain at all stages of the cell cycle (Anamthawat-Jónsson et al. 1993).

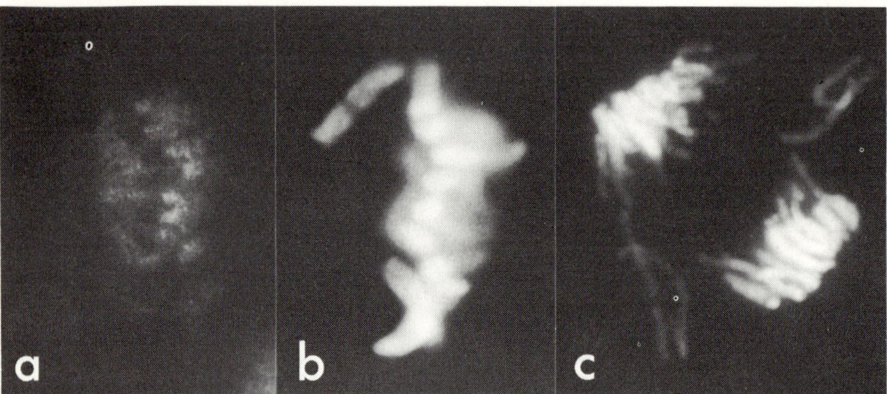

Figure 3. Whole mount preparation of a metaphase from a root tip of barley x *H. bulbosum* after (**a**) immunocytochemical detection using a monoclonal antibody (YOL1/34) against tubulin and rhodamine labelled secondary antibodies (method as in Doonan and Clayton 1986) and (**b**) simultaneous DAPI staining. One chromosome (arrow), probably of *H. bulbosum* origin, has not congressed onto the metaphase plate and shows less fluorescence after antitubulin immunolabelling. Bar 10μm.**c**. Reconstruction from deblurred optical sections of DAPI stained whole mount chromosome preparations.At anaphase two chromosomes, presumably of *H. bulbosum* origin, are lagging although the chromatids have separated Bars 10μm.

Interestingly, despite the similar genome separation shown in the two hybrids, the hybrids showed contrasting behaviour at anaphase (Fig. 2b and 3c). In the *H. bulbosum* hybrid, the *H. bulbosum* origin chromosomes are mostly lagging behind during division (Fig. 3c), while in the *S. africanum* hybrid, chromosomes from the two species move simultaneously to the poles (Fig. 2b). Thus there is a genotype dependent control of behaviour of the sets of chromosomes, and it seems unlikely that the differential behaviour has evolved only for functions in unusual, although important, interspecific hybrids. Using whole mount preparations labelled with anti-microtubule antibodies we were able to show that some chromosomes, of *H. bulbosum* origin, did not congress onto the metaphase plate and had very few microtubules attached to their centromeres (Fig. 3a and b). The ultrastructure of the centromeres from the two parental species was also contrasting (Schwarzacher et al. 1992a).

4. Detection of alien chromosomes and chromosome arms

While the combination of characters from two species in hybrids has occurred both naturally and in plant breeding programmes, crossing programmes can be designed to transfer chromosomes, chromosome segments or cytoplasms between different species. Many generations of inbreeding and selection have reduced the genetic base of many crops, but it can be expanded by crossing the crop with alien species, backcrossing, and selecting lines with recombinant chromosomes (e.g., Gale and Miller 1987). Sexual wide-hybrids provide the starting point for the transfer of smaller numbers of agronomically desirable traits from a wild species into a crop species.

In situ hybridization is proving valuable for identifying the origin of chromosomes in first generation and backcrossed hybrids between wheat and its wild and cultivated relatives, for screening lines for the presence of alien chromosomes, and, particularly, small alien segments that may carry important characters but are difficult to detect by other methods (Schwarzacher et al. 1992b). The genomic probing method enables easy identification of the translocated chromosome segment and characterization of its size. Fig. 4 shows a wheat line which includes a chromosome arm from *Hordeum chilense* translocated onto a wheat chromosome. The highest yielding wheat varieties currently grown in Europe include one pair of chromosome arms originating from rye, translocated onto a pair of wheat chromosome arms (a 1BL/1RS translocation; e.g., Heslop-Harrison et al. 1990 and Fig. 5a).

The genomic probing method allows the identification of particular chromosome arms not only at metaphase, but throughout all stages of the cell cycle and in non-dividing tissues. The homologous chromosomes lie in discrete domains. Chromosome arms usually run from the centromeric end of the nucleus to the telomeric end, displaying the typical Rabl orientation. In disomic lines, the homologous alien chromosome arms often show parallel coiling and condensation patterns, but are not spatially associated (Schwarzacher et al. 1992b and Fig. 5b).

5. Genetic and physical maps

Having a saturated gene map is important for cloning and manipulation of genes. The knowledge of where genes lie both physically and genetically is valuable for breeding, and clones are useful for functional and genetic analysis. In the future, knowledge of gene positions is likely to be useful for transformation or directed genetic manipulation of cereals and other crops. *In situ* hybridization, using both single copy probes and by breakpoint mapping in translocation lines is a vital tool in physical gene mapping. Some types of genes can be mapped only by *in situ* hybridization - in particular those that are present in many copies, including some slight variants, within the genome (Maluszynska and Heslop-Harrison 1991, Leitch and Heslop-Harrison 1992). Hence, the molecular cytogenetic approaches uniquely complement other gene mapping strategies which are being applied to crops.

The genetic map of the short arm of chromosome 1RS, which is involved in the translocation described above, shows that many DNA and isozyme markers and genes map distal of the secondary constriction at the NOR (nucleolus organizer region, the site of the rRNA genes) and that this distal segment comprises most of the genetic length of the chromosome arm (Baum and Appels 1991, Heslop-Harrison 1991, Wang et al 1992a; see Fig. 6). At metaphase, the rDNA locus is about 25% from the telomere (Schlegel et al. 1986). The satellite, distal to the rDNA locus, has a segment of genetically inactive telomeric heterochromatin, and a euchromatic segment comprising about 15% of the total euchromatin of the short arm (Fig. 6) which is available to contain most mapped markers and genes. In the wheat variety Beaver, with a 1BL/1RS translocation, the NOR of 1RS is inactive, and remains, together with the telomeric heterochromatin, condensed throughout the cell cycle (Fig. 5c). Both strongly fluorescent regions provide markers along metaphase and interphase chromosomes, the telomeric heterochromatin distally and the r-DNA proximally of the euchromatic segment of the satellite. At metaphase this euchromatic segment represents about 15% of the euchromatic length of the arm (see above), while at interphase it can be up to 45% (Fig. 5c). Thus, the euchromatin of the satellite is relatively more decondensed at interphase than is

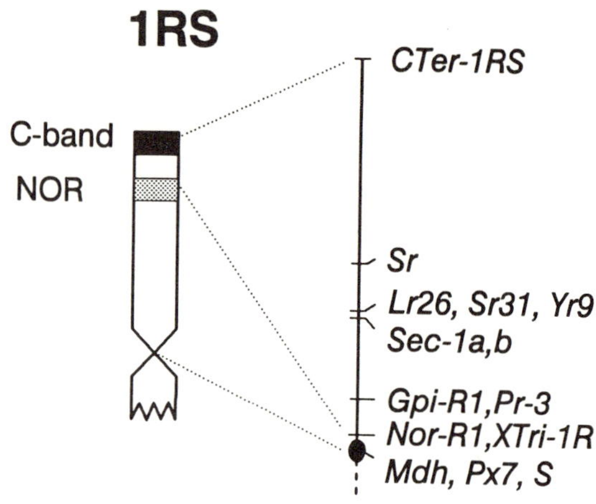

Fig. 6. Diagram of the short arm of chromosome 1R at somatic metaphase (left) and its genetic map showing genes and a telomeric heterochromatic clone only (right). Mapping data are taken from Baum and Appels (1991) and Wang et al. (1992a).

expected from its physical length at metaphase. It contains many genes and its large proportion of the genetic length of the chromosome arm indicates that most recombination occurs there (Fig. 6). Interestingly, in pachytene surface spreads of rye, the location of the r-DNA on chromosome 1RS corresponds to the one found at metaphase indicating that the satellite region is not represented by a relatively extended synaptonemal complex (Albini and Schwarzacher 1992). The description of the modulated decondensation during development, and perhaps the ability for manipulation to optimize gene expression in new varieties of crop plants, will be important research areas over the next few years.

6. Conclusions

Techniques which have come to the fore since the last Chromosome Conference include the development of routine non-radioactive DNA:DNA *in situ* hybridization methods for single copy and repetitive gene mapping, sometimes with multiple labels, and flow cytometry for chromosome sorting, isolation of chromosome specific libraries and karyotype analysis (evidenced by many papers in this volume). Genome mapping has been advanced by the development of methods for large DNA fragment cloning in yeast artificial chromosomes, and amplification of sequences from complex DNAs of specific segments by the polymerase chain reaction (PCR). Many of these techniques are applicable in crops and are now rapidly advancing genome organization studies (see e.g. Heslop-Harrison 1991; Moore et al. 1993; Wang et al. 1992b).

We hope that this paper has shown the many ways research in molecular cytogenetics is impinging on plant breeding. Cytogenetic analysis is valuable for research, and perhaps has applications, for a wide range of different needs in understanding the genomes of crop plants. The manipulation and understanding of chromosomes and their behaviour continues to be a vital part of ensuring that our food supply remains secure, and in working with plants to improve our environment.

7. Acknowledgements

We thank M. Shi and G.E. Harrison for technical assistance, K. Anamthawat-Jónsson, A.R. Leitch, I.J. Leitch, N. Neves and A. Castilho for continuing collaboration and helpful discussion, T.E. Miller, J.W. Rogers, and I.P. King for generation of many of the alien lines. We acknowledge support from BP Venture Research, the Royal Society and AFRC plant molecular biology grants.

8. References

Albini, S.M. and Schwarzacher, T. (1992) *In situ* localization of two repetitive DNA sequences to surface-spread pachytene chromosomes of rye. **Genome** 35, 554-563.

Anamthawat-Jónsson, K. and Heslop-Harrison, J.S. (1990) Centromeres, telomeres and chromatin in the interphase nucleus of cereals. **Caryologia** 43, 205-213.

Anamthawat-Jónsson, K. et al. (1990) Discrimination of closely related Triticeae species using genomic DNA as a probe. **Theor. Appl. Genet**. 79, 721-728.

Anamthawat-Jónsson, K., Schwarzacher, T. and Heslop-Harrison, J.S. (1993) Behavior of parental genomes in the hybrid *Hordeum vulgare* x *H. bulbosum*. **J. Heredity** 84(1) (in press).

Baum, M. and Appels, R. (1991) The cytogenetic and molecular architecture of chromosome 1R - one of the most widely utilized sources of alien chromatin in wheat varieties. **Chromosoma** 101, 1-10.

Bennett, M.D. and Smith, J.B. (1976). Nuclear DNA amounts in angiosperms. **Phil. Trans. R. Soc. Lond. B** 274, 227-274.

Doonan, J.H. and Clayton, L. (1986) Immunofluorescent studies on the plant cytoskeleton, in **Immunology in Plant Science** (ed T.L. Wang) Cambridge University Press, Cambridge, pp. 111-136.

Finch, R.A., Smith, J.B. and Bennett, M.D. (1981) *Hordeum* and *Secale* mitotic genomes lie apart in a hybrid. **J. Cell Sci.** 52, 391-403.

Gale, M.D. and Miller, T.E. (1987) The introduction of alien genetic variation into wheat, in **Wheat Breeding, Its Scientific Base** (eds F.G.H. Lupton), Chapman and Hall, London, pp. 173-210.

Gleba, Y.Y. et al. (1987) Spatial separation of parental genomes in hybrids of somatic plant cells. **Proc. Natl. Acad. Sci. U.S.A.** 84, 3709-3713.

Greilhuber, J., Volleth, M. and Loidl, J. (1983) Genome size of man and animals relative to the plant *Allium cepa*. **Canad. J. Genet. Cytol.** 25, 554-560.

Heslop-Harrison, J.S. (1991) The molecular cytogenetics of plants. **J. Cell Sci**. 100, 15-21.

Heslop-Harrison, J.S. and Bennett, M.D. (1984) Chromosome order - possible implications for development. **J. Embryol. Exp. Morph.** 83, Suppl, 51-73.

Heslop-Harrison, J.S. and Bennett, M.D. (1990) Nuclear architecture in plants. **Trends Genet.** 6, 401-405.

Heslop-Harrison, J.S. et al. (1990) Detection and characterization of 1B/1R translocations in hexaploid wheat. **Heredity** 65, 385-392.

Heslop-Harrison, J.S. et al. (1988) A method of identifying DNA sequences in chromosomes of plants. **European Patent Application** Number 8828130.8. December 2.

Kasha, K.J. and Kao, K.N. (1970) High frequency haploid production in barley (*Hordeum vulgare* L.). **Nature** 225, 874-876.

Law, C.N. (1981) Chromosome manipulation in wheat. **Chromosomes Today** 7, 194-205.

Leitch, A.R. et al. (1990) Genomic *in situ* hybridization to sectioned nuclei shows chromosome domains in grass hybrids. **J. Cell Sci.** 95, 335-341.

Leitch, A.R. et al. (1991) Parental genomes are separated throughout the cell cycle in a plant hybrid. **Chromosoma** 101, 206-213.

Leitch, I.J. and Heslop-Harrison, J.S. (1992) Physical mapping of the 18S-5.8S-26S rRNA genes in barley by *in situ* hybridization. **Genome** (in press).

Linde-Laursen, I. and Jensen, J. (1991). Genome and chromosome disposition at somatic metaphase in a *Hordeum* x *Psathyrostachys* hybrid. **Heredity** 66, 203-210.

Maluszynska, J. and Heslop-Harrison, J.S. (1991) Localization of tandemly repeated DNA sequences in *Arabidopsis thaliana*. **Plant J.** 1, 159-166.

Moore, G. et al. (1993) Key features of cereal genome organisation as revealed by the use of cytosine methylation-sensitive restriction endonucleases. **Genomics** (accepted).

Ordartschenko, N. and Keneklis, T. (1973) Localization of paternal DNA in interphase nuclei of mouse eggs during early cleavage. **Nature** 241, 528-529.

Schlegel, R., Melz, G. and Mettin, D. (1986) Rye cytology, cytogenetics and genetics - current status. **Theor. Appl. Genet.** 72, 721-734.

Schwarzacher, T. et al. (1992a) Parental genome separation in reconstructions of somatic and premeiotic metaphases of *Hordeum vulgare* x *H. bulbosum*. **J. Cell Sci.** 101, 13-24.

Schwarzacher, T. et al. (1992b) Genomic *in situ* hybridization to identify alien chromosomes and chromosome segments in wheat. **Theor. Appl. Genet.** 84/8 (in press).

Schwarzacher, T. et al. (1989) *In situ* localization of parental genomes in a wide hybrid. **Ann. Bot.** 64, 315-324.

Wang, M.L. et al. (1992a) Comparative RFLP-based genetic maps of barley chromosome 5 (1H) and rye chromosome 1R. **Theor. Appl. Genet.** 84, 339-344.

Wang, M.L. et al. (1992b) Construction of a chromosome-enriched HpaII library from flow-sorted wheat chromosomes. **Nucleic Acid Res.** 20, 1897-1901.

Zelesco, P.A. and Marshall Graves, J.A. (1988) Chromosome segregation from cell hybrids. IV. Movement and position of segregant set chromosomes in early-phase interspecific cell hybrids. **J. Cell Sci.** 89, 49-56.

16 Molecular organization of genes and repeats in the large cereal genomes and implications for the isolation of genes by chromosome walking

R.B. FLAVELL, M.D. GALE, M. O'DELL, G. MURPHY, G. MOORE
John Innes Centre, UK
and H. LUCAS *INRA Laboratoire de Biologie Cellulaire, Versailles, France*

1   INTRODUCTION

The chromosomes of the small grain cereals in the *Triticeae* (wheat, barley, rye etc.) have received much attention from cytogeneticists because they are of a relatively large size and especially because so much of wheat cytogenetics has been based upon the manipulation of chromosome number using aneuploids (Review, Law *et al*, 1987). For the molecular biologist these chromosomes have offered both advantages and handicaps. Many years ago it was established that over 85% of these cereal genomes consisted of repetitive DNA families with many thousands of copies (Flavell *et al*, 1977). It was also shown that the sequences in this fraction turned over at a relatively high rate during evolution to create major structural differences between homoeologous chromosomes carrying the same genes in closely related species (Flavell, 1982). These studies were relatively easy technically because of the large amount of repetitive DNA. In such genomes the 50,000 or so different genes probably constitute less than 1% of the total DNA (Flavell, 1980). This statistic raises a number of issues and problems. In this paper we review current results on how the genes and repeats are organised in the chromosomes, the discovery that retrotransposons are probably very abundant among the repeats in the chromosomes and the recent observations on sites containing unmethylated cytosine that are conserved in the genomes.

Why is there a special need to know in fine detail how genes and repetitive DNA are organised with respect to one another in these cereal species? There is, of course, the general reason that it is desirable to understand chromosome structure in as much detail as possible. However, a further and practical reason is that to manipulate plants for plant improvement via the introduction of single genes, it is necessary to first purify the desired genes. Many such genes will come from non *Triticeae* species but some will necessarily come from the cereal species because these are the only species in which they have been recognised. Genes can be recognised and isolated by "transposon tagging" as is well established in maize and antirrhinum, (Review, Balcells *et al*, 1991; Bhatt and Dean, 1992) by T DNA tagging in species in which it is possible to insert DNA directly into plant genes via infection by *Agrobacterium tumefaciens* or by "walking" in isolated DNA from known isolatable markers resident nearby to the desired gene. At present there is no endogenous transposable element system to achieve gene tagging in the *Triticeae* and procedures to generate fertile transgenic plants from single cells are so difficult that attempts to insert autonomously active elements from maize have not been successful. Also these species are not infected efficiently by *Agrobacterium* (Dale *et al*, 1989). Therefore, it is necessary to consider very carefully the potential for gene isolation in cereals by "chromosome walking". Walking would be extremely difficult if (i) the desired genes lay a long way away from the single copy isolatable markers and (ii) the DNA between the genes was all highly repetitive. It is therefore highly desirable to

Chromosomes Today   Volume 11. Edited by A.T. Sumner and A.C. Chandley. Published in 1993 by Chapman & Hall, London. ISBN 0 412 47670 3

know more about how the single copy/few copy genes and repetitive DNA sequences are organised with respect to one another.

At the end of this paper a new strategy for identifying and isolating genes in cereal genomes is presented. It is based upon the premise that modern cereal species have evolved from a common ancestor and therefore gene synteny will be very extensive between the modern species. Integration of known genes from all related species should therefore produce a representative fine scale map for any current species. Also, chromosome walking within any species, especially in rice which has a very small genome and relatively little repetitive DNA, should facilitate the isolation of homoeologous genes in homoeologous positions in any modern cereal species.

2   GROSS ORGANISATION OF SINGLE COPY SEQUENCES IN THE CHROMOSOMES

Over 750 cDNA and other single copy sequences have now been mapped with respect to each other on wheat chromosomes by determining the frequency of recombination between them (Chao *et al*, 1989; Devos *et al*, 1992 a & b; Gale, M. D. unpublished). Wheat is an allohexaploid so approximately 250 have been positioned on each of the three constituent genomes. The order and position of most of the markers is approximately the same on each of the genomes indicating the common evolutionary origin of the three parental species. To illustrate some of the findings the genetic maps of the homoeologous chromosomes 2A, 2B and 2D are shown in Figure 1. Many single copy cDNA clones map in the regions proximal to the centromeres. A similar situation is found for all of the homoeologous groups of chromosomes analyzed (Chao *et al*, 1989; Devos *et al*, 1992 a & b; Gale, M. D. unpublished). This raises the question whether these genes are concentrated in the regions around the centromeres or whether the genetic maps based on recombination frequencies are distorted because recombination occurs non randomly along the chromosomes. This is almost certainly the case for the following reasons.

Firstly, for wheat chromosomes 6B and 1B (and rye chromosome IR) the NOR loci containing the ribosomal DNA have been located approximately two thirds of the way from the centromere to the telomere, as judged from *in situ* hybridization of ribosomal DNA to condensed metaphase chromosomes. However, on the recombination map the NOR loci lie much closer to the centromeres (Dvorak and Chen 1984; Snape *et al*, 1985; Lawrence and Appels 1986; Wang *et al*, 1991).

Secondly, the pericentromeric regions are known to consist of very large arrays of repeated sequences, especially in the B genome chromosomes, condensed in heterochromatin and therefore likely to have a very low density of genes (Flavell *et al*, 1987; Gerlach & Peacock, 1980). These facts argue against the conclusion that most of the genes lie clustered close to the centromeres and thus it must be concluded that recombination events occur preferentially in the distal parts of the chromosome arms, thus expanding the recombination map distances between the telomere and at least the nucleolus organisers. It has long been recognised that the positions of chiasma in the metaphase chromosomes of wheat are invariably relatively distal (Holm, 1986). While there is not direct evidence that recombination is very rare around the centromeres this is likely. There is probably a high to low gradient of recombination from distal to proximal regions and therefore the majority of the cDNAs (genes) which are positioned close to the centromeres on the recombination map lie much more distal on the arms.

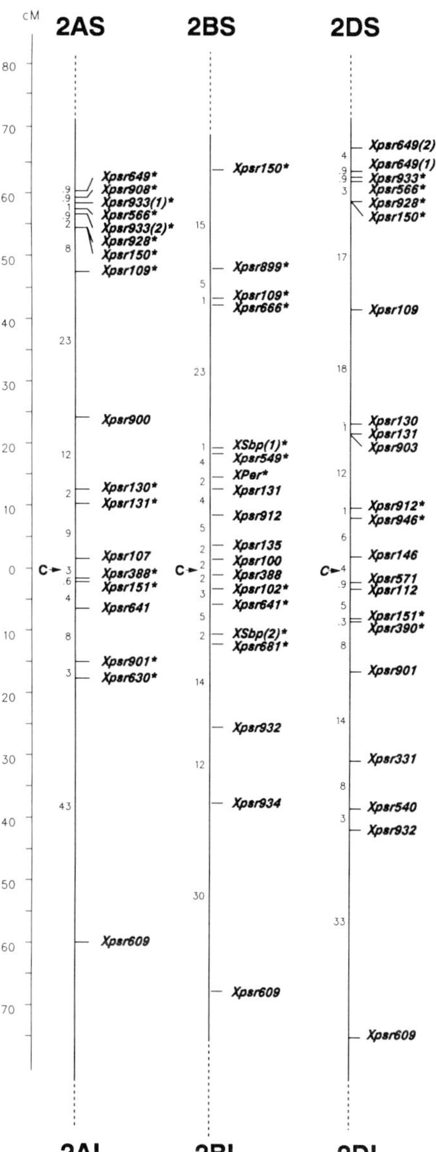

FIGURE 1. The current RFLP maps of wheat homoeologous chromosomes 2A, 2B and 2D.

The short arms are denoted 2AS, 2BS and 2DS while the long arms are denoted 2AL, 2BL and 2DL. C=Centromere. A centimorgan scale is shown on the left and distances between markers are given in Centimorgans on the maps.
* = cDNA clone markers. (see Devos *et al* 1992b)

The predisposition of some regions to be more susceptible to recombination than others could be due to the distribution of DNA sequences that are targets for an initiating endonuclease or because chromatin conformation makes such regions much more accessible to recombination enzymes.

The cDNA clones localised on the recombination maps of the three chromosomes of each homoeologous set of hexaploid wheat map at very similar locations with only few exceptions (for example, wheat homoeologous group 7, Chao *et al*, 1989). The exceptions are due to reciprocal translocations of chromosomal segments or where deletions have occurred. Thus during divergence of the progenitor species within the *Triticeae* gene order has been conserved as expected from long established results on other loci determining plant characters (Review Worland *et al*, 1987). The gene order has also been conserved between wheat, rye and barley chromosomal segments (for example homoeologous group 2; Devos *et al*, 1992 b) although more rearrangements in the form of translocations have occurred (Liu *et al*, 1992; Devos *et al*, 1992 c). There is evidence for conservation of gene order between these small grain cereals and maize (Gale, M. D., unpublished results). Extensive conservation between tomato and potato has already been published (Bonierbale *et al*, 1988; Gebhardt *et al*, 1991). This conservation of gene order over very long evolutionary times matches that discovered between mammalian species and needs to be contrasted with the fluidity and non conservation of the repetitive parts of the genomes. Further comments on this important point are made near the end of this paper.

In a model of sequence organisation published in 1978 it was shown that about 25% of the wheat DNA comprised single or few copy sequences when sheared into small fragments and assayed by hybridisation kinetics under reasonably stringent hybridization conditions (Rimpau *et al*, 1978). These single or few copy sequences were presumed to consist of genes and very diverged members of families of more highly repeated sequences. Because the proportion of single copy sequences increased considerably when the hybridisation conditions did not allow mismatched sequences to form stable duplexes (Smith & Flavell, 1975), it is reasonable to assume that a major proportion of the 25% of the total DNA that behaved as single copy DNA comprised diverged members of major repetitive families. As stated earlier perhaps only 1% of the total DNA would be expected to comprise the genes.

It was estimated that about 19% of the genome comprised short single or few copy sequences (<1000bp) interspersed with repeated DNA sequences, while approximately 6% consisted of single or few copy sequences several thousand base pairs long interspersed with repeats. Two percent of the genome comprised much longer single copy sequences. At that time it was not known whether the genes lay predominantly interspersed with repetitive DNA or whether they resided in the long regions of non repetitive DNA. It now appears, but from a very small sample size, that genes isolated from hexaploid wheat often have some repetitive DNA relatively close to them. Thus it can be concluded that many of the genes are likely to lie in the domains where short single copy and repetitive DNA segments are interspersed. These domains occupy a minimum of 46% of the genome from calculations of the proportions of DNA consisting of short non repeats interspersed with short repeats (Rimpau *et al*, 1978). However, it cannot be concluded from this information whether DNA fragments containing genes interspersed with repetitive DNA, which will occupy only a small fraction of the DNA consisting of interspersed single copy and repetitive DNA, are clustered together in the chromosomes or distributed at random. The issue is discussed towards the end of this paper.

3   GROSS ORGANISATION OF THE REPEATED DNA SEQUENCES

It was established sometime ago that long tandem arrays of a relatively simple sequence (Gerlach & Peacock, 1980; Flavell *et al*, 1987) are clustered around the centromeres of many wheat chromosomes, especially those of the B genome. Tandem arrays of a more complex repeat are localised near most of the telomeres and at a few interstitial sites, most of which coincide with the positions of C bands (Jones & Flavell, 1982). More recently, evidence for repeats typical of those conserved at telomeres have been detected in wheat chromosomes (Cheung, Gale and Moore unpublished). The ribosomal RNA genes are, of course, tandemly arranged at the NOR loci. From these results one can guess that perhaps up to 10% of the total DNA of the cereal genome may consist of tandemly arrayed short repeat units.

From the experiments carried out in the 1970's on the organisation of repeats (Rimpau *et al*, 1978, 1980; Flavell *et al*, 1981), it was concluded that a large fraction of the genome consists of short repetitive sequences interspersed in many permutations such that (i) the neighbouring sequences of individual members of the same repeat family are different (ii) there is substantial heterogeneity within families of repeats (iii) there is gain and loss of such repetitive elements during evolution and (iv) much structural heterogeneity is created between related chromosomes of related species during evolution due to the changes in these repetitive elements.

These conclusions led to the concept of the dynamic nature of the plant genome with respect to its repetitive components and implied that short segments of DNA were evolving, multiplying, moving into new locations and being lost during evolution (Flavell 1982, 1985). These conclusions have now been reinforced by the knowledge that retrotransposons are common in cereal genomes. Two particular families Wis-2, and Wire-1 have been studied in detail (Moore *et al*, 1991a; Flavell and Harris unpublished results). The initial member of Wis-2 was isolated from within a seed storage protein gene in a particular variety in which the gene is not expressed (Harberd *et al*, 1987). Like retrotransposons in mammalian genomes, Drosophila and yeast, it has long terminal repeats that presumably contain a promoter and terminator for transcription of mRNA. These transcripts are converted into double stranded DNA molecules via transposon-encoded reverse transcriptase and can be reintegrated possibly at random into the genome at new sites. The Wis-2 copies characterised have sequences that would give rise to proteins with characteristic resemblances to the reverse transcriptases and integrases of other retrotransposons (Moore *et al*, 1991a; Murphy *et al*, 1992). Active copies of such elements would lead to similar repeats being localised at many different locations in the genome, as found (Rimpau *et al*, 1978, 1980). The Wis-2 element sequenced is 8.1 Kbp long (Murphy *et al*, 1992). The Wire-1 copy characterised is 12.5 Kbp long (Flavell & Harris unpublished results). If such retrotransposon elements occupied much of the cereal genomes, then the interspersed repetitive elements making up much of the DNA would each be 8 to 12 thousands of base pairs long. In the intra and interspecies DNA hybridization studies (Rimpau *et al*, 1978, 1980) the lengths of renatured duplexes were however much less than this, implying that the retrotransposon families are heterogeneous and consequently frequently form mismatched duplexes on renaturation within and between species. Studies on Wis-2 and Wire-1 have shown this to be the case for these families of sequences. Wis-2-2, a subfamily probably containing more copies than the subfamily Wis-2-1 whose member was sequenced, possesses an internal fragment very different from that in Wis-2-1 (Moore *et al*, 1991a). Renaturation of members of these subfamilies would therefore produce relatively short perfect duplexes.

Most segments of Wis-2-1 are present in rye and barley genomes although the major variants are substantially different in the different species. The segment containing the reverse transcriptase gene is found conserved even in the distantly related oats genome, although somewhat diverged. The other segments of Wis-2-1 are highly diverged or absent from the oats genome. If one were to hypothesize that the cereal genomes were constructed from accumulated diverged retrotransposon elements, such as Wis-2 elements, then one would predict when comparing the oats and wheat genomes that short regions of repetitive DNA in common between the species (reverse transcriptase components) would be interspersed with longer regions of repetitive DNA that were very dissimilar or unrelated. This is what was found (Rimpau *et al*, 1978, 1980) and therefore one can speculate that the conserved repeats between the distantly related cereal genomes represent those repetitive elements that encode proteins such as reverse transcriptase important in repeat propagation.

Studies on Wire-1 have revealed similar features to those of Wis-2. One internal segment characteristic of most copies of the element in most wheat-related species, is absent from the elements in *Aegilops squarrosa*, another close relative of wheat. Therefore elements from wheat and *Aegilops squarrosa* would hybridise *in vitro* to form short and not full length heteroduplexes.

The long terminal repeats of Wis-2 are particularly long and the presence of many inverted repeats within them has led to the hypothesis that the inverted repeats arose as errors during the replication process in which different RNA molecules are used as templates for copying by reverse transcription (Lucas *et al*, 1992). This model also implies that sequence mismatches in the LTRs are created frequently, which would lead to divergence in the LTRs of a retrotransposon family during evolution. Divergence of elements within a retrotransposon family, both within and between species would also result if recombination between retrotransposons occurs at a high frequency, as in retroviruses.

It appears then that a large fraction of the cereal genomes consists of complex permutations of diverse copies or fragments of copies of repetitive elements, many of which are likely to be or to have evolved from retrotransposons. Their complex arrangement is the result of movement of the elements. Since the initial Wis-2 element was isolated from within a gene there is no reason to suppose that such repetitive elements have failed to integrate into and survive in domains containing genes. Therefore, it is reasonable to assume that genes lie in domains containing repeats as concluded earlier.

4   THE GROSS ORGANISATION OF UNMETHYLATED CpG SITES IN GENES AND REPEATS

More than 80% of the cytosine residues in CpG or CpXpG motifs are methylated in wheat (Gruenbaum *et al* 1981). This raises the question of why the other 20% are not and what might determine such a methylation pattern. In mammalian species many "housekeeping" genes lie adjacent to CpG rich segments (islands) that are unmethylated (Bird, 1987) and recently Saccone *et al* (1992) have reported that the highest concentration of genes in the human genome lie in a GC rich fraction (isochore) that also has the highest concentration of CpG islands, transcriptional and recombinational activities. Unlike mammalian genomes, the bulk of the cereal genomes exhibit little suppression of CpG dinucleotides (Swartz *et al*, 1962, Russell *et al*, 1971). Both repetitive and single copy sequences possess near expected CpG content and therefore contain many recognition sites for CpG methylation-sensitive restriction enzymes.

We have recently investigated the distribution of unmethylated CpG sites in single copy and repetitive DNA, and in genes, using restriction endonucleases that do not cleave DNA at specific recognition sites containing CpG or CpXpG bases unless the cytosine is unmethylated. Eighty percent of short DNA fragments (<500bp) released by HpaII which cleaves at CCGG sites were found to hybridise to only single or few copy DNA sites in the genome (Cheung *et al*, 1992). Similarly 60% of short fragments released by PstI hybridised to single or few copy sequences (Harcourt, 1992). This is a major departure from what is expected for a genome consisting of 80% repetitive DNA and exhibiting little CpG suppression, on the hypothesis that unmethylated CpG or CpXpG motifs are randomly distributed. We thus conclude that unmethylated CpG regions are preferentially associated with single copy sequences (and possibly genes).

The endonuclease NotI has a comparatively long and very CpG rich recognition sequence (GCGGCCGC) and requires the CpGs to be unmethylated before it cleaves. It therefore cuts cereal DNA very infrequently compared with HpaII. When short DNA segments released by NotI from genomic DNA i.e. segments created by two unmethylated NotI sites located close together were assayed, only 1 out of 56 hybridised to highly repetitive DNA. Ten such sequences taken at random all hybridised only to single or low copy DNA. When sequences in domains containing only a single NotI site were assayed, 90% were localised within repetitive DNA but frequently within 10 to 20 kilobase pairs of a cDNA hybridization site. These experiments imply that clustered unmethylated CpG sites are associated with single copy sequences, and genes lie close to unmethylated NotI sites (Cheung *et al*, 1992; Moore *et al*, 1992).

An example of the presence of clusters of unmethylated CpGs around genes is provided by the α Amy B1 locus in wheat (Cheung *et al* 1992) shown in Figure 2. Sites for NotI, Mlu and Nru-I, all methylation-sensitive enzymes whose recognition sites contain CpGs, are shown.

The gross organisation of unmethylated NotI, Mlu-I (cleavage site ACGCGT) or Nru-1 (cleavage site TCGCGA) sites in the fraction of the genome containing sequences that hybridise to dispersed repetitive elements (Wis-2, or Bis-I) (Moore *et al*, 1991b) was investigated by treating genomic DNA with one of the enzymes, fractionating them using a CHEF pulse field system, transferring the DNA to a membrane and hybridizing the genomic fragments with the dispersed repeat. The results when using barley and wheat DNA are shown in Figure 3. Several surprises were revealed in these experiments. Firstly, the dispersed repetitive elements revealed discrete hybridization bands rather than the smears expected if Nru I and Mlu I sites were distributed randomly through the genome. Secondly, the signal intensity of the hybridization bands suggested that each consisted of multiple copies

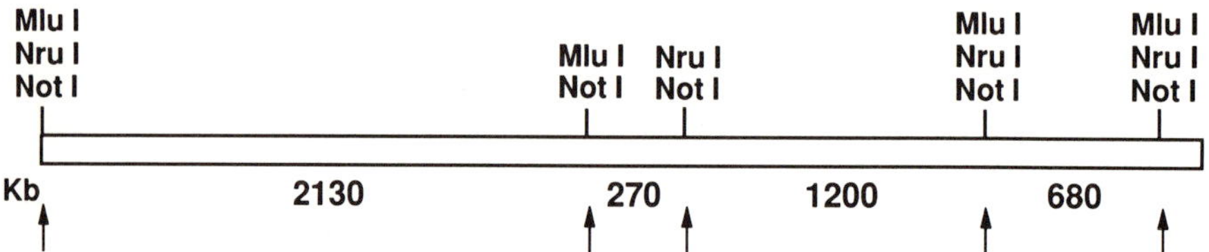

FIGURE 2.   A long range of map of the α-Amy-B1 locus.
Clusters of unmethylated recognition sites for methylation-sensitive restriction
enzymes are shown. Each marks the location of genes in this locus. Single
unmethylated Not I, NruI and MluI sites are not shown on this map. Distances
between clusters of sites are given in Kilobase pairs. (See Cheung *et al* 1992)

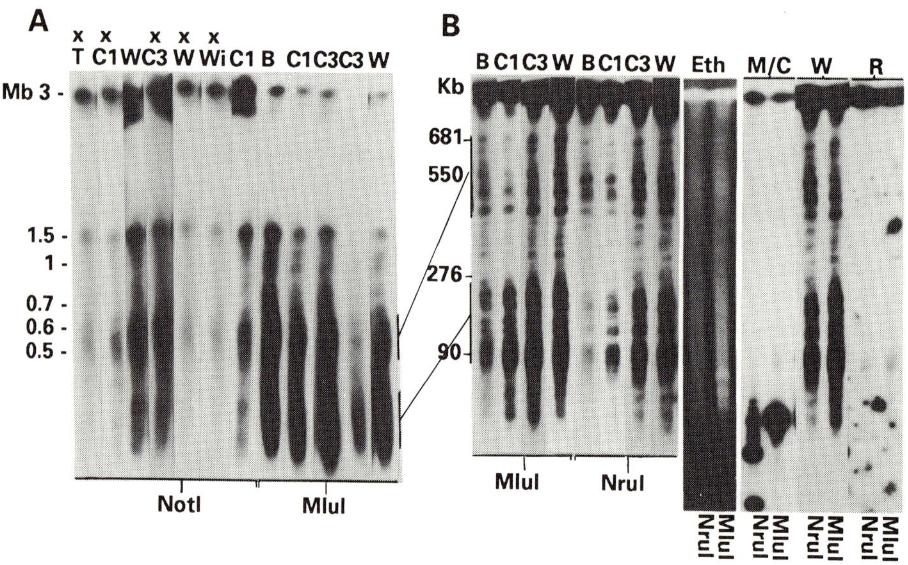

**FIGURE 3.   Long range organisation of Not I, MluI and NruI sites in the Barley
and Wheat Genomes**
5ug of barley and wheat DNA were digested with restriction enzymes as indicated,
separated on a CHEF gel, to resolve restriction fragments of 100 kb-3Mb (A) and
50-750 kb (B and C) and probed with either total wheat DNA (T), ribosomal DNA
pTA71 (R) mitochondrial/chloroplast DNA (M/C) and a number of different
repetitive sequences (B, W, C1, C3 and Wi). Ethidium bromide staining of the
Nru I and MluI digests is shown (Eth). Tracks marked by an x are wheat DNA
digests. Exposure on film varied from 2 hours to 5 days.

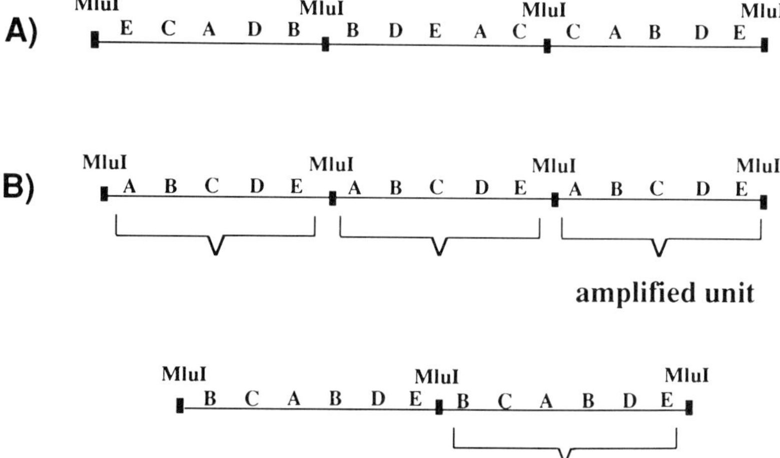

FIGURE 4. **Models for Genome Organisation of Repetitive Sequences**
A) Multicopy MluI (and NruI) restriction fragments arise from cleavages at GC rich "motifs" which are regularly spaced and which are unmethylated because they correspond to a position-specific function. The restriction fragments will contain random distributions of repetitive sequences (A, B, C, D, E).
B) Multicopy Mlu I (and Nru I) restriction fragments arise from cleavage of regularly spaced GC rich "motifs" in amplified regions. The restriction fragments will contain non-random distributions of repetitive sequences.

of fragments of similar if not identical lengths. This conclusion was supported by the observation that the intensity of hybridization correlates with the amount of DNA present revealed by ethidium bromide fluorescence. Thirdly, the Mlu-I and Nru-I digestion patterns are extremely similar implying that unmethylated sites for each of these enzymes frequently lie close together (see also figure 2). Fourthly, the digestion patterns were similar for wheat and barley even though the repeated DNA sequences and their organisation are very different between the species (Flavell *et al*, 1977; Rimpau *et al*, 1978, 1980). The discrete hybridization bands were greater than 90 kbp in length and other much longer fragments contained members of the Wis-2 or Bis-1 dispersed repeat families but no unmethylated sites for Mlu-I or Nru-I. The multicopy fragments of specific lengths after restriction of both barley and wheat DNAs with Mlu-I and Nru-I could reflect the conservation of unmethylated sites at conserved reiterated intervals through a major fraction of the genome as schematically drawn in Figure 4A. Alternatively, the multicopy fragments could result from amplification of long repeats of DNA (some greater than 440kbp) containing at least two conserved sites as shown in Figure 4B. This latter scheme predicts that the fragments of similar length within a species will be essentially identical. Which of these explanations is correct is unknown at present but is being investigated. They are not mutually exclusive.

Whichever is correct, the finding that a major fraction (approximately 60%) of cereal DNA contains unmethylated CpG sites at a series of evolutionary conserved, reiterated intervals raises the questions "what could be the function of such sites and how are they maintained unmethylated?" They might represent scaffold binding sites (Gasser and Laemmli 1987) that are preferentially maintained by natural selection at fixed distances apart for reasons of maintaining essential chromatin condensation patterns. Alternatively they might be replication initiation sites. It will be important to do further research to determine their function. The sites would be maintained unmethylated by, presumably, the binding of proteins to them which interfere with the normal CpG and CpXpG methylation processes. If these multicopy fragments have resulted from amplification, then it is unlikely that they will carry genes and single copy DNA. This would imply that long regions of repetitive DNA exist, devoid of single copy DNA and thus the segments of DNA consisting of interspersed single copy and short stretches of repetitive DNA would be concentrated in clusters. Such clusters would account for about 46% of the wheat genome (Rimpau *et al*, 1978).

The distribution of unmethylated NotI and Mlu-I sites along a single rye chromosome arm has been investigated recently (Moore *et al* 1992). They are more frequent towards the telomere, excluding the tandem arrays of repeats localised at the ends (Jones and Flavell 1982). This is consistent with a chromosome model in which there is a gradient of single copy/genes from centromere proximal to distal regions, with more single copy/genes sequences being distal. As described earlier, recombination is much more frequent in the distal regions, presumably where the single copy/gene sequences are localised. A schematic model of a chromosome based on all these observations is given in Figure 5. The clusters of single copy sequences interspersed with repeats and associated with unmethylated islands are concentrated in distal regions while long stretches of repeated DNA punctuated by regularly spaced unmethylated regions and other unmethylated islands are more concentrated in proximal regions. This is obviously over simplified but is designed to emphasize that our results to date imply that the single copy sequences/genes and long arrays of repeats are not distributed at random. While providing valuable information on the gross organisation of the genome structure, much more detailed information will be required to describe the organisation of specific genes with respect to one another and to repeats.

5   SEQUENCE ORGANISATION AND CHROMOSOME WALKING

The preceding sections indicate that uncertainties still remain about the distribution of genes within the DNA segments consisting of interspersed single copy and repetitive DNA (including diverged repeats that behave as single copy DNA under hybridisation conditions) and the long stretches of repetitive DNA.If it is assumed that genes do not exist in say 10% of the diploid cereal genome (the known long tandem arrays) and there are 50,000 genes, then on a random distribution of genes throughout the remaining DNA, a gene should occur approximately every 100,000 base pairs. However as described above, and in figure 5 if there are very much longer regions of repetitive DNA then the genes will lie much closer together.

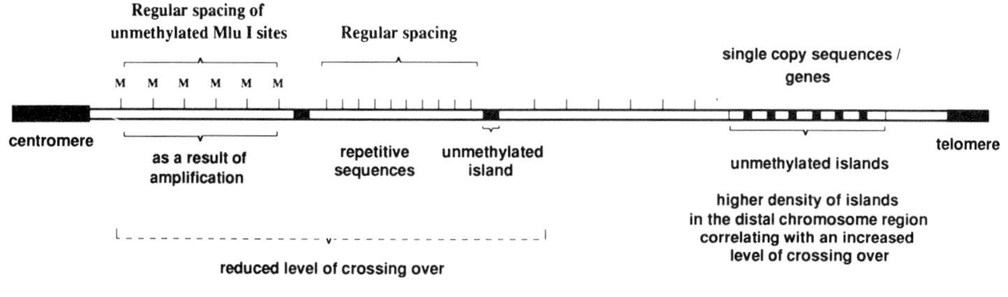

FIGURE 5.   **A Model for the distribution of genes and blocks of repetitive sequences along a wheat/rye Chromosome**

A chromosome is depicted showing a preponderance of single copy sequences/genes distal, associated with unmethylated islands and in locations where recombination is most frequent. Other regions more proximal are where much longer stretches of repetitive DNA sequences defined by regular spacing of unmethylated Mlu-1 sites are concentrated. The diagram is an over simplification of the distribution of single copy sequences along the chromosome arm. A gradient from distal to proximal regions is likely to be a more accurate description. Further details are described in Moore *et al* 1992.

For "chromosome walking" to genes from closely linked markers it is necessary to find single copy markers that are <u>physically</u> close to the desired gene. This requires optimally that recombination occurs frequently between markers and the desired gene so that close <u>genetic</u> linkage is correlated with close <u>physical</u> linkage. If recombination is frequently distal in the chromosomes, it is important to know which genes lie in these regions and in particular the location of the gene to be isolated. To establish the position of a marker sequence relative to a gene a detectable polymorphism is required in the vicinity. The frequency of restriction site polymorphism to serve as markers is unfortunately low in small grain cereals compared with for example maize (Burr and Burr, 1991: Gale, *et al* 1990).

These issues of uncertainty over (i) the distances apart of the genes (ii) the sites and frequencies of recombination with respect to the genes and (iii) the reduced levels of polymorphism, implies that chromosome walking is going to be difficult or at least very inefficient in these large genomes. The probability of having a marker and the desired gene in the same YAC up to 200,000 base pairs seems remote. Walking from one YAC to another demands in principle the same single copy sequence in at least two YACs. Finding a single copy sequence in a YAC consisting mainly of repetitive DNA is time consuming and so walking will be additionally inefficient in these large genomes unless the genes are clustered in regions where single copy DNA is frequent.

6 A STRATEGY FOR IDENTIFYING AND ISOLATING GENES IN CEREALS WITH LARGE GENOMES

The major difficulties of using "genome walking" to identify and isolate genes from large genomes raise the issue of whether a species with a small genome can be used as a "model" or an "aid" in gene isolation. The dicot plant *Arabidopsis* has been adopted as a model plant species with the intention of generating complete knowledge of a plant genome and its complement of genes. The genes will be isolated by transposon tagging, T-DNA tagging, chromosome walking from a complete library of overlapping YACS and cosmids and by the eventual sequencing of a vast number of cDNAs and the entire genome. This will greatly aid knowledge of and the isolation of homologous genes in monocots. However for the foreseeable future many of the desired cereal genes will not be readily recognisable from the *Arabidopsis* source. Therefore substantial resources have been committed by Japan and China to rice genome research and by USA to maize genome research. In the Japanese programme large numbers of rice cDNAs are being sequenced and the eventual aim is to sequence the entire rice genome.

To augment these studies we propose a different strategy for identifying and isolating genes from cereal species with large genomes. This strategy does not focus on one species as a model and is emerging from the knowledge that modern cereal monocot species are descendants of a single ancestral genome. The proposal is based on the following: if gene order has been retained within chromosome segments as indicated earlier, even though many translocations and rearrangements have taken place during evolution, then it should be possible by studying gene order in a range of present day cereal species to reconstruct the gene order of ancestral species. Recognition of all the different genes by mutation or molecular mapping in any of the species and their integration into a common map should be very instructive to the position and function of homologous genes in all present day genomes. This should mean that (i) a more comprehensive catalogue of putative loci controlling any trait could be accumulated for any cereal genome and (ii) such genes could be isolated by walking within the chromosome of any species and their homologues in other species retrieved by sequence homology. As it happens the rice genome is very small compared to the genomes of barley, rye and wheat. Rice YACs could therefore be a major tool in gene isolation strategies for large cereal genomes. It may be possible to walk within the small rice genome either to discover markers very close to the desired gene, to complete a walk initiated in the large genome species, or to discover and isolate the homologous gene in rice. This strategy appears very attractive for overcoming the substantial problems associated with the large complex cereal genomes. It is dependent on the extent of synteny between these cereal species which needs therefore to become a major focus of research. If synteny is extensive and any modern cereal genome can be represented as a composite from multiple species, then a major step forward in fundamental and applied plant science will have been accomplished.

Balcells, L., Swinburne, J., and Coupland, G., (1991) Transposons as tools for the isolation of plant genes. **Tibtech** 9, 31-36.

Bhatt, A. M., and Dean, C. (1992) Development of tagging systems in plants using heterologous transposons. **Current opinion in Biotechnology** 3, 152-158.

Bird, A. P. (1987) CpG islands as markers in the vertebrate nucleus. **Trends in Genetics** 3, 342-347.

Bonierbale, M. W., Plaisted, R. L. and Tanksley, S. D., (1988) RFLP maps based on a common set of clones reveal modes of chromosomal evolution in potato and tomato. **Genetics** 120, 1095-1103.

Burr , B. and Burr, F. A., (1991) Recombinant inbreds for molecular mapping in maize. **Trends In Genetics** 7, 55-60.

Chao, S., *et al* (1989) RFLP based genetic maps of wheat homoeologous group 7 chromosomes. **Theor. Appl. Genet.** 78, 495-504.

Cheung, W. Y., *et al* (1992) Hpa II library indicates methylation free islands in wheat and barley. **Theor. Appl. Genet.** 84, 739-746.

Dale, P. J., *et al* (1989). Agroinfection of wheat: inoculation of in vitro grown seedlings and embryos. **Plant Science** 63, 237-245.

Devos, K., *et al* (1992a) RFLP based genetic map of the homoeologous group 3 chromosomes of wheat and rye. **Theor. Applied Genetics.** 83, 931-939.

Devos, K. Miller, T. and Gale, M. D., (1992b) Comparative RFLP maps of the homoeologous groups 2 chromosomes of wheat, rye and barley. **Theor. Appl. Genet.** in press.

Devos, K., *et al* (1992c). Chromosome rearrangments in the rye genome relative to that of the wheat genome. **Theor. Appl. Genet.** in press.

Dvorak, J. and Chen, K. C. (1984) Distribution of nonstructural variation between wheat cultivars along chromosome arm 6bp: Evidence from the linkage map and physical map of the arm. **Genetics** 106, 325-333.

Flavell, R. B., (1980). The molecular characterisation and organisation of plant chromosomal DNA sequences. **Annual Review of Plant Physiology**, 31, 569-596.

Flavell, R. B., O'Dell, M. and Hutchinson, J. (1981). Nucleotide sequence organisation in plant chromosmes and evidence for sequence translocation during evolution. Cold Spring Harbour Symp. **Quant. Biol.**, 45, 501-508.

Flavell, R. B. (1982) Sequence amplification deletion and rearrangement: major sources of variation during species divergence, in **Genome Evolution.** (eds. G. A. Dover, and R. B. Flavell) Academic Press. 301-323.

Flavell, R. B. (1985). Repeated sequences and genome change. in **Genetic Flux in Plants.** (eds. B. Hohn, and E. S. Dennis), Springer Verlag, Wien, New York, 129-156.

Flavell, R. B., Bennett, M. D. and Hutchinson-Brace, J. (1987) Chromosome structure and organisation. in **Wheat.** (ed. F. Lupton) Chapman and Hall. 211-268.

Flavell, R. B., Rimpau, J. and Smith, D. B. (1977) Repeated sequence DNA relationships in four cereal genomes. **Chromosoma** 63, 205-222.

Gale, M. D., Chao, S. and Sharp, P., (1990) RFLP mapping in wheat progress and problems. in **Gene Manipulation in Plant Improvement II** (ed J P Gustafson), Plenum Press, New York.

Gasser, S. M. and Laemmli, U. K. (1987) A glimpse at chromosomal order. **Trends in Gentics** 3, 16-22.

Gebhardt, C., Ritter, E., Barone, A., *et al.* (1991) RFLP maps of potato and their alignment with the homoeologous tomato genome. **Theor. Appl. Genet.** 83, 49-57.

Gerlach, W. L. and Peacock, W. J. (1980) Chromosome locations of highly repeated DNA sequences in wheat. **Heredity** 44, 269-276

Gruenbaum, U., *et al* (1981) Sequence specificity of methylation in higher plant DNA. **Nature** 292, 860-862.

Harberd, N. P., Flavell, R. B. and Thompson, R. D.. (1987) Identification of a transposon like element in a Glu-I allele of wheat. **Mol. Gen. Genet.** 209, 326-332.

Harcourt, R. (1992) DNA sequence polymorphisms in Triticeae species. Ph.D thesis Cambridge University.

Holm, P. B. (1986) Chromosome pairing and chiasma formation in allohexaploid wheat, Triticum aestivum analysed by spreading of meiotic nuclei. Carslberg Res. **Commun.** 51, 239-294.

Jones, J. and Flavell, R. B. (1982). The mapping of highly repeated DNA families and their relationship to C bands in chromosomes of Secale cereale. **Chromosoma**, 86, 595-612.

Law, C. N., Snape, J. W. and Worland, A. J. (1987) Aneuploidy in wheat and its uses in genetic analysis. in **wheat** (ed Lupton) Chapman and Hall. p71-108.

Lawrence, C. J. and Appels R. (1986) Mapping the nucleolus organiser, seed protein loci and isozyme loci on chromosome IR in rye. **Theor. Appl. Genet.** 71, 742- 749.

Liu, C., *et al* (1992) Non-homoeologous translocations between group 4, 5 and 7 chromosomes in wheat and rye. **Theor. Appl. Genet.** 83, 305-312.

Lucas, H., *et al* (1992) The nucleotide sequence of the wheat WIS-2-1A retrotransposon-like element LTRs suggests that their large size is partly due to palindromic structures. **Mol. Biol. and Evol** 9, 716-728.

Moore, G., *et al* (1991a) A family of retrotransposons and associated genomic variation in wheat. **Genomics.** 10, 461-468.

Moore, G., *et al* (1991b) BIS1, a major component of the cereal genome and a tool for studying genomic organisation. **Genomics.** 10, 469-476.

Moore,G., *et al* (1992) Key features of cereal genome organisation as revealed by the use of cytosine methylation sensitive restriction endonucleases. **Genomics** in press.

Murphy, G., *et al* (1992) Sequence analysis of WIS-2-1A, a retrotransposon. **Plant Mol. Biol.** in press.

Rimpau, J. Smith, D. B. and Flavell, R. B. (1978) Sequence organisation analysis of the wheat and rye genomes by interspecies DNA/DNA hybridisation. **J. Mol. Biol.**, 123, 327-359.

Rimpau, J. Smith, D. B. and Flavell, R. B. (1980) Sequence organisation in barley and oats chromosomes revealed by DNA/DNA hybridisation. **Heredity**, 44, 131-149.

Russell, C. J., *et al* (1971). The double standed DNA of Cauliflower mosaic virus. **J. Gen. Virol** 11, 129-138.

Saccone, S., *et al* (1992). The highest gene concentrations in the human genome are in telomeric bands of metaphase chromosomes. **Proc. Natl. Acad. Sci.** (USA) 89, 4913-4917.

Smith, D. B. and Flavell, R. B. (1975) Characterisation of the wheat genome by renaturation kinetics. **Chromosoma**, 50, 223-242.

Snape, J. W., *et al* (1984). Intrachromosomal mapping of the nucleolar organiser region relative to three marker loci on chromosome 1B of wheat (*Triticum aestivum*). **Theor. Appl. Genet.** 69, 263-270.

Swartz, M. N., Trautner, T. A. and Kornberg, A. (1962). Enzymatic synthesis of dcoxyribonucleic acid. **J. Biol. Chem.** 237, 961-1967.

Wang, M. L., Atkinson, M. D., Chinoy, C. N., Devos, K. M., Liu, C., Rogers, J. and Gale, M. D. (1991) RFLP based map of rye (*S. cereale*) chromosome IR. **Theor. Appl. Genet.** 82, 174-178.

Worland, A. J., Gale. M. D. and Law, C. N., (1987) Wheat genetics in **Wheat** (ed Lupton) Chapman and Hall.

17 Fragile sites and unstable elements

R.I. RICHARDS and G.R. SUTHERLAND

Adelaide Children's Hospital, Australia

1. Introduction

1.1 Fragile Sites

The term fragile site was introduced to describe gaps or
discontinuities in chromatids or chromosomes seen by
cytogeneticists (see Sutherland, 1991a for review).
Specific culture conditions are required for fragile sites
to be observed reproducibly. Fragile sites are always at
exactly the same point on the chromosome in any particular
patient or kindred, are inherited in a Mendelian fashion
and appear under particular in vitro conditions. They are
grouped into two major classifications based on relative
frequency. Rare fragile sites are carried by less than one
in twenty individuals, whereas all individuals are probably
homozygous for common fragile sites. Both groups are
further divided according to the specific culture
conditions required for their expression. For example of
the rare fragile sites the majority are folate sensitive,
some are distamycin-A inducible and two require
bromodeoxyuridine (Sutherland, 1991a).
   Only the rare fragile site at Xq27.3 referred to as
FRAXA is of any known clinical significance. Indeed its
association with the most common familial form of mental
retardation (consequently referred to as fragile X
syndrome) has been a major stimulus to research into
fragile sites (Richards and Sutherland, 1992a).

1.2 Fragile X Syndrome

Fragile X syndrome has an incidence of about one in 2000
children. It has no discernible ethnic predilection,
having been found in all racial groups. It is the most
common form of familial mental handicap, and is second in
frequency only to Down syndrome as a cause of mental
handicap. For many years the diagnosis of the disorder has
been difficult, requiring the cytogenetic demonstration of

Chromosomes Today   Volume 11. Edited by A.T. Sumner and A.C. Chandley. Published in
1993 by Chapman & Hall, London. ISBN 0 412 47670 3

the fragile X chromosome. Detection of carrier individuals, who show no cytogenetic abnormalities, has been possible in families only by using DNA polymorphisms linked to the fragile site. The syndrome can be clinically indistinct and the genetics are bizarre (Sutherland, 1985).

The clinical symptoms of the fragile X syndrome in essence constitute mental handicap of variable severity (more marked in males than females), minor dysmorphic facial features, behaviour disturbances, macroorchidism in the males and a generalized disorder of connective tissues. One of the notable aspects of the syndrome is its great variability, with some retarded males showing few of the physical components. The syndrome can be very difficult to diagnose clinically and all mentally retarded individuals should be considered as candidates for it and have appropriate diagnostic studies performed.

Diagnosis still depends on cytogenetic demonstration of the fragile X - although this could change in the future. This demonstration is not straightforward. Cells need to be cultured under conditions of thymidine (dTTP) or deoxycytidine (dCTP) starvation to induce the fragile site. Even under optimum conditions the fragile site is usually seen in only 10-40% of metaphases in retarded males, less frequently in retarded females, in only a proportion of asymptomatic carrier females and almost never in asymptomatic carrier males.

Fragile X syndrome has a unique place in human genetics because of the existence of the normal transmitting male - males who are cytogenetically and intellectually normal but transmit the fragile X. The label 'X-linked dominant disorder with incomplete penetrance' has been used to describe its inheritance pattern. In addition, detailed segregation analysis of affected pedigrees (Sherman, 1991) has revealed several other non-mendelian aspects of fragile X genetics.

(1)   The mothers of all fragile X males are fragile X carriers - no males are afflicted as a result of mutation in the egg of their mother. A mutation frequency of 7.2 x10^{-4} per male gamete per generation was calculated but all mutations had to pass through a female before having any phenotypic effect.

(2)   The segregation ratio (proportion of affected male children of carrier mothers) is 0.4, not 0.5 as expected for an X-linked mutation. The segregation ratio for sons of affected fragile X females is 0.5.

(3)   Affected females receive the fragile X from their mothers, not their fathers. The fragile X carrier daughters of normal transmitting males are unaffected. The fragile X carrier mothers of normal transmitting males are normal themselves

and have a low risk of having fragile X syndrome
children.
This last property means that the daughters of normal
transmitting males are more likely than the mothers of
normal transmitting males to have affected children. This
observation is known as the Sherman paradox.

2.   Molecular Genetics of Fragile X Syndrome

In the absence of an understanding of the biochemical basis
for the disorder, mapping and positional cloning were the
main techniques used to define the fragile X syndrome
mutation. A number of laboratories cooperated in the
isolation and mapping of DNA probes and somatic cell hybrid
breakpoints in the vicinity of Xq27.3 (Suthers et al.,
1990; Rousseau et al., 1991a; Poutska et al., 1991).
Several of these DNA probes detect or contain polymorphic
sequences and allowed linkage analysis to be performed on
affected pedigrees. This improved the accuracy of
diagnosis and confirmed previous segregation analyses.
   Using pulsed-field gel electrophoresis (PFGE) and probes
closely linked to the fragile site (2-34 and Do33), Mandel
and colleagues (Vincent et al., 1991) showed that the
methylation of certain restriction enzyme sites differed
between individuals with fragile X syndrome and normal
individuals, a result confirmed in other laboratories (Bell
et al., 1991; Yu et al., 1992; Pieretti et al., 1991).
   Physical isolation of the fragile X site was undertaken
cooperatively (and competitively) by many of the groups
working in this area (Kremer et al., 1991a; Heitz et al.,
1991; Dietrich et al., 1991; Hirst et al., 1991; Verkerk et
al., 1991). Schlessinger, in collaboration with our
laboratory, used VK16 as a probe to isolate a yeast
artificial chromosome (YAC), XTY26, that was subsequently
shown by in situ hybridization to span the fragile site and
to contain both 2-34 and Do33, which flank the fragile site
(Kremer et al., 1991a). To localize the fragile site
itself we primarily used in situ hybridization of a contig
of lambda subclones of the YAC to fragile X chromosomes,
supported by Southern blot mapping to two somatic cell
hybrids. These hybrids were constructed by Warren et al
(1990) from a fragile-X-expressing somatic cell hybrid in a
way designed to achieve breakage at the fragile site. The
hybrids had complementary parts of the fragile X chromosome
and were found to have breakpoints localizing to the same 5
kb EcoRI restriction fragment (Yu et al., 1991). This
fragment was also found to vary in length in fragile X
males. The instability associated with the fragile X
genotype was subsequently localized to a 1 kb PstI fragment
that could be readily detected by pfxa3, a probe adjacent
to the instability.
   Mandel's group focused on the CpG island associated with
the fragile X methylation (Vincent et al., 1991), when

characterizing their YACs that spanned the fragile X site (Heitz et al., 1991). They found that the probes selected to detect the methylation differences at higher resolution on normal Southern blots (instead of PFGE blots) also detected unstable sequences adjacent to the CpG island (Oberlé et al., 1991).

Sequence analysis and further mapping demonstrated that the unstable element at the fragile X site was a p(CCG)n repeat sequence (Kremer et al., 1991b) that varies in length, probably as a result of amplification, although insertion of foreign sequences is still a possible explanation. Once the amplification reaches a critical length - about 600 bp, or 200 copies of the repeat - then methylation of an adjacent CpG island is observed. The unstable repeat also contains CpG dinucleotides, but it is not known whether this sequence is methylated or whether such methylation is directly associated with the increased methylation in the adjacent CpG island. Amplification of the repeat may itself hinder efficient transcription and render the region liable to methylation as a consequence of this inactivity. Whether methylation has an active or a passive role in the elaboration of the phenotype of fragile X syndrome is yet to be determined.

3.   Fragile Sites and Unstable Elements

The fragile X unstable element belongs to a class of sequences known as microsatellite repeats. The level of polymorphism of these sequences is proportional to their repeat length (Weber, 1990). In some instances new mutations have been observed in large pedigrees (Dracopoli et al., 1991). Analysis of flanking genetic markers has shown that recombination does not occur at the site of mutation during meiosis, so variation in copy number of repeat sequences is not a simple process of unequal crossover during meiotic recombination (Morral et al., 1991).

The fragile X p(CCG)n repeat shows these same properties. Instability of the p(CCG)n repeat appears to be proportional to length (Richards and Sutherland, 1992a); the sequence is stable in length in normal pedigrees, although at least 12 alleles have been detected with an observed heterozygosity in the CEPH pedigrees of 65%. Normal alleles have from approximately six to 58 copies of the repeat (Fu et al., 1991), whereas asymptomatic carriers appear to have about 60-230 copies; within this range the p(CCG)n repeat is meiotically unstable within the pedigree, particularly when transmitted by females (Yu et al., 1992; Fu et al., 1991; Rousseau et al., 1991b). Another hypermutable satellite also shows a gender bias and, again, copy number tends to increase rather than decrease (Vergnaud et al., 1991). Finally, fragile X individuals with more than about 230 copies of the repeat frequently exhibit mitotic instability, which is reflected as somatic variation of repeat length in different cells, even in one tissue.

Mutations in microsatellites have been termed dynamic (Richards and Sutherland, 1992a, b, c); each alteration in copy number affects the mutability of the product of the mutational event. Other forms of mutation are 'static' that is, the mutation does not alter the mutability of the product. Analysis of other fragile sites will be necessary to demonstrate whether a similar mechanism is responsible for their expression.

4.   Segregation and the Sherman paradox

The segregation of the unstable element through affected pedigrees can account for many (if not all) of the bizarre genetic properties of the disorder. In an affected pedigree no two individuals who carry the fragile X mutation need be genetically identical at the fragile X locus. Amplification of the repeat is far more likely when transmitted by females. In addition, the length of the repeat in a fragile X mother has a direct relationship to the length of the repeat in her fragile X offspring (Yu et al., 1992; Fu et al., 1991). These properties combine to account for the Sherman paradox: the general increase in size of the repeat with successive generations of an affected pedigree correlates with increasing phenotypic manifestation of the disorder. The mothers and daughters of transmitting males are at different stages of progression of the mutation. This situation is analogous to 'anticipation' in myotonic dystrophy, where increasing severity and earlier onset of the disorder reflects an increase in the copy number of a similar repeat sequence, p(AGC)n, with successive generations of the affected pedigree (Brook et al., 1992).

Increase in copy number for a repeat of this composition is also responsible for Kennedy's disease, although in this instance the amplification is limited and no unusual inheritance characteristics have been reported (La Spada et al., 1991).

5.   Molecular pathway from genotype to phenotype

Before the molecular cloning of the fragile X site the biochemical basis of the disorder was unknown. From this perspective the relationship between the fragile X mutation and the FMR1 protein is one of the most important issues in understanding the molecular basis of the disorder. The transcribed sequence of the FMR1 gene was identified by screening human fetal brain cDNA libraries with cosmid clones that flank the fragile X mutation (Verkerk et al., 1991). The FMR1 transcript contains the fragile X repeat near its 5' end. According to Verkerk et al. (1991) the mutation leads to the expansion of a tract of arginine residues at the amino-terminal region of FMR1, but if the number of consecutive arginines ranges from six to >200 in phenotypically normal individuals, the protein would have to show an extraordinary degree of flexibility. An alternative interpretation (Yu et al., 1992; Richards and Sutherland, 1992a) is that the repeat sequence is located in the 5′ untranslated region of the FMR1 gene. In this case the fragile X mutation would have a quantitative (rather than a qualitative) effect on the FMR1 protein. Consistent with this view, most affected males have no FMR1 mRNA (and therefore none of its encoded protein, Pieretti et al., 1991).

The function of FMR1 is far from clear, although its apparently ubiquitous expression and extraordinary cross-species conservation suggest that it may have a 'house-keeping' role. Rousseau et al., (1991c) demonstrated an inverse correlation between the age of fragile X female carriers and the proportion of cells whose fragile site is on the active X chromosome. Such a relationship suggests that selection has occurred for an essential autonomous cellular function (presumably conferred by FMR1), although this is at odds with the fact that fragile X males do not express FMR1 mRNA (Pieretti et al., 1991).

6.   Myotonic dystrophy

The observation that a heritable unstable DNA sequence accounted for anticipation in FMR led to the prediction (Sutherland et al., 1991) that this type of sequence, and its unusual molecular properties, might account for several other ill-understood genetic phenomena. The most notable was anticipation as seen in myotonic dystrophy (DM) (Harper et al., 1992).

The molecular basis of DM has been determined as an increase in copy number of a trinucleotide repeat, p(AGC) (Harley et al., 1992; Buxton et al., 1992; Aslandis et al., 1992; Brook et al., 1992; Fu et al., 1992; Mahadevan et al., 1992). This is the same repeat sequence found in the androgen receptor gene and amplified in Kennedy disease. Transcription is from the opposite strand of DNA (La Spada et al., 1991) from that in DM. The DM repeat is located in the 3′ untranslated region of a gene encoding a member of the protein kinase family (Brook et al., 1992; Fu et al., 1992; Mahadevan et al., 1992). The normal repeat length of the DM p(AGC)n repeat is 5 to 27 copies with the majority of normal individuals having five copies. Carriers have 50 or more copies and a correlation exists between the length of the repeat and the phenotype of an affected individual. The instability of the DM unstable element extends beyond meiotic instability in affected pedigrees to mitotic instability, manifest as somatic variation - a smear of bands evident in most affected individuals. These

properties are common to both the FMR and DM (Table 1)
unstable sequences.

Table 1.   Properties of Heritable Unstable Elements
           associated with Fragile X Syndrome and Myotonic
           Dystrophy

<u>Similarities</u>

Trinucleotide repeat

Normally polymorphic in
copy number

Disorder correlates with
amplification

Instability of copy number
in - pedigree (anticipation)
   - individual (somatic mosaicism)

Amplification not by meiotic recombination

Founder chromosomes indicative of
low 'initial' mutation frequency

Transcribed

<u>Differences</u>

Composition of repeat
- p(AGC)n is not methylated
- is methylation a cause or a consequence?

Location of repeat in transcript -
independent of translation?

Autosomal (DM;19) versus sex (FraXA;X)
chromosome location

No observed fragile site for DM

---

7.   Common Properties Of Disorders Caused By Dynamic

Mutation

Unifying concepts help to limit descriptive aspects of
phenomena. For the quality of <u>dynamic</u> mutation to be a
useful new concept for the mechanism of heritable unstable
DNA sequence mutation it must exhibit at least some

properties which are common and therefore predictable.
Fortunately this has already occurred. The atypical
segregation patterns of fragile X syndrome and DM (referred
to as the Sherman paradox and anticipation respectively)
allowed for the prediction that since a heritable unstable
element was the molecular basis for fragile X syndrome,
then a similar element and molecular mechanism might
underlie the molecular basis for DM (Sutherland et al.,
1991).

Similarly, Harley et al. (1992) were able to predict
that since fragile X syndrome and DM both have a heritable
unstable element as their molecular basis and DM exhibits a
founder chromosome (linkage disequilibrium) effect (Harley
et al., 1991), then fragile X syndrome might well exhibit a
similar phenomenon. This was indeed found to be the case
(Richards et al., 1992), with only a few (perhaps three)
founder mutations being able to account for the majority of
fragile X syndrome cases in the populations studied.

Analysis of normal individuals with one of the
haplotypes common to fragile X individuals revealed a high
incidence of high (normal) copy number (n > 39) (Richards
et al., 1992). This suggests that there might be a pool of
individuals whose descendants are at high risk of
subsequent mutation due to the relationship between copy
number and mutability.

Another similarity of fragile X syndrome and DM is that,
despite their variable phenotypes, they exhibit a
remarkable degree of homogeneity in the site of their
mutation. Many human genetic disorders have an array of
different insertions, deletions and point mutations
(sometimes in multiple genes, e.g. thalassaemias) as their
molecular basis and exhibit a variable phenotype as a
consequence. Variable phenotype in fragile X syndrome and
DM does not (in the vast majority of cases) arise from the
degree to which different types of mutation affect the
function of the gene but from the degree of amplification
of a single site of mutation (a trinucleotide repeat).
In   fragile   X   syndrome   the   amplification   blocks
transcription of the FMR-1 gene (Pieretti et al., 1991).
There does not appear to be an abnormal gene product.   The
location of the repeat in DM-1 in the 3′ untranslated
region suggests that a similar quantitative (rather than
qualitative) defect forms the molecular basis for the
pathway from genotype to phenotype.

8.   Prospects

Despite the recent major advances in our understanding of
the molecular basis for fragile sites and heritable
unstable elements much remains to be learned. For the
disorders caused by dynamic mutation the pathways from
genotype to phenotype now need to be elucidated. The
relationship between the chemistry of fragile site

expression and the sequence composition of the unstable trinucleotide repeat may become clear through the analysis of additional fragile sites.

9.   Acknowledgements

Research conducted in our laboratories was supported in part by grants from the National Health and Medical Research Council of Australia, The Channel 7 Children's Medical Research Foundation of South Australia and the Adelaide Children's Hospital Research Trust. R.I.R. thanks Shelley Richards for support and encouragement during the preparation of this chapter.

10.   References

Aslanidis, C., et al., (1992) Cloning of the essential myotonic dystrophy region and mapping of the putative defect. Nature 355:548-551.

Bell, M.V., et al., (1991) Physical mapping across the fragile X: hypermethylation and clinical expression of the fragile X syndrome. Cell 64:861-866.

Brook, J.D., et al., (1992) Molecular basis of myotonic dystrophy: Expansion of a trinucleotide (CTG) repeat at the 3' end of a transcript encoding a protein kinase family member. Cell 68:799-808.

Buxton, J., et al., (1992) Detection of an unstable fragment of DNA specific to individuals with myotonic dystrophy. Nature 355:547-548.

Dietrich, A., et al., (1991) Molecular cloning and analysis of the fragile X region in man. Nucleic Acid Res 19:2567-2572.

Dracopoli, N.C., et al., (1991) The CEPH consortium linkage map of human chromosome 1. Genomics 9:686-700.

Fu, Y.-H., et al., (1991) Variation of the CGG repeat at the fragile X site results in genetic instability: resolution of the Sherman Paradox. Cell 67:1-20.

Fu, Y.-H., et al., (1992) An unstable triplet repeat in a gene related to myotonic muscular dystrophy, myotonin protein kinase. Science 255:1256-1258.

Harley, H.G., et al., (1991) Detection of linkage disquilibrium between the myotonic dystrophy locus and a new polymorphic DNA marker. Am J Hum Genet 49:68-75.

Harley, H.G., et al., (1992) Expansion of an unstable DNA region and phenotypic variation in myotonic dystrophy. Nature 355:545-546.

Harper, P.S., et al., (1992) Anticipation in myotonic dystrophy: New light on an old problem. Am J Hum Genet 51:10-16.

Heitz, D., et al., (1991) Isolation of sequences that span the fragile X and identification of a fragile X-related CpG island. Science 251:1236-1239.

Hirst, M.C., et al., (1991) A YAC contig across the fragile X site defines the region of fragility. Nucleic Acid Res 19:3283-3288.

Kremer, E.J., et al., (1991a) Mapping of DNA instability at the fragile X to a trinucleotide repeat sequence p(CCG)n. Science 252:1711-1714.

Kremer, E.J., et al., (1991b) Isolation of a human DNA sequence which spans the fragile X. Am J Hum Genet 49:656-661.

La Spada, A.R., et al., (1991) Androgen receptor gene mutations in X-linked spinal and bulbar muscular atrophy. Nature 352:77-79.

Mahadevan, M., et al., (1992) Myotonic dystrophy mutation: an unstable CTG repeat in the 3′ untranslated region of a candidate gene. Science 255:1253-1255.

Morral, N., et al., (1991) CA/GT microsatellite alleles within the cystic fibrosis transmembraner conductance regulator (CFTR) gene are not generated by unequal crossing over. Genomics 10: 692-698.

Oberlé, I., et al., (1991) Instability of a 550-base pair DNA segment and abnormal methylation in fragile X syndrome. Science 252:1097-1102.

Pieretti, M., et al., (1991) Absence of expression of the FMR-1 gene in fragile X syndrome. Cell 66:817-822.

Poustka, A., et al., (1991) Physical map of human Xq27-qter: Localizing the region of the fragile X mutation. Proc Natl Acad Sci USA 88: 8302-8306.

Richards, R.I. and Sutherland, G.R. (1992a) Fragile X syndrome: The molecular picture comes into focus. Trends in Genetics 8:249-255.

Richards, R.I. and Sutherland, G.R. (1992b) Heritable unstable DNA sequences. Nature Genetics 1:7-9.

Richards, R.I. and Sutherland, G.R. (1992c) Dynamic mutations: A new class of mutations causing human disease. Cell 70: 709-712.

Richards, R.I., et al., (1992) Evidence of founder chromosomes in fragile X syndrome. Nature Genetics 1:257-260.

Rousseau, F., et al., (1991a) Four chromosomal breakpoints and four new probes mark out a 10-cM region encompassing the fragile-X locus (FRAXA). Am J Hum Genet 48:108-116.

Rousseau, F., et al., (1991b) Direct diagnosis by DNA analysis of the fragile X syndrome of mental retardation. N Engl J Med 325:1673-1681.

Rousseau, F., et al., (1991c) selection in blood cells from female carriers of the fragile X syndrome: inverse correlation between age and proportion of active X chromosome carrying the full mutation. J Med Genet 28:830-836.

Sherman, S. In Fragile X Syndrome, Edited by Hagerman, R.J. and Silverman, A.C., The John Hopkins University Press, Baltimore USA (1991).

Sutherland, G.R. (1985) Fragile X syndrome. Trends in Genetics 1:108-112.

Sutherland, G.R. (1991a) Chromosomal fragile sites. GATA 8:161-166.

Sutherland, G.R. (1991b) The enigma of the fragile X chromosome. Trend Genet 1:108-111.

Sutherland, G.R., et al., (1991) Hereditary unstable DNA: a new explanation for some old genetic questions? Lancet 338:289-292.

Suthers, G.K., et al., (1990) Physical mapping of new DNA probes near the fragile X mutation (FRAXA) by using a panel of cell lines. Am J Hum Genet 47:187-195.

Vergnaud, G., et al., (1991) The use of synthetic tandem repeats to isolate new VNTR loci - cloning a human hypermutable sequence. Genomics 11:135-144.

Verkerk, A.J.M.H., et al., (1991) Identification of a gene (FMR-1) containing a CGG repeat coincident with a breakpoint cluster region exhibiting length variation in fragile X syndrome. Cell 65:905-914.

Vincent, A., et al., (1991) Abnormal pattern detected in fragile-X patients by pulsed-field gel electrophoresis. Nature 349:624-626.

Warren, S.T., et al., (1990) Isolation of the human chromosomal band Xq28 within somatic cell hybrids by fragile X site breakage. Proc Natl Acad Sci USA 87:3856-3860.

Weber, J.L. (1990) Informativeness of human (dC-dA)n(dG-dT)n polymorphisms. Genomics 7:524-530.

Yu, S., et al., (1992) Fragile-X syndrome unique genetics of the heritable unstable element. Am J Hum Genet 50: 968-980.

Yu, S., et al., (1991) Fragile X genotype characterized by an unstable region of DNA. Science 252:1179-1181.

18 The 'AZF'-function of the human Y chromosome during spermatogenesis

P. VOGT, R. KEIL and S. KIRSCH
University of Heidelberg, Germany

Introduction

First evidence that the human Y chromosome must have a function in spermatogenesis was given by Tiepolo and Zuffardi (1976). Analysing a terminal deletion in the Y chromosome of 6 sterile males with a normal phenotype but azoospermia, they suggested the presence of a male fertility gene complex, called "azoospermia factor: AZF", at the distal region of the euchromatic part of the long Y arm (Yq11). This mapping position has been confirmed by different laboratories at the cytogenetic level (Yunis et al., 1977; Hartung et al., 1988; Chandley et al., 1989) and at the molecular level (Disteche et al., 1986; Gal et al., 1987; Andersson et al., 1988; Beverstock et al., 1989; Bardoni et al., 1991; Ma et al., 1992; Vogt et al., 1991a,b; 1992).

In the testicles of males with different karyotypes but a common aberration of the Yq11 chromosome region, generally, a premeiotic disruption of spermatogenesis is observed in testis tissue sections. Their histological phenotypes can be, however, variable in males with the same karyotype and similar in males with a different karyotype. If one address the function of "AZF" to the expression of one Y gene, the complexity observed may be explained by different mutations of this gene causing pleiotropic phenotypes. Other possibilities, however, may be, that the function of "AZF" is expressed by more than one Y gene or that azoospermia is not only caused by disruption of a Y gene coding a protein, but also by a general disruption of the chromosome structure in certain chromatin domains located in distal Yq11 (Limon and Gibas, 1985). As a first step towards a molecular understanding of the "AZF" function of the human Y chromosome, we, therefore, put the question, whether it is possible to reduce the complexity of its mutant phenotypes by comparing the testis histology of sterile males with a proposed dysfunction of AZF in more detail.

Testis histology of azoospermic males with a monocentric deleted Yq–chromosome: 46,XYq–

A partial deletion of distal Yq11 in six azoospermic males was first described by Tiepolo and Zuffardi (1976). Their external genitalia were normal and their testicles hypoplastic. Analysing the testis tissue sections of three patients, it was revealed that one (case 1) had a low

Chromosomes Today Volume 11. Edited by A.T. Sumner and A.C. Chandley. Published in 1993 by Chapman & Hall, London. ISBN 0 412 47670 3

number of spermatogonia and Sertoli cells, the second (case 3) had only Sertoli cells and the third (case 4) neither germ cells nor Sertoli cells in his seminiferous tubules (Tiepolo and Zuffardi, 1976). In two patients (cases 1 and 4) a testicular fibrosclerosis was observed.

These variable phenotypes have been confirmed by analysing testis tissue sections of a series of other azoospermic males with a 46,XYq-karyotype (Yunis et al., 1977; Hartung et al., 1988; Chandley et al., 1989; Bardoni et al., 1991). Some authors describe additionally an abnormal thickness of the basal lamina of the patient's testis tubules. Only if the thickness of the lamina is normal, Hartung et al (1988) observed spermatogonia and spermatocytes in the testis tubules.

Testis histology of azoospermic males with a dicentric Yp isochromosome: 46,X,idic(Yp)

A karyotype with a dicentric Yp isochromosome in azoospermic males is often associated with a 45,X cell line. Its proportion is highly variable and often different in the patient's fibroblasts and lymphocytes (Daniel, 1985). Since the breakpoint and fusion point of the two associated Y chromosomes lies in the distal Yq11 region, the rearranged Y chromosome has a size like the normal Y chromosome. But after staining the heterochromatic block in the long arm of the normal human Y chromosome by quinacrine, it can be recognized that the rearranged "normal looking" iso-Y chromosome usually has lost the fluorescent heterochromatic Yq12-block. It was therefore called "nonfluorescent" Y chromosome (Ynf). The 46,X,Ynf karyotype sometimes includes more than two cell lines with "different looking" Y chromosomes, suggesting that the dicentric "Ynf" chromosome is unstable in vivo or at least in "in vitro" cell cultures (Chandley et al., 1986; Fryns et al., 1978; Diekmann et al., 1992). In all cases, where a testis biopsy has been performed, spermatogenesis is blocked before meiosis. Only Sertoli cells are observed in a case of Chandley et al. (1986) and Affara et al (1986). Degenerating premeiotic germ cells and Sertoli cells are observed in the case of Kaluzewski et al. (1988) and an arrest at the pachytene stage is described for the second case of Affara et al. (1986).

Testis histology of azoospermic males with a ring-Y chromosome: 46,X,r(Y)

Sterile males with a ring-Y chromosome are phenotypically often quite normal, but azoospermic (Daniel, 1985). This has been surprising, since in most cases the karyotype of those males were a mosaic containing the 45,X cell line in their lymphocyte cultures. It is known that cells with ring chromosomes will generate "in vitro" cells with 45 chromosomes minus a ring because of media-induced sister chromatid exchanges (Daniel, 1985). Since also sterile males with a nonmosaic ring-Y karyotype have been described (Wilson et al., 1976), the question is raised whether a mosaic karyotype with a ring-Y chromosome exist in vivo, at all (Steinbach et al., 1979; Daniel, 1985).

Usually it is difficult to analyse a **ring form** in the small dotlike Y chromosome of this patient class. Only in a meiotic chromosome study, the ring constitution of the abnormal Y chromosome has been established

(Chandley and Edmond, 1971). In this patient, disruption of spermatogenesis takes place during meiosis. In another azoospermic patient with the same karyotype disruption of spermatogenesis takes place before the development of spermatogonia (Steinbach et al., 1979). The patient with a spermatogenic disruption at meiosis had a higher amount of 45,X cells in his lymphocytes (30% vs. 6.6%).

Testis histology of azoospermic males with a Y/Y translocation chromosome: 46,X,t(Yp11;Yq11)

Translocations between two Y chromsomes are difficult to examine and can be only differentiated from Yq–isochromosomes by a different length of both Y arms. Whether the function of AZF is distorted in the translocated Yq11 chromosome at all, is still an open question, since no studies have been done to determine its breakpoint in Yq11. The second Y chromosome has a breakpoint in distal Yp and most probably a normal Yq11 region. It is therefore assumed that AZF is functional on this Y–chromosome.

The seven cases described are azoospermic or oligospermic (Wahlström, 1985). Their testicles are small and soft. An analysis of the testis histology in one case reveals a maturation arrest at the primary spermatocyte stage. Often their karyotype is a mosaic with a 45,X cell line.

Testis histology of azoospermic males with an autosomal translocation of the Y chromosome: 46,X,t(Y;A)

Autosomal translocations of the Y chromosome are observed in a balanced or non–balanced karyotype. In most familial cases with a balanced Y–A translocation the distal heterochromatic part of the Y chromosome is translocated to the short arm of an acrocentic chromosome (Smith et al., 1979). This chromosomal preference may be explained by the observed nonrandom association of those chromosomes to the XY pairing complex in the sex vesicle (Stahl et al., 1984). The Y–heterochromatin, attached to the autosome, can be easily detected by fluorescent staining.

Balanced or reciprocal Y–A translocations are observed in fertile and sterile males (Fryns et al., 1985). It is assumed that in fertile males the breakpoint of the Y chromosome is in the heterochromatic Yq12 region and in sterile males adjacent in the euchromatic distal Yq11 region. A family with the same balanced Y–A translocation and fertile and sterile males in its pedigree (Reitalu, 1973) suggests that a breakage event on the Y chromosome in Yq12 can be unstable and may shift to the flanking euchromatic Yq11 region.

Sterile males with a balanced Y–A translocation chromosome have normal external genitalia or hypogonadism. The histology of their testis, shows an arrest of spermatogenesis at the spermatocyte stage before or during meiosis (Laurent et al., 1982; Viguie' et al., 1982; Fryns et al., 1985) or during the formation of spermatids (Faed et al., 1982). In two cases where an analysis of meiotic chromosomes has been performed, the condensation of the X and Y chromosome during their premeiotic pairing process was, partly, inhibited (Gonzales et al., 1981; Viguie' et al. 1982).

In the unbalanced 46,X,t(Y;A) karyotype the euchromatic "Yq12-" chromosome part is translocated to an autosome. The acentric heterochromatic chromosome part (Yq12) is lost. Azoospermic males with this karyotype are rare. They were described earlier as 45,X males including a rare mosaic with a 46,XY cell line, necessary to explain their sexual phenotype (de la Chapelle et al., 1986). Molecular analysis of the presence of the Y chromosome in those males had revealed however, that the male determining Yp chromosome part is present and translocated to an autosome. The Y-breakpoint is in the Yq11 region (Disteche et al., 1986; Gal et al., 1987; Andersson et al., 1988) and the Y centromere assumed to be inactive.

Phenotypically, their testis can be normal but also hypoplastic and small. Testis histology reveals an early disruption of germ cell development, only showing Sertoli cells and immature seminiferous tubules.

Testis histology of azoospermic males with an interstitial deletion in Yq11

Interstitial deletions in Yq11 can be visible or invisible at the cytogenetic level. In the latter case, the patient has a normal 46,XY karyotype and the deletion can only be revealed by molecular deletion mapping as described in the following chapter.

A cytogenetically visible interstitial deletion in Yq11 of an azoospermic male has been described by Skare et al. (1990). His testicles were small. A testicular biopsy was declined. From molecular hybridization analysis it was concluded that part of the Yq12 heterochromatin is present and probably the total G band Yq11.23 deleted.

Small interstitial deletions ("microdeletions") in Yq11 of two azoospermic males have been described by Vogt et al. (1991a;1992). Both patients had a normal phenotype but small testicles. One patient had a deletion in distal Yq11.22/proximal Yq11.23. The second patient had a deletion in the distal part of Yq11.23. According to the molecular deletion map of Ma et al. (1992) both microdeletions do not overlap. A testis biopsy of the patient with the proximal microdeletion revealed only Sertoli cells in his seminiferous tubules. A testis biopsy of the second patient has been declined.

Phenotypic expression picture of the genetic "AZF" function during early spermatogenesis

To summarize the observations of the testis histology seen in azoospermic males with a proposed dysfunction of "AZF" two common major stages of "AZF activity" during spermatogenesis can be distinguished.

The first stage (stage "a") is, before or during the development of spermatogonia. The second stage (stage "b") is during the development of primary or secondary spermatocytes. Since in spermatogenesis the second stage is timely fixed to the onset of puberty, it seems reasonable to split the function of AZF into a prepubertal function at stage "a", called "AZFa" and a pubertal function at stage "b", called "AZFb". The question now raised, whether both spermatogenic Y-functions are coded and/or regulated by Y chromosomal gene(s) located in distal Yq11.

The cytogenetic resolution of the distal Yq11 chromosome region is inherently low and it is difficult or even impossible to distinguish the position of the Y-breakpoint in one patient from the Y-breakpoint of another patient at this resolution level. It is therefore an open question, whether patients with a spermatogenic disruption at stage "a" and stage "b" may have different Y breakpoints at the molecular scale.

It can be shown, that patient "JOLAR" with a small interstitial deletion in distal Yq11.22/proximal Yq11.23 reveals a spermatogenic disruption at stage "a" (Fig.1). Therefore, if the function of "AZFa" is expressed by a Y gene, part of its coding DNA structure should be located in distal Yq11.22/proximal Yq11.23. If the function of "AZFb" is expressed by a different Y gene its molecular position should be distinguishable from the Y gene expressing "AZFa". The second azoospermic patient with a microdeletion in distal Yq11.23, not overlapping with the microdeletion of "JOLAR" (Ma et al., 1992). suggested that a Y gene expressing "AZFb" is located in distal Yq11.23 (Vogt et al., 1992). To prove this hypothesis a testis biopsy of the second patient needs to be analysed. Unfortunately, this is not yet possible. Another possibility, however, should be deletion mapping of the Y breakpoints of sterile males with a known disruption of their spermatogenesis at stage "a" and stage "b" at a common molecular scale.

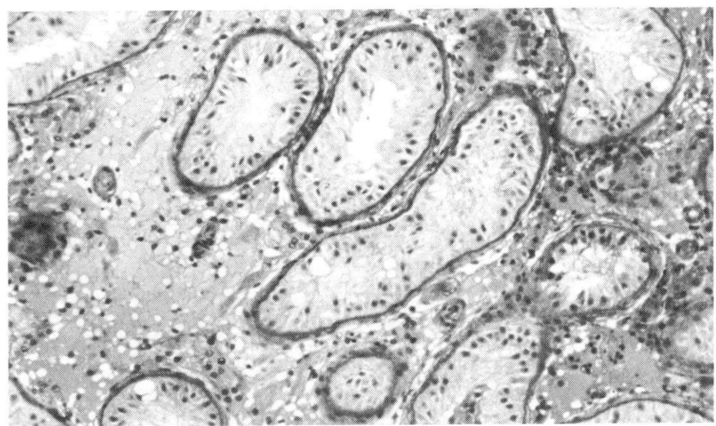

Fig. 1. Testis histology of azoospermic patient "JOLAR" with a small interstitial deletion in distal Yq11.22/proximal Yq11.23.

Deletion mapping of "AZFa" and "AZFb" in Yq11.22-23

A series of molecular breakpoints in distal Yq11 of azoospermic males has been described recently from different laboratories (Affara et al., 1986; Ferguson-Smith et al., 1987; Bardoni et al., 1991; Ma et al., 1992; Vogt et al., 1991b). The position of those breakpoints were mapped by deletion mapping according to Vergnaud et al. (1986). This mapping procedure is very powerful in mapping the molecular position of Y breakpoints in relation to each other, but it gives no information about the molecular distance between them. This is only possible after the establishment of a molecular PFGE map for this chromosome region. Towards the production of a PFGE map, a detailed Vergnaud-deletion map with a lot of Y specific DNA probes is a necessary prerequisite.

The most detailed deletion maps of the Yq11 chromosome region have been published by Ferguson-Smith et al. (1987) and Ma et al. (1992). The Ferguson-Smith map includes 12 subintervals along the total euchromatic Yq11 region. The Ma map includes 15 subintervals along the same chromosome region but is concentrated to chromosome region Yq11.22-23, dividing it into 14 subintervals. The Y breakpoint of two azoospermic males (863-829; 863-830) with a 45,X/46,Xidic(Yp) karyotype, described

PROBES	M188044	732	87SFB	JOWAL	BITRA	WSM184	863829	LGL2172	L1078	FRABO	863712	DM	863830	MN	870740	DA	FW	F443	GM2103	GM2469	GM118	M	FS	B
pJA4402	-	+	+	+	+	+	+	+	N	+	+	+	+	+	+	+	+	N	-	-	-	0	1/2/3	A-G
12f3	-	+	+	+	+	+	+	+	+	+	+	+	+	+	+	+	+	-	-	-	-			
pY6HP35	-	-	+	+	+	+	+	+	+	+	+	+	+	+	+	+	+	-	-	-	-	I		
118e	-	-	+	+	+	+	+	+	+	+	+	+	+	+	+	+	+	-	-	-	-			
Y-253	-	-	-	+	+	+	+	+	+	+	+	+	+	+	+	+	+	+	-	-	-	II		
Y-198	-	-	-	+	+	+	+	+	+	+	+	+	+	+	+	+	+	+	-	-	-			
Y-221	-	-	-	-	+	+	+	+	+	+	+	+	+	+	+	+	+	+	-	-	-	III		
Y-216a/A	-	-	-	-	-	+	+	+	+	+	+	+	+	+	+	+	+	+	-	-	-	IV	4	H
Y-294	-	-	-	-	-	+	+	+	+	+	+	+	+	+	+	+	+	+	-	-	-			
pY6BS64	-	-	-	-	-	+	+	+	+	+	+	+	+	+	+	+	+	+	-	-	-			
pY6BS65/B	-	-	-	-	-	-	+	+	+	+	+	+	+	+	+	+	+	+	-	-	-	V		
Y-202	-	-	-	-	-	-	-	+	+	+	+	+	+	+	+	+	+	+	-	-	-	VI		
pY6PHc54	-	-	-	-	-	-	-	+	+	+	+	+	+	+	+	+	+	+	-	-	-			
pY6D14-1	-	-	-	-	-	-	-	+	+	+	+	+	+	+	+	+	+	+	+	-	-	VII	5/6	
pY6Ba34	-	-	-	-	-	-	-	+	+	+	+	+	+	+	+	+	+	+	+	-	-			
Fr25-II/B	-	-	-	-	-	-	-	+	+	+	+	+	+	+	+	+	+	+	+	-	-			
pY6BS65/C	-	-	-	-	-	N	-	+	+	+	+	+	+	+	+	+	+	+	+	-	-			
LLY22g/A	-	-	-	-	-	-	-	-	-	-	+	+	+	+	+	+	+	+	+	-	-	VIII	7	I
50f2/E	-	-	-	-	-	-	-	-	-	-	+	+	+	+	+	+	+	+	+	-	-			
52d/A	-	-	-	-	-	-	-	-	-	-	+	+	+	+	+	+	+	+	+	-	-			
Y-367/A	-	-	-	-	-	-	-	-	-	-	+	+	+	+	+	+	+	+	+	+	-	IX		
GMGXY10	-	-	-	-	-	-	-	-	-	-	-	+	+	+	+	+	+	+	+	+	-	X	8	
GMGY18	N	-	-	-	-	N	-	N	-	-	-	N	N	N	+	+	N	N	N	+	N	XI	9	J
Y-216a/B	-	-	-	-	-	-	-	-	-	-	-	-	+	+	+	+	+	+	+	+	-			
Fr25-II/A	-	-	-	-	-	-	-	-	-	-	-	-	+	+	+	+	+	+	+	+	-			
pY6BS65/D	-	-	-	-	-	-	-	-	-	-	-	-	+	+	+	+	+	+	+	+	-			
LLY22g/B	-	-	-	-	-	-	-	-	-	-	-	-	+	+	+	+	+	+	+	+	-			
p69/6	-	-	-	-	-	-	-	-	-	-	-	-	+	+	+	+	+	+	+	+	-			
pY6HP52	-	-	-	-	-	-	-	-	-	-	-	-	-	-	-	+	+	+	+	+	-	XII		
50f2/C	-	-	-	-	-	-	-	-	-	-	-	-	-	-	-	+	+	+	+	+	-			
p116/21	-	-	-	-	-	-	-	-	-	-	-	-	-	-	-	+	-	+	+	+	-			
pY6BS65/E	-	-	-	-	-	-	-	-	-	-	-	-	-	-	-	-	-	+	+	+	-	XIII	10	K
Y-367/B	-	-	-	-	-	-	-	-	-	-	-	-	-	-	-	-	-	+	+	+	-			
Fr15-II	-	-	-	-	-	-	-	-	-	-	-	-	-	-	-	-	+	+	+	+	-			
RBF8	-	-	-	-	-	-	-	-	-	-	-	-	-	-	-	-	-	+	+	+	-			
49f	-	-	-	-	-	-	-	-	-	-	-	-	-	-	-	-	-	+	+	+	-			
GMGY5	-	-	-	-	-	-	-	-	-	-	-	-	-	-	-	-	-	+	+	+	-	XIV	11	
GMGY1	-	-	-	-	-	-	-	-	-	-	-	-	-	-	-	-	-	+	+	+	-		12	
poxY1	-	-	-	-	-	-	-	-	-	-	-	-	-	-	-	-	-	+	+	+	-			
pHY2.1	-	-	-	-	-	-	-	-	-	-	-	-	-	-	-	-	-	+	+	+	+		13	

Fig. 2. Comparison of three molecular deletion maps of the long Y arm with emphasis on the Yq11.22-23 region. M=Ma et al.(1992); F-S=Ferguson-Smith et al. (1987); B=Bardoni et al. (1991)

as panel member 6 and 7 by Affara et al. (1986) were used in both
mapping experiments as subinterval-marker. Since additionally, 4 other
patients with a 45,X,46,Xidic(Yp) karyotype but unknown sexual
phenotype were used similarly in both deletion maps (JC; DM; MN; FW), it
is possible to relate the subintervals of both maps to each other (Fig.2).
Obviously the Ma map (M) is more differentiated in the distal part of
Yq11.22 and proximal part of Yq11.23. The Ferguson-Smith map (F-S) is
somewhat more detailed in distal Yq11.23.
The deletion map of Bardoni et al. (1991) defines 11 subintervals in Yq11
but only 6 intervals (F, G, H, I, J, K) in the Yq11.22-23 chromosome
region. Since the Bardoni map (B) used a subset of probes from the Ma
map it is again possible to relate both deletion maps to each other
(Fig.2). The relative position of the Y-DNA probes used in the Ma map
has been confirmed by other research groups (Donlon and Müller, 1991;
Kotecki et al., 1991; Cheng et al., 1992). The composite map in Fig.2 may
therefore be useful, to compare the Y breakpoint of azoospermic males
from different laboratories using the same set of Y-DNA probes.

Using this composite map the positions of the Y breakpoint of five
azoospermic males known to have a disruption of their spermatogenesis at
stage "a" and three azoospermic males known to have a disruption of
their spermatogenesis at stage "b" are compared schematically in Fig.3.
The detailed clinical backgound of all patients has been described
previously (Affara et al., 1986; Bardoni et al., 1991; Ma et al., 1992). For
patients of stage "a" the most proximal Y breakpoint is defined by
deletion of the 12f3 probe in "JOLAR". The most distal Y breakpoint must
be located distal to subinterval "C" of the repetitive pY6BS65 probe in
subinterval VII as shown in LGL2172 (Fig.2 u.3). The position of a Y gene
expressed during stage "a" is therefore suggested to be between
subinterval I and VII (Fig.4a).

The Y breakpoint of 863-830 in subinterval XI supports the position of a
different Y gene expressed during stage "b" in distal Yq11.23. This

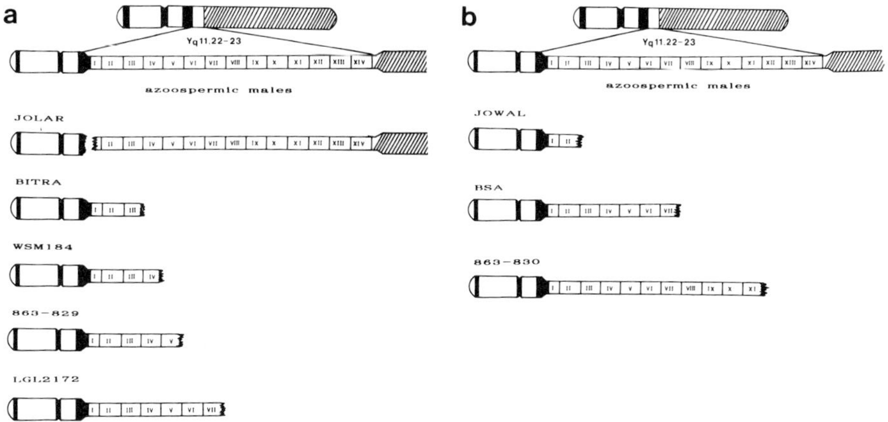

Fig. 3. Schematic comparison of Y breakpoints of AZF⁻ males
with a known spermatogenic disruption at stage "a" (a) and at
stage "b" (b). Subintervals are defined by the Ma map in Fig.2.

subinterval is in the molecular neighbourhood of the proximal Y
breakpoint of the second patient with a microdeletion in distal Yq11.23
(Ma K, unpublished results).

The position of the Y-breakpoint of JOWAL, in subinterval II but with
a spermatogenic disruption at stage "b" is surprising in this context.
Following the concept of two distinguishable Y spermatogenesis functions
"AZFa" and "AZFb" in distal Yq11 expressed by different Y genes, this
case is only explained, if one assumes in this patient a leaky dysfunction
of "AZFa" by special rearrangments of his Y chromosome in proximal
Yq11.23, due to its fusion to the distal part of the Y short arm. Another
possibility may be that secondary deletion events in distal Yq11.23 have
happened in the Y ring during growing the lymphocytes repeatedly in cell
cultures after the original analysis.

Molecular extension of "AZFa"

The description of the testis histology of BSA is somewhat confusing
stating the absence of germ cells beside a severe reduction of pachytenes
(Bardoni et al., 1991). Since the Y breakpoint of patient LGL2172 has
been mapped to the same interval as BSA, and LGL2172 has a distinct
spermatogenic disruption at stage "a", one may assume that the function
of "AZFa" is "leaky" in BSA and that in between subinterval VII the Y
breakpoint of LGL2172 is proximal to the Y breakpoint of BSA. The distal
end of Y DNA sequences related functional to "AZFa" is then supposed to
be in subinterval VII.

Although a complete PFGE map of chromosome region Yq11.22-23 is not
yet available, a large molecular distance (greater than 2Mb) between the
12f3 probe and pY6BS65/C is already indicated (Keil et al., manuscript in
preparation). The molecular position of the distal breakpoint in the Y
chromosome of JOLAR and the proximal breakpoint in the Y chromosome of
LGL2172 isn't yet known on this PFGE map. But one can assume that DNA
sequences functional during spermatogenesis at stage "a" should be
located in subinterval I and VII. This may indicate that the function of
"AZFa" is coded and/or regulated by more than one Y gene or that the Y
gene expressing "AZFa" is large.

General discussion

In summary, it is obvious that the genetically defined spermatogenesis
function "AZF" of the human Y chromosome is complex. Observations of
the testis histology of sterile males with different Yq11 anomalies
suggest that the "AZF" function is expressed during the development of
spermatogonia (stage "a") and during the development of spermatocytes at
puberty (stage "b"). Since these AZF-functions are active at two different
times during spermatogenesis, the definition of two distingiushable
functions "AZFa" and "AZFb" is suggested as a first step towards the
resolution of the complex functional phenotype of AZF.

With aid of the analysis of a small interstitial deletion in distal
Yq11.22/proximal Yq11.23 of the azoospermic male "JOLAR" and the
knowledge of his testis histology (Fig.1), it has been possible, to map

functional DNA sequences, coding and/or regulating the AZFa-function, to distal Yq11.22/proximal Yq11.23 (Vogt et al., 1991a; 1992). Comparing the Y breakpoints of different azoospermic males with a similar spermatogenic disruption stage at stage "a" or "b" it is shown that "AZFa" can be mapped between subinterval I and subinterval VII of the Ma map (Fig.4a and 4b), and that this map position can be distinguished from the map position of "AZFb" in distal Yq11.23.

It is not yet possible to map the distal extension of "AZFb" by subinterval mapping, since the mapping of Y breakpoints in distal Yq11.23 of fertile males are not included in the Ma map, but only in the map of Bardoni et al. (1991). Since the probe 49f detects a highly polymorphic DNA region (Weissenbach and Goodfellow, 1991) one may assume that "AZFb" ends proximal to this Y DNA region. The proximal extension of "AZFb" should be not larger than to subinterval VII. The proximal extension of "AZFa" is not known. This result of two different mapping positions for "AZFa" and "AZFb" confirms our view of two genetically different Y spermatogenesis functions in Yq11.22-23, earlier summarized as "AZF" (Tiepolo and Zuffardi, 1976) and now split in "AZFa" and "AZFb".

It may be argued, that this statement is too preliminary, since it is mainly based on only one microdeletion in distal Yq11.22/proximal Yq11.23 and only two patients with a spermatogenic disruption at stage "b" and a Y breakpoint distal to the Y breakpoints of five patients with a spermatogenic disruption at stage "a" (Fig.4a;b). Obviously, the analysis of more interstitial microdeletions in proximal and distal Yq11.23 is needed to proof our concept of two distinguishable Y functions during spermatogenesis in more detail.

Similarly, only the analysis of interstitial microdeletions in distal Yq11.23 found in azoospermic males with a spermatogenic disruption at stage "b" will help to reveal, whether the function of "AZFb" is an intrinsically genetic function of DNA sequences in distal Yq11.23 or only a secondary effect of X-Y pairing problems before and/or during meiosis due to the deletion of terminal Y euchromatin.

It is well known, that during early pachytene the Y chromosome pairs to the X chromosome, transiently, along its total euchromatic length, most probably including distal Yq11 (Chandley et al., 1984). In this context, it is interesting to note that a specific chromatin structure in distal Yq11.23 has been described by Limon and Gibas (1985). If one accept the view, that the genetic function of a given DNA sequence is, not only expressed by its coding potential for a specific protein, but also by its ability to adopt, or to initiate a functional important chromatin structure (Trifonov, 1989; Vogt, 1990) the locus-specific chromatin structure in distal Yq11.23 may well be an intrinsic part of the AZFb-function.

The variable phenotypes observed in patients with a proposed dysfunction of "AZFb" indicate, however, that the function of "AZFb" cannot be only related to the X-Y pairing process. An arrest at stage "b" is observed at pachytene (Affara et al., 1986; Smith et al., 1979: Bardoni et al., 1991) but also during meiotic divisions (Chandley and Edmonds, 1971; Laurent et al., 1982) and during the formation of spermatids (Faed et al., 1982). Males with a balanced Y-A translocation can also be fertile. Only in sterile males the Y breakpoint is assumed to be in distal Yq11 (Fryns et al., 1985). A molecular confirmation of this assessment is, however, still missing.

Mapping the extension of "AZFa" over at least 2Mb of DNA between subinterval I and VII suggests that more than one Y gene are responsible for the expression of "AZFa". Variable histological pictures of spermatogenic arrest at stage "a", observed in different patients may support this assumption, but it may also mean that different mutations in one large AZFa gene can cause pleiotropic phenotypes.

Another possibility may be, that the observed variability represents different stages of a degenerative process in the patients' testes tubules, which occurs generally, when the function of "AZFb" at puberty is missing.

It is interesting to note that six members of the pY6H sequence family are located between subinterval I and VII and that one sequence member, pY6H35, is deleted in the small interstitial microdeletion of "JOLAR" (Vogt et al., 1992). This sequence family has been isolated by crosshybridization to a male fertility gene sequence on the Y chromosome of Drosophila hydei (Vogt et al., 1991c). Based on this result it has been speculated that the spermatogenesis function of the Y chromosome is evolutionarily very old and that the molecular structure of the AZF gene may be similar to the structure of the Y-fertility genes of Drosophila (Vogt, 1990). If this holds true, one may speculate that the function of "AZFa" is not based on protein(s) at all, but on a gene coding for a large structural RNA molecule, as in Drosophila (Hennig et al., 1987).

Acknowledgements
We would like to thank all our friends and colleagues, especially Ann Chandley, Tim Hargreave, Mike Köhler, Ma Kun, Andrew Sharkey, Bob Speed, Jochen Süß and Professor R.A.Pfeiffer, who have contributed with their invaluable clinical material, ideas and discussions so much to our view on the function of AZF as presented in this manuscript. Mrs. A. Wiegenstein is thanked for her excellent photographic assistance. We are grateful to Professor F. Vogel for his invaluable support and encouragement. The research on AZF is supported by the Deutsche Forschungsgemeinschaft (Vo403/1-4).

References

Affara, N. A. et al. (1986) Regional assignment of Y-linked DNA probes by deletion mapping and their homology with X-chromosome and autosomal sequences. **Nucl. Acid Res.**, 14, 5253-5373.

Andersson, M. et al. (1988) Y;autosome translocations and mosaicism in the aetiology of 45,X maleness: assignment of fertility factor to distal Yq11. **Hum. Genet.**, 79, 2-7.

Bardoni, B. et al. (1991) A deletion map of the human Yq11 region: Implications for the evolution of the Y chromosome and tentative mapping of a locus involved in spermatogenesis. **Genomics**, 11,443-451.

Beverstock, G.C. et al. (1989) Y chromosome specific probes identify breakpoint in a 45,X/46,X,del(Y)(pter-q11.1) karyotype of an infertile male. **J. Med. Genet.**, 26, 330-342.

Chandley, A.C. et al. (1984) On the nature and extent of XY pairing at meiotic prophase in man. Cytogenet. Cell Genet., 38, 241-247.

Chandley, A.C. and Edmond, P. (1971) Meiotic studies on a subfertile patient with a ring Y chromosome. **Cytogenetics**, 10, 295–304.

Chandley, A.C. et al. (1986) Short arm dicentric Y chromosome with associated statural defects in a sterile man. **Hum. Genet.**, 73, 350–353.

Chandley, A.C. et al. (1989) Deleted Yq in the sterile son of a man with a satellited Y chromosome (Yqs). **J. Med. Genet.**, 26, 145–153.

Cheng, S–D. Gasparini, R. Müller, U. (1992) Molecular analysis of aberrations of Xp and Yq. **Hum. Genet.**, 88, 379–382.

Daniel, A. (1985) Y isochromosomes and rings, in **The Y chromosome, Part B: Clinical Aspects of Y chromosome abnormalities** (ed A.A. Sandberg), Alan R. Liss, Inc., New York, pp. 105–135.

De la Chapelle, A. et al. (1986) The origin of 45,X males. **Am. J. Hum. Genet.**, 38, 330–340.

Diekmann, L. et al. (1992) Multiple minute marker chromosomes derived from Y identified by FISH in an intersexual infant. **Hum. Genet.**, 90, 181–183.

Disteche, C.M. et al.(1986) Molecular detection of a translocation (Y;15) in a 45,X male. **Hum. Genet.**, 74: 372–377.

Donlon, T.A. and Müller, U. (1991) Deletion mapping of DNA segments from the Y chromosome long arm and their analysis in an XX male. **Genomics**, 10: 51–56.

Faed, M.J.W. Lamont, M.A. Baxby, K. (1982) Cytogenetic and histological studies of testicular biopsies from subfertile men with chromosome anomaly. **J. Med. Genet.**, 19, 49–56.

Ferguson–Smith, M.A. Affara, N.A. Magenis, R.E. (1987) Ordering of Y specific sequences by deletion mapping and analysis of X–Y interchange males and females. **Development**, 101 (Supplement), 41–50.

Fryns, P. Cassiman, J.J. Van den Berghe, H. (1978) Unusual in vivo rearrangments of the Y chromosome with mitotic instability in vitro. **Hum. Genet.**, 44, 349–355.

Fryns, P. Kleczkowska, A. Van den Berghe, H. (1985) Clinical manifestations of Y/autosome translocations in man, in **The Y chromosome, Part B: Clinical Aspects of Y chromosome abnormalities** (ed A.A. Sandberg), Alan R. Liss, Inc., New York, pp. 213–243.

Gal, A. et al. (1987) A 45,X male with Y–specific DNA translocated onto chromosome 15. **Am. J. Hum. Genet.**, 40, 477–488.

Gonzales, J. Lesourd, S. Dutrillaux, B. (1981) Mitotic and meiotic analysis of a reciprocal translocation t(Y;3) in an azoospermic male. **Hum. Genet.**, 57, 111–114.

Hartung, M. et al. (1988) Yq deletion and failure of spermatogenesis. **Ann. Genet.**, 31, 21–26.

Hennig, W. et al. (1987) Structure and function of Y chromosomal genes in Drosophila, in **Chromosomes Today, vol.9** (eds A. Stahl, J.M. Luciani, A.M. Vagner–Capodano), Allen & Unwin, London, pp. 48–58.

Kaluzewski, B. et al. (1988) Two mosaic cases with nonfluorescent Y chromosome analysed with Y–specific DNA probes. **Am. J. Med. Genet.**, 31, 489–503.

Kotecki, M. et al. (1991) Deletion mapping of interval 6 of the human Y chromosome. **Hum. Genet.**, 87, 234–236.

Laurent, C. et al. (1982) The use of surface spreading in the pachytene analysis of a human t(Y;17) reciprocal translocation. **Cytogenet. Cell Genet.**, 33, 312–318.

Limon, J. and Gibas, Z. Lateral asymmetry of the Y chromosome and its significance, in **The Y chromosome, Part A: Basic Characteristics of the Y chromosome** (ed A.A. Sandberg), Alan R. Liss, Inc., New York, pp. 317–326.

Ma, K. et al. (1992) Towards the molecular localization of the AZF locus: mapping of microdeletions in azoospermic men within 14 subintervals of interval 6 of the human Y chromosome. **Hum. Mol. Genet.**, 1, 29–33.

Reitalu, J. (1973) A familial Y–22 translocation in man. **Hereditas**, 74, 155–160.

Skare, J. et al. (1990) Interstitial deletion involving most of Yq. **Am. J. Med. Genet.**, 36, 394–397.

Smith, A. Fraser, I.S. Elliot, G. (1979) An infertile male with balanced Y;19 translocation. Review of Y;autosome translocations. **Ann. Genet.**, 22, 189–194.

Stahl, A. et al. (1984) The association of the nucleolus and the short arm of acrocentric chromosomes with the XY pair in human spermatocytes: its possible role in facilitating sex–chromosome acrocentric translocations. **Hum. Genet.**, 68, 173–180.

Steinbach, P. Fabry, H. Scholz, W. (1979) Unstable ring Y chromosome in an aspermic male. **Hum. Genet.**, 47, 227–231.

Tiepolo, L. and Zuffardi, O. (1976) Localization of factors controlling spermatogenesis in the nonfluorescent portion of the human Y chromosome long arm. **Hum. Genet.**, 34, 119–124.

Trifonov, E.N. (1989) The multiple codes of nucleotide sequences. **Bull. Math. Biol.**, 51, 417–432.

Turleau, C. Chavin-Colin, F. de Grouchy, J. (1980) A 45,X male with translocation of euchromatic Y chromosome material. **Hum. Genet.**, 53, 299–302.

Vergnaud, G. et al. (1986) A deletion map of the human Y chromosome based on DNA hybridization. **Am. J. Hum. Genet.**, 38, 109–124.

Viguie', F. Romani, F. Dadoune J.P. (1982) Male infertility in a case of (Y;6) balanced reciprocal translocation. Mitotic and meiotic study. **Hum. Genet.**, 62, 225–227.

Vogt, P. (1990) Potential genetic functions of tandem repeated DNA sequence blocks in the human genome are based on a highly conserved "chromatin folding code". **Hum. Genet.**, 84, 301–336.

Vogt, P. et al. (1991a) Towards the molecular localization of AZF, a male fertility gene on the human Y chromosome by comparative mapping of microdeletions in the Y chromosome of men with idiopathic infertility. **Am. J. Hum. Genet.**, 49, Supplement, vol.4.

Vogt, P. et al. (1991b) Towards a molecular map of the human Y chromosome in Yq11.23 with view to the potential extension of the functional DNA structure of the AZF locus in this chromosome region, in **Genome analysis: From Sequence to Function, series: Adv. in Mol. Gen. vol. 4** (eds J. Collins and A.J. Driesel), Hüthig Buch Verlag Heidelberg, pp. 277–280.

Vogt, P. et al. (1991c) Selection of DNA sequences from interval 6 of the human Y chromosome with homology to a Y chromosomal fertility gene sequence of Drosophila hydei. **Hum. Genet.**, 86, 341–349.

Vogt, P. et al. (1992) Microdeletions in interval 6 of the Y chromosome of males with idiopathic sterility point to disruption of AZF, a human spermatogenesis gene. **Hum. Genet.**, 89, 491–496.

Wahlström, J. (1985) Y/Y translocations and their cytologic and clinical manifestations, in **The Y chromosome, Part B: Clinical Aspects of Y chromosome abnormalities** (ed A.A. Sandberg), Alan R. Liss, Inc., New York, pp. 207–212

Weissenbach, J. Goodfellow, P.N. (1990) Report of the committee on the genetic constitution of the Y chromosome. **Cytogenet. Cell Genet.**, 51, 438–449.

Wilson, M.G. Stein, R.B. Towner, J.W. (1976) Ring Y chromosome without mosaicism. **Birth Defects,** 12, 105–112.

Yunis, E. et al. (1977) Yq deletion, aspermia, and short stature. **Hum. Genet.**, 39, 117–122.

Sex chromosomes

19 Unpaired sex chromosomes and gametogenic failure

P.S. BURGOYNE and S.K. MAHADEVAIAH

MRC Mammalian Development Unit, London, UK

1 Introduction

It has long been appreciated that there is a correlation between the presence of unpaired sex chromosomes and spermatogenic failure, but there is as yet no general agreement as to the reasons for this relationship. In 1974 Miklos put forward an intriguing hypothesis which proposed that there are meiotic 'pairing sites' distributed over the chromosomes, which must be 'saturated' by pairing during meiotic prophase. If chromosome regions remain unpaired at pachytene, these pairing sites are 'activated' and set in motion a process that eventually leads to the destruction of the nucleus and hence the elimination of the cell with the meiotic pairing error. The hypothesis relates the efficiency of this elimination process to the number of 'unsaturated' pairing sites. In the special case of the XY bivalent, the requirement for pairing is restricted to the region of pairing and homology. In 1984 Burgoyne & Baker, in the light of their studies on the etiology of oocyte deficiency in XO mice (Burgoyne and Baker, 1981; 1985), extended Miklos' concept of a 'meiotic quality control' to include meiosis in females. The present paper focusses on studies of male and female mice with sex chromosome pairing anomalies, which provide data that can be used to test the main tenets of Miklos' hypothesis.

2 General considerations

Miklos' model suggests a causal relationship between the extent of pairing failure at pachytene and the extent of gametogenic failure; furthermore, it predicts that the gametogenic failure will occur during or subsequent to pachytene. Thus, in order to test the model in the context of specific cases where there are unpaired sex chromosomes, we need information on the extent of pairing failure, the extent of germ cell loss, and the stage at which this loss occurs.

The extent of pairing failure should ideally be assessed by pachytene synaptonemal complex (SC) analysis, since this allows the frequency of cells affected, and the extent of the pairing failure in the affected cells, to be determined. Frequently however, the extent of sex chromosome pairing failure is described in terms of the frequency of univalent sex chromosomes at diakinesis and MI. The

Chromosomes Today Volume 11. Edited by A.T. Sumner and A.C. Chandley. Published in 1993 by Chapman & Hall, London. ISBN 0 412 47670 3

justification for using MI data is that there is apparently a mechanism in most species which ensures a minimum of one chiasma per synapsed bivalent, even (as proposed by Burgoyne, 1982) within the short region of synapsis between the X and Y of eutherian mammals. It is the presence of this 'obligatory chiasma' which in most species is responsible for maintaining the bivalent association through to MI, thus ensuring the orderly disjunction of the chromosomes. There are two major problems with using MI data in the present context. First, although univalence at MI is a strong indication that there has been a primary failure to pair during pachytene, this does not necessarily mean there was no synapsis - as we shall see, univalent sex chromosomes can fold and undergo 'self-synapsis' to form hairpin structures, but the absence of chiasmata in these non-homologously synapsed structures means that the hairpins are not maintained through to MI. Second, if there is selection against cells with unsynapsed sex chromosomes prior to MI, the frequency of univalence at MI will underestimate the primary incidence of pairing failure.

The extent and stage of germ cell loss should ideally be assessed from germ cell counts before, during and after the pachytene stage. In female mice this means obtaining oocyte counts during late fetal and early postnatal life, while in males it is important to obtain the germ cell counts during puberty, since there may be secondary damage to the spermatogenic epithelium in older males due to the continual presence of degenerating cells. In the absence of such direct information it is possible to obtain some indication as to the degree of gametogenic impairment from some other criteria. In female mice, a shortened reproductive lifespan is a good indication of a reduced oocyte pool. On the other hand, adult ovarian weight, the number of eggs ovulated per cycle, and litter size, are all very little affected, even when there are major reductions in the oocyte pool. Ovarian histology can also be misleading, because the number of oocytes in growing follicles is also little affected by large reductions in the pool, and it requires serial sections to judge whether there is a deficiency of oocytes in primordial follicles. By contrast, reproductive lifespan in males is not a good guide to the degree of spermatogenic impairment, whereas testis weight, measures of sperm output, and testicular histology all provide a useful guide.

Having defined the information needed to test the main predictions of Miklos' hypothesis, we can now examine specific examples of male and female mice in which unpaired sex chromosomes are present during meiosis.

3 Unpaired sex chromosomes in male meiosis

3.1 XSxr^{a}O males
It is possible to produce XO male mice by mating XY males carrying the sex reversal factor 'Sxr' (see section 3.3) to XO females (Cattanach et al., 1971) or to females heterozygous for the large X inversion In(X)1H (Levy & Burgoyne, 1986). The Sxr factor originates from the short arm of the mouse Y (McLaren et al., 1988; Roberts et al., 1988) and confers maleness because it includes the testis determining gene *Sry* (Gubbay et al., 1990). Sxr^a is the original form of Sxr and in

XSxr^aO males it is attached to the X chromosome distal to the pseudoautosomal region (see XSxr chromosome in Fig.2).

XSxr^aO males have testes which are about half the normal size, and are typically aspermic, although a few sperm are sometimes found in testicular preparations. Any sperm that are produced have abnormal heads and the majority of spermatids and sperm are diploid (Levy & Burgoyne, 1986). Burgoyne & Baker (1984) proposed that the spermatogenic impairment was due to the presence of an unpaired sex chromosome at pachytene (Miklos' model), and suggested that diploid sperm were produced when cytodifferentiation continued despite nuclear damage and arrest just prior to MII. In agreement with the requirements of Miklos' model, we have now established from a study of pubertal XSxr^aO males that the spermatogenic deficiency arises subsequent to chromosome synapsis, there being an almost total block during the meiotic metaphases (Sutcliffe et al., 1991). In adult XSxr^aO males there is some deficiency of earlier stages, including spermatogonia (Kot & Handel, 1990), which we interpret as secondary damage resulting from continual spermatocyte arrest and degeneration.

One might think that SC analysis is of little relevance when all the pachytene cells must necessarily have an XSxr^a univalent chromosome. However, SC analysis has shown that the univalent X sometimes partially or completely self-synapses to form a 'balloon' or 'hairpin' (Chandley & Fletcher, 1980; Mahadevaiah et al., 1988) just as it does in XO females (see section 4.1). Levy & Burgoyne (1986) suggested that this non-homologous self-synapsis might protect the cells from meiotic arrest between MI and MII, thus accounting for the presence of some haploid spermatids and sperm.

Having established that the stage at which the spermatogenic failure occurs in XSxr^aO mice is consistent with Miklos' model, we then sought to establish that it is the lack of X-Y pairing, rather than the deficiency of Y-specific DNA, that is responsible for the spermatogenic block. To test this we have produced XSxr^aY*^X males (Burgoyne et al., 1992a) in which the Y*^X chromosome (a recombinant product of XY* males - see Fig.4B) provides the pseudoautosomal pairing region without any Y-specific DNA (Eicher et al., 1991). The addition of this tiny chromosome to the XSxr^aO complement had a dramatic effect: testis weights improved from a mean of 48mg in the XSxr^aO 'controls' to 113mg in the XSxr^aY*^X males, sperm counts in the epididymis improved from zero in XSxr^aO males to a mean of 1.7 million in the XSxr^aY*^X males, and a quantitative histological analysis showed that the meiotic block had been removed. Although the sperm counts were near normal and within the range expected for fertile XYSxr^a carrier males (see section 3.2), the mice were nevertheless completely sterile. This is due to the fact that the sperm produced are grossly abnormal, although, unlike XSxr^aO males, these abnormal sperm are haploid.

We can conclude from this experiment, that the meiotic block and diploid sperm seen in XSxr^aO males are due to the lack of the Y-chromosomal pseudoautosomal region which mediates X-Y pairing. This provides very strong support for Miklos' model. The sperm abnormality, on the other hand, must be due to the deficiency of Y specific DNA which encodes information needed for the normal development of the sperm head.

3.2 $X^{Y*}O$ males

In addition to the $Y*^X$ chromosome, XY* males produce a second type of recombinant, X^{Y*}, which is essentially an X chromosome attached to a Y chromosome via a shared pseudoautosomal region (see Fig.4B). $X^{Y*}O$ males are similar to XSxr[a]O males in that they have a single sex chromosome, but differ in that they almost certainly have a full complement of Y-specific genes. We would therefore expect to see severe spermatogenic impairment, but if self-synapsis of the univalent sex chromosome occurred in some spermatocytes, this would allow some potentially functional sperm to be produced. We have tested three such males; all were sterile with testes less than half the size of control testes, and all lacked sperm in the epididymis, although one did have a few sperm present in air-dried testis preparations. Pachytene SC analysis revealed the expected univalent X^{Y*}, and this never achieved full self-synapsis (Fig.1). Eicher & Washburn (1986)

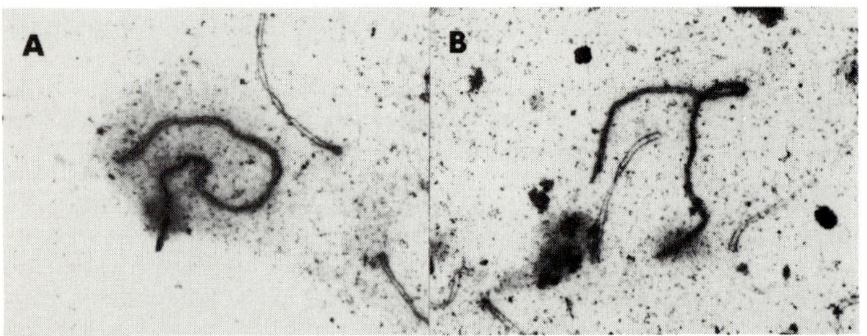

Fig. 1. Pachytene synaptonemal complexes from an $X^{Y*}O$ male. From 32 cells analysed, 29 of the X^{Y*} axes showed no synapsis (A), and 3 showed unconvincing partial self-synapsis (B).

on the other hand report that some $X^{Y*}O$ males are fertile, confirming that these males (unlike XSxr[a]O males) are able to produce functional sperm. It would be very interesting to see if the rare fertile $X^{Y*}O$ males have a much higher level of X^{Y*} self-synapsis than we have observed in our sterile examples.

3.3 XYSxr[a] males

The XYSxr[a] males used in standard Sxr[a] stock matings carry Sxr[a] attached distally to the pseudoautosomal region of the Y chromosome; the Y chromosome originates from the RIII inbred strain. These males transmit Sxr[a] on the Y to XYSxr[a] carrier sons and on the X to sterile XXSxr[a] sons, Sxr[a] having been transferred to the X by crossing over in the pseudoautosomal region (Fig.2B). In 1982, it was reported that T(X;16)16HXSxr[a] mice could develop as females due to the inactivation of Sxr[a] on the X chromosome (Mclaren & Monk, 1982; Cattanach et al., 1982) and this enables the production of XSxr[a]Y males which carry Sxr[a] on the X chromosome (section 3.3), and also allows the introduction of a different strain of Y chromosome.

XYSxr[a] males with an RIII Y have been the most extensively studied, and these have usually been fertile males, Sxr[a] carrier status having been established from breeding tests. Nevertheless, these fertile males bear the hallmarks of spermatogenic impairment, in that they have reduced testis weights (Lyon et al., 1981) and reduced numbers of sperm in the caput epididymis (Mahadevaiah et al., 1988). Severely affected males may be sterile, and sterility is in fact the norm in XYSxr[a] males in which the RIII Y chromosome has been replaced by a 101 strain Y (Tease & Cattanach, 1989). Histologically, some tubules in XYSxr[a] testes show the deficiency of spermatids and sperm which is characteristic of XSxr[a]O testes, and in these tubules (as in XSxr[a]O testes) the spermatids are predominantly diploid (Hannappel & Drews, 1979; Hannappel et al., 1980).

As early as 1978, Winsor et al. reported that XYSxr[a] males had elevated levels of sex chromosome univalence at MI. It was subsequently established from SC analysis that this univalence is largely due to a primary failure to pair (Chandley & Fletcher, 1980; Evans et al., 1980; Mahadevaiah et al., 1988; Tease and Cattanach, 1989); probably because the presence of Sxr[a] interferes with synaptic initiation (Tease & Cattanach, 1989). Burgoyne & Baker (1984) interpreted the spermatogenic impairment and presence of diploid spermatids in XYSxr[a] testes as being consequences of this sex chromosome univalence. Significantly, the XYSxr[a] males carrying the 101 Y chromosome, which have particularly small testes and are usually sterile, have particularly high levels of sex chromosome univalence (Tease & Cattanach, 1989).

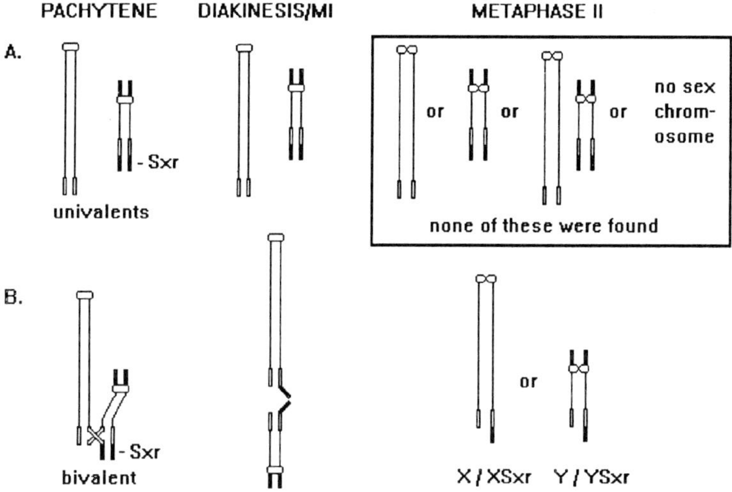

Fig.2. Meiosis in XYSxr[a] males. Although many cells at pachytene have univalent sex chromosomes (A), many of which survive through to diakinesis/MI, all metaphase II cells have Sxr present on only one X or Y chromatid, showing that they are derived from pachytene cells in which the X and Y paired and recombined (B).

In conjunction with our quantitative analysis of the spermatogenic block in XSxr[a]O males, we also analysed testes from pubertal XYSxr[a] males, and this established that the spermatogenic deficency in XYSxr[a] testes first appears during late pachytene (Sutcliffe et al., 1991).

Although this is consistent with the requirements of Miklos' model, it is an earlier onset than that seen in XSxr^aO males. We have argued that this earlier onset is a consequence of having two unpaired sex chromosomes rather than one in the affected cells, so that in terms of Miklos' model there are twice as many 'unsaturated pairing sites'. Although in these pubertal males, much of the spermatocyte loss occurs in late pachytene, in mature XYSxr^a males spermatocyte losses seem to be concentrated during the meiotic metaphases (Hannappel & Drews, 1979; Kot & Handel, 1990).

In expert hands, Sxr^a can be visualised in G-banded meiotic cells; analysis of the disposition of Sxr^a on the X and Y chromosomes at diakinesis (i.e. just prior to MI) and MII has fully substantiated the view that it is the spermatocytes in which the sex chromosomes failed to pair, that are eliminated. Thus, Evans et al., (1982) reported that many, if not all, diakinesis spermatocytes with X-Y univalence carried Sxr^a on both Y chromatids, while those spermatocytes with an X-Y bivalent always showed evidence of transfer of one copy of Sxr^a to an X chromatid. Since MII cells always had Sxr^a on only one chromatid of the X or Y, they concluded that X-Y univalence 'appears to act as a cell lethal'. This conclusion is further supported by the fact that XYSxr^a males do not produce large numbers of aneuploid XXY and XO offspring (univalents would be expected to assort randomly at MI leading to the production of XY and O gametes – cf. XY females, section 4.3). These findings are summarised in Fig.2.

We can summarise the situation in XYSxr^a males as follows: (a) The presence of Sxr^a inhibits X-Y synapsis, so that in many spermatocytes the X and Y fail to pair. (b) Spermatocytes in which the X and Y fail to pair are usually eliminated in late pachytene or between MI and MII. (c) Diploid spermatids are formed from cells which complete the first meiotic division, and then continue to cytodifferentiate despite nuclear arrest prior to MII. The importance of these findings in the context of Miklos' hypothesis, is that they show the presence of two pseudoautosomal regions is not sufficient in itself to prevent spermatocyte elimination – these regions must also pair.

3.4 XSxr^aY males

Cattanach et al. (1990) have described some unusual features in the transmission and meiotic behaviour of Sxr^a in XSxr^aY males carrying a 101 strain Y chromosome. While XYSxr^a males produce equal numbers of Sxr^a-recombinant and non-recombinant progeny (due to the presence of the 'obligatory crossover' in the X-Y bivalent), these XSxr^aY males were significantly deficient in recombinant offspring. Meiotic analysis revealed that just under half the cells at diakinesis had formed an X-Y bivalent, but 62% of the bivalents showed an association between the Y-short arm and Sxr^a located at the distal end of the X, rather than between the two pseudoautosomal regions. Furthermore, in contrast to XYSxr^a males, MII cells were found in which both X chromatids carried Sxr^a or both Y chromatids lacked Sxr^a, showing that the expected ('obligatory') recombination in the pseudoautosomal region had not taken place. These non-recombinant MII cells provide the explanation for the deficiency of recombinant offspring.

How are we to interpret these findings? We are sure Cattanach et al. (1990) are right to conclude that the association between the Y

short arm and Sxrᵃ at diakinesis is maintained by a chiasma occurring
between these regions of homology. However, they go on to conclude
that it is the presence of this chiasma which has enabled the cells to
survive, despite the absence of crossing over (and, they assume, a
lack of pairing) between the X and Y pseudoautosomal regions. Thus,
they view chiasma formation, rather than X-Y pairing *per se*, as
essential for spermatocyte survival; X-Y pairing is only required in
that it is a prerequisite for chiasma formation. This is certainly at
variance with Miklos' model, since it was originally constructed on
the basis of data from male *Drosophila*, where there is synapsis but no
crossing over. The flaw in Cattanach et al.'s logic, is that they
take the lack of crossing over in the pseudoautosomal region as
evidence that the regions have not paired. However, if there was
pseudoautosomal pairing *and* Sxrᵃ-Y short arm pairing at pachytene, a
crossover between Sxrᵃ and the Y short arm would satisfy the
requirement for one crossover per bivalent (the 'obligatory' chiasma),
while the pseudoautosomal pairing would satisfy the pairing
requirements (Fig.3). With no chiasma in the pseudoautosomal region,

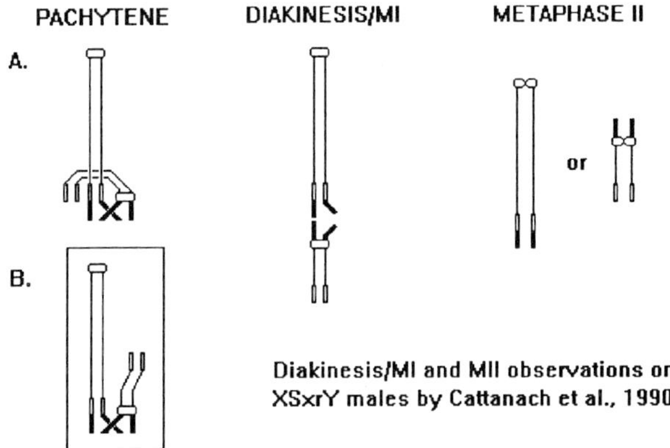

Fig.3. Meiosis in XSxrᵃY males. In addition to the expected form
of sex bivalent, 62% of bivalents at diakinesis/MI showed a Y
short arm to X long arm association, indicating the presence of
a chiasma between the Y short arm and Sxr. The presence of MII
cells which were non-recombinant for Sxr demonstrates that a
chiasma in the pseudoautosomal region is not a prerequisite for
spermatocyte survival. Cattanach et al. took the lack of pseudoautosomal exchange as evidence for a lack of pseudoautosomal
pairing (B), but we favour a double pairing configuration with a
single crossover (A).

the association in this region would not be maintained through to
Diakinesis/MI, in agreement with observations of Cattanach et al. By
showing that a chiasma in the pseudoautosomal region is not essential
for spermatocyte survival, we would argue that these observations on
XSxrᵃY mice provide support for Miklos' view that the meiotic quality
control is directly monitoring pairing.

3.5 XY* and $X^{Y*}Y*^{X}$ males

XY* males were first mentioned in a review by Eicher (1982) as having a Y chromosomal rearrangement which generated large and small recombinant chromosomes by crossing over. An elegant paper giving details of the rearrangement has recently been published (Eicher et al., 1991); Y* and its recombinant products are illustrated in Fig.4. $X^{Y*}Y*^{X}$ males carry the two recombinant chromosomes, and as such have the same complement of sex chromosomal DNA as XY* males. However, XY* males are usually fertile (although, their testes are reduced in size – Hale et al., 1991), while the 5 $X^{Y*}Y*^{X}$ males we have tested, together with three tested by Hunt & LeMaire (1992a), have all been sterile. Why should this be? Once again, pachytene SC analysis has proved very instructive. The XY* males, like XYSxr[a] males, have increased levels of sex chromosome univalence which we believe accounts for the reduced testis size, but 44% of early/mid pachytene nuclei in the study of Hale et al. (1991) had formed bivalents. By contrast, we found no bivalents among the 39 $X^{Y*}Y*^{X}$ cells we have examined at pachytene (Burgoyne et al., 1992a). Thus, the difference in fertility correlates with a difference in the extent of sex chromosome pairing.

The meiotic data for XY* males also bear out conclusions drawn from the meiotic data for XYSxr[a] males. At diakinesis/MI the chiasmate transfer of most of one Y* chromatid onto one of the X chromatids was clearly visible in nearly all the XY* bivalents, while the 80% of cells at this stage which had X and Y* univalents showed no sign of this transfer. However, at MII all the cells had recombinant sex chromosomes. Thus, once again there is clear evidence that the cells with sex chromosome univalents at MI do not contribute products to MII (Hale et al., 1991).

Hale et al. go on to discuss the question as to whether the cells with univalents are being eliminated because the chromosomes failed to pair, or because they failed to recombine, and (like Cattanach et al., 1990 – see section 3.4) they conclude that it is recombination which is essential for spermatocyte survival. The main points of their argument are: (a) There is an increase in the frequency of univalents between pachytene (64%) and diakinesis/MI (80%) which must be due to the dissociation of bivalents. (b) All the bivalents, but none of the univalents at diakinesis/MI have undergone prior recombination, so the dissociating bivalents must have failed to recombine. (c) All MII cells show evidence of prior recombination, so those diakinesis/MI cells in which the univalent X and Y* chromosomes had previously *paired* but had not *recombined,* must have been eliminated.

This is a very cogent argument, but some of the details regarding the nature of the bivalent associations at pachytene, lead us to an alternative explanation (Fig.4). Crucial to our explanation, is the observation by Hale et al. (1991), that only 4.5% of the early/mid pachytene cells with X-Y* bivalents showed the 'staggered' configuration predicted from the deduced position and orientation of the Y* pseudoautosomal region. The remaining 95.5% of bivalents showed a 'parasynapsed' sex bivalent similar in appearance to that in normal males. Of the 191 parasynapsed bivalents analysed, four were said to have the X and Y* centromeres involved in the paired segment. In most of the remainder, the X centromere could be distinguished at the unpaired end of the X, and in four of these the Y* centromere could be

seen within the paired segment. However, in the majority of the
parasynapsed bivalents the Y* centromere either could not be resolved,
or, was seen to be *at the unpaired end of the Y* chromosome* (Fig.4).

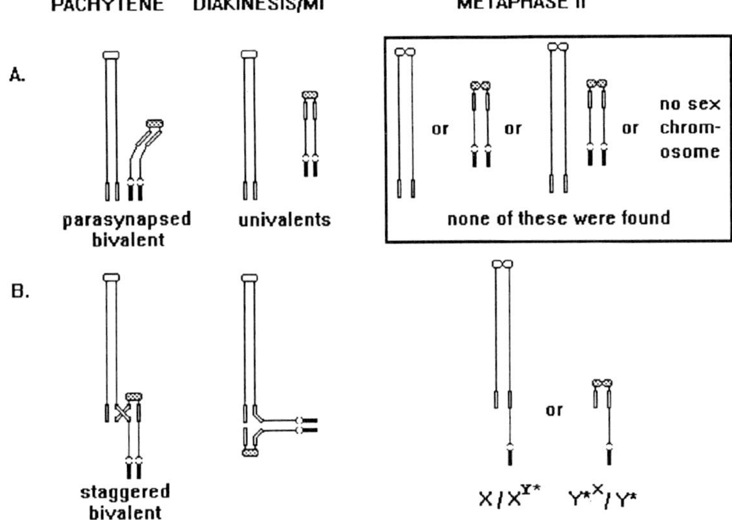

Fig.4. Meiosis in XY* males. The Y* chromosome is in essence a
  normal Y which has been hijacked by a non-Y centromere attached
  to the Y 'pseudoautosomal' region. The Y centromere is presumed
  to be present but inactive. Pachytene SC analysis suggests
  that the majority of bivalents undergo non-homologous 'para-
  synapsis' of the form illustrated in A. Due to the absence of a
  chiasma these will become univalent at diakinesis/MI, adding to
  those univalents which were already present at pachytene. All
  cells with univalents are eliminated prior to MII; this is
  consistent with Miklos' model because even in the parasynapsed
  configuration the Y* pseudoautosomal region remains unpaired.
  A minority of pachytene cells have staggered bivalents (B) and
  these generate diakinesis/MI cells with a clear chiasma. Although
  only 20% of diakinesis/MI cells have this chiasma, all MII cells
  are recombinant.

In the latter case, the pairing taking place would be non-homologous
so that there would be no chiasma to maintain the association through
to Diakinesis/MI, and because the pseudoautosomal region located at
the centromeric end of the Y* chromosome had not been paired during
pachytene, the cells would be eliminated prior to MII.

3.6 XY interspecific hybrid males

There are other instances where there is an increase in sex chromosome
univalence at MI, and this is particularly true of interspecific mouse
hybrids. An illuminating study of male hybrids between *Mus musculus*
and *Mus spretus* has recently been published by Matsuda et al. (1991).
The F$_1$ males were severely oligospermic and sterile due to an almost
total arrest during the meiotic metaphases, and this was associated

with 95% X-Y univalence at MI. Mating males from either of the
parental stocks to F_1 females, produced male offspring ranging from
sterile to fully fertile, and there was a strong correlation between
the extent of the X-Y univalence and the degree of spermatogenic
impairment, as is predicted by Miklos' model. Moreover, by using a
linked gene marker, it was found that the level of X-Y pairing was
high when the two pseudoautosomal regions came from the same species,
and low when one was *musculus* and one was *spretus*.

3.7 XYY males

Reports of sporadic cases of XYY mice have indicated that they are
almost invariably sterile (Cattanach & Pollard, 1969; Rathenberg &
Müller, 1973; Evans et al., 1978; Das & Kar, 1981; Tease, 1990),
although the sample is biassed in that sterility was usually the
reason for obtaining a karyotype. All had clear signs of spermatogenic impairment, but the reduction in testis weight was very
variable, and paradoxically the only fertile male had particularly
small testes (19 & 20mg). We suspect that this may have been an
undisclosed XO/XY/XYY mosaic, since they can have very small testes,
yet retain fertility. Miklos (1974) suggested that the spermatogenic
impairment in XYY mice might be a consequence of one of the sex
chromosomes failing to pair, and when Burgoyne & Biddle (1980) used
the available meiotic data to test various models of spermatocyte
loss, they found that the model in which all MI spermatocytes with a
univalent sex chromosome are eliminated before MII, gave the best fit
to the data. Under this model, spermatocytes which satisfy the
pairing needs of all three sex chromosomes by forming XYY trivalents,
are able to generate viable sperm, so fertility is not precluded.

We have recently been able to assess the fertility of a large
series of XYY males (Burgoyne et al., 1992b) produced by mating XXY
females (section 4.4) to XY males, or to XY-del males. The maternally
derived Y is normal except for a small 11kb deletion which has removed
the testis determining gene *Sry*; the paternally derived Y-del is about
half normal size due to a large deletion in the long arm, enabling it
to be distinguished in meiotic preparations. Most of the information
collected has been for XYY-del males. Of 33 such males tested, 5
showed immediate fertility and half their offspring had inherited two
paternal sex chromosomes, demonstrating that 'XYY' spermatocytes can
produce functional products. We believe these spermatocytes must have
satisfied pairing requirements by forming 'XYY' trivalents, as
suggested by Burgoyne & Biddle (1980), and have confirmed by SC
analysis that trivalents are formed. However, after two or three
litters, these males either stopped breeding or ceased producing
aneuploid offspring. Meanwhile, some of the 28 initially sterile
males began to breed, and by 9 months of age 90% of the males had
become fertile; none of the offspring from these males were aneuploid.
Burgoyne et al. (1992b) interpreted this later phase of breeding as
being due to an age-related delay in the process which eliminates
cells with unpaired sex chromosomes. It was envisaged that this delay
allowed spermatocytes with unpaired sex chromosomes to complete MII,
and then the spermatids carrying a chromosome which had been unpaired
(i.e. the aneuploid spermatids) were eliminated. However, in
collaboration with Dr. E.P. Evans, we have now carried out an MII

analysis in some of these males, and this has led to the conclusion that this later phase of fertility is supported by small numbers of XY-del and XY,*Sry⁻* cells. It is not clear whether this represents progressive Y loss with age or selection over time in favour of 'XY' cells present at a very low level from the outset. What is clear, is that there is strong selection against XYY cells between MI and MII in these older fertile males. In a previous study of XO/XY/XYY mosaics we observed a similar strong selection against XYY cells in late pachytene (Palmer et al., 1990).

3.8 XYY*^{X} males

Hunt and Eicher (1991) have used XY*^{X} females (see section 4.6) to generate XYY*^{X} males. These males differ from XYY males, in that they are fertile from the outset, and regularly produce sex-chromosomally aneuploid offspring. If we make the assumption that this improved fertility is a consequence of having the tiny Y*^{X} chromosome (Fig.4) in place of a normal Y, then the fact that the Y*^{X} chromosome lacks any Y-specific DNA immediately suggests that the improved fertility is due to the restoration of the correct dosage of Y-linked genes. However, in the context of Miklos' model, it is necessary to consider whether the Y*^{X} might be more efficient at pairing so that more trivalents are being formed. Hunt & Eicher reject this possibility because the percentage of trivalents they observed at diakinesis/MI (19%) fell within the range for the previously reported sporadic males. In fact, only two of these published cases had trivalent frequencies above 19%, and judging from their testis weights and sperm counts, these two males would have proved fertile if they had ever been mated (they were unusual in that infertility was not the reason for obtaining a karyotype). Furthermore, Hunt & Eicher's own data do suggest that the Y*^{X} chromosome is a more efficient pairing partner than a normal Y, because it was paired in 80% of XYY*^{X} cells at diakinesis/MI as compared to only 45% for the normal Y. However, all these comparisons at diakinesis/MI only give an indirect indication of the incidence of trivalent formation. In males with three sex chromosomes it is particularly important to obtain SC data, because a single chiasma in a trivalent may satisfy the 'obligatory chiasma' requirement, and in the absence of a second chiasma many of the trivalents may resolve into a bivalent + univalent by MI; indeed, in one of the XYY males studied by Evans et al. (1978) the frequency of trivalents fell from 24% in early diakinesis to only 6% by MI.

3.9 Unpaired sex chromosomes in males – conclusions

The information from these examples of males with unpaired sex chromosomes leads to the firm conclusion that a failure of sex chromosomes to pair during pachytene is causally related to spermatogenic failure. While some have argued that it is chiasma formation rather than pairing *per se*, which is essential for spermatocyte survival, we have shown here that the data on which their arguments are based are in fact compatible with Miklos' model; that is to say that the meiotic surveillance system is monitoring chromosome synapsis ('pairing') rather than chiasma formation. As we shall see, this is further supported by meiotic data obtained from females with unpaired sex chromosomes.

4 Unpaired sex chromosomes in female meiosis

4.1 XO females

The first indication that XO mice might have a reduced number of oocytes came from evidence that they have a shortened reproductive lifespan (Lyon & Hawker, 1973). Burgoyne & Baker (1981; 1985) subsequently showed that XO mice have less than half the normal number of oocytes, and that this oocyte deficiency is a consequence of elevated levels of oocyte atresia during late pachytene. This juxtaposition of sex chromosome univalence with a postsynaptic gametogenic failure, led Burgoyne and Baker (1985) to invoke Miklos' hypothesis as a basis for explaining the oocyte deficiency in XO mice. However, this raised the question as to why nearly half the oocytes in XO mice survive despite having a single X chromosome.

In 1986, Speed published the results of a detailed SC analysis for XO pachytene oocytes from the 17th-20th day *post coitum* (d*pc*), and this revealed that the single X chromosome was able to fold to form a fully synapsed 'hairpin' structure; he suggested that this self-synapsis might satisfy 'pairing requirements' and protect the cells from elimination. This is another way of saying that the excess pachytene oocyte loss in XO mice is due to the elimination of oocytes in which the X failed to achieve synapsis. How can the SC data be used to test this hypothesis? If there is selection against cells with no X chromosome synapsis, the proportion of cells with a self-synapsed X should increase as pachytene proceeds. This is borne out by Speed's data, there being a marked increase in the frequency of cells showing self-synapsed X axes as you move from the 17th day (mostly early pachytene) to the 19th day (mostly mid/late pachytene). On the 20th day there is a small decrease in the frequency of self-synapsed X chromosomes, but this can be attributed to the desynapsis which occurs at the pachytene to diplotene transition, since the majority of the unsynapsed X axes lacked the thickening associated with asynapsis.

The importance of these studies on XO mice is that they provided the first direct evidence that unpaired sex chromosomes might be causing gametogenic impairment in females as well as males. In the context of Miklos' model, the XO data are also important in that they introduce the concept that non-homologous synapsis can 'saturate pairing sites'. Since this non-homologous synapsis rarely if ever leads to crossing over, as judged from the lack of ring X chromosomes at MI (Dr E.P. Evans, personal communication), this supports our view that it is pairing *per se*, rather than recombination, which is being monitored by the 'meiotic quality control'.

4.2 Females heterozygous for the X inversion In(X)1H

Burgoyne & Baker (1985) used females heterozygous for the large paracentric X inversion In(X)1H to generate the females for their study of the etiology of oocyte deficiency in XO mice, and they found that these In(X)/X females also had an increased level of pachytene oocyte loss as compared to the XX controls; after birth the oocyte pool was found to be reduced by about 35%. They suggested that the

inversion might be causing this increased oocyte loss by interfering
with X chromosome synapsis.

Tease & Fisher (1986) have carried out an SC analysis in In(X)1H
heterozygous fetuses from the 16th to 20th d*pc*, and have made a number
of intriguing observations. A surprising finding was that the
majority of pachytene cells showed no sign of the expected inversion
loop; loops were most frequent on the 16th d*pc* (29%) falling to 8% on
the 20th d*pc*. Since 31% of late zygotene cells also showed no signs
of loop formation, it was concluded that in many cases the two X axes
must be synapsing non-homologously from the outset. Of importance in
the present context, was the observation that half of the loops
present on the 16th day showed incomplete synapsis. In a further
paper Tease & Fisher (1988) carried out a C-banded analysis of
pachytene spreads from In(X)/X fetuses, and found that nearly half of
the oocytes with no inversion loop showed reverse pairing of the X
chromosomes (i.e. the two centromeres were at opposite ends of the
bivalent, allowing homologous pairing of the inverted segment).

Once again, the SC findings lead to a specific prediction under
Miklos' model: namely, that the increased pachytene atresia seen in
In(X)/X ovaries is due to the elimination of the pachytene cells with
incompletely paired loops. If this is so, then the frequency of loops
should decrease as pachytene proceeds, in particular there should be a
reduction in the frequency of loops showing partial pairing. This is
borne out by the data: at early pachytene the frequency of loops was
37%, in mid pachytene - 24%, and late pachytene - 6%; moreover, it was
the loops showing partial pairing which were specifically depleted.
[It should be said that Tease & Fisher (1988) interpreted the
reduction in loop frequency as being due to synaptic adjustment, with
the loops converting into bivalents with non-homologous pairing across
the inversion. This interpretation is not supported by their data,
since the ratio of bivalents with non-homologous pairing to those with
reverse pairing, remained unchanged between the 16th and 18th d*pc*,
although the frequency of loops was halved during this period.]

4.3 XY females
The first opportunity to systematically study XY female mice came when
Eicher et al. (1982) discovered an inherited form of incompletely
penetrant XY sex reversal. This sex reversal occurred when a Y chromosome derived from some wild caught mice from the Val Poschiavo in
Switzerland was backcrossed into the C57BL/6 inbred strain; the sex
reversal is due to some incompatibility between this Y^{POS} chromosome
and the C57BL/6 background (Eicher & Washburn, 1983;1986; Burgoyne &
Palmer, 1991; Palmer & Burgoyne, 1991). These XY^{POS} females proved to
be almost invariably sterile, but the fact that some C57BL/6 XY^{POS}
individuals are male, allows the production of further XY females.

From the start it was apparent that the sterility was associated
with severe oocyte deficiency, although some oocytes were found to be
present in prepubertal females (Eicher et al., 1992). Taketo-Hosotani
et al. (1989) have traced the origins of the oocyte deficiency in
C57BL/6 XY females (carrying a different, but related, feral Y
chromosome) and, just like XO females, the deficiency is due to excess
atresia during late pachytene. Recently Hunt & LeMaire (1992b) have
claimed that the pattern of loss in these XY females differs from that

in XO females, in that there is a second wave of germ cell loss around
the time of puberty. However, we interpret this second wave of
failure as representing the atresia that occurs in XO (and XX) mice
when their pool of primordial follicles is exhausted, and regular
cyclicity ceases. The severe oocyte deficiency in these XY females
simply means that the initial establishment of the population of
growing follicles often exhausts the pool of primordial follicles, so
that they move straight into the phase of irregular cyles (Taketo-
Hosotani et al., 1989) and atresia.

An indication that the oocyte loss in these XY females might be
associated with the presence of unpaired sex chromosomes first came
from an abstract (Kundu et al., 1983) describing the results of an SC
analysis. In 67% of the pachytene cells analysed the X and Y were
present as univalents. More recently Hunt & LeMaire (1992b) have
reported that 80% of MI cells have X and Y univalents.

In 1990, Lovell-Badge & Robertson described some XY female mice
resulting from a mutation in the Y-chromosomal testis determining
gene. It was subsequently established that 11kb of Y-chromosomal DNA,
including the testis determining gene (*Sry*), is deleted in these
females (Gubbay et al., 1990;1992). We have assessed the breeding
performance of these XY females (Mahadevaiah et al., 1993) and
although the majority are sterile, they do appear to be more often
fertile than C57BL/6 XY females. This may simply be a consequence of
having an outbred rather than an inbred genetic background. At 5 days
after birth these *Sry*-negative XY females have ca. 29% of the number
of oocytes found in XX litter mates (S.J. Palmer & P.S. Burgoyne,
unpublished); we presume that this oocyte deficiency is once again due
to increased atresia during pachytene.

We have also carried out a detailed SC analysis, which revealed
that 89% of pachytene cells had the X and Y present as univalents.
However, just as in XO females, univalence does not necessarily mean
no synapsis – the X and the Y both have the capacity to fold back and
self-synapse. Following the approach we took with the XO data, we can
ask whether the frequency of cells showing some form of sex chromosome
synapsis increases during pachytene, as would be the case if those
showing no sex chromosome synapsis are being eliminated. This is
indeed the case, the incidence of cells with sex chromosome synapsis
rising from 10% in early pachytene, to 55% in mid pachytene, and to
71% in late pachytene (Mahadevaiah et al., 1993).

4.4 XXY females
When the X and Y chromosomes in XY females are present as univalents
at MI (Hunt & LeMaire, 1992b) it means that there is no mechanism to
ensure the regular segregation of the X from the Y. It is therefore
no surprise that the few *Sry*-negative XY females that are fertile
produce a number of sex-chromosomally aneuploid offspring, including
some *Sry*-negative XXY females (Mahadevaiah et al., 1993). All XXY
females we have tested so far have been fertile, but at 5 days after
birth they have as few oocytes as *Sry*-negative XY females (Palmer &
Burgoyne, unpublished).

SC analysis has shown that the two X axes regularly synapse, while
the Y almost always remains as a univalent (we observed only one XXY
trivalent among 69 cells analysed). As in XY females, the univalent Y

sometimes self-synapses to form a hairpin. The prediction from
Miklos' model is that the cells in which the Y fails to self synapse
will be preferentially eliminated, and this is supported by our
finding that the proportion of cells with an unsynapsed Y decreases
markedly during pachytene (Mahadevaiah et al., 1993). We attribute
the improved fertility of XXY females over XY females to the better
oocyte quality associated with two X chromosomes (Burgoyne & Biggers,
1976) and to the fact that XY females produce a high proportion of
lethal (YO, YY) or developmentally compromised (X^PO, X^PY) zygotes.

4.5 XYY females

Mahadevaiah et al. (1992) also produced some *Sry*-negative XYY females,
and obtained only one offspring from five matings, indicating that XYY
females are more severely affected than XY females. However, this is
not borne out by oocyte counts, which at 5 days after birth are
similar to those of *Sry*-negative XY females (Palmer & Burgoyne, unpublished). The poorer fertility may be due to the increased likelihood
of producing lethal (YO, YY, YYY) zygotes as compared to XY females.

   SC analysis revealed that despite the identity of the two Y
chromosomes, both Y chromosomes remained unpaired in the majority of
cells. As in XO, XY and XXY females, some of the univalent sex
chromosomes were self-synapsed. The overall frequency of cells
showing at least one unsynapsed sex chromosome, was equivalent to the
frequency in XY females.

4.6 XY*^{X} females

XY*^{X} females are found among the progeny of males carrying the Y*
rearrangement (Eicher et al., 1991). The Y*^{X} chromosome consists of
most of the 'pseudoautosomal region' (which mediates pairing between
the X and Y of normal males) attached to a centromere which is known
to be derived from either an X chromosome or an autosome. The mode of
generation of the Y*^{X} chromosome (Fig.4) means that it cannot carry
any Y-specific DNA. The feature that distinguishes XY*^{X} females from
XY females is their fertility - even on a C57BL/6 inbred background 8
out of 9 females tested produced some offspring, and in terms of
reproductive lifespan they appeared to be roughly equivalent to XO
females on the same genetic background (Hunt, 1991).

   Unfortunately, there is as yet no pachytene SC data available for
these females, so we can only make inferences as to the level of sex
chromosome pairing from MI data. Significantly, Hunt and LeMaire
(1992b) found a much higher level of 'X-Y' association at MI in XY*^{X}
oocytes (63%) than they found in XY^{POS} oocytes (20%). If this is a
reflection of an increased level of pairing at pachytene, then under
Miklos' model this will reduce pachytene oocyte loss and improve
fertility. Support for the view that the Y*^{X} chromosome is a more
efficient pairing partner than a complete Y chromosome, comes from the
study by Hunt & Eicher (1991) of XYY*^{X} males (section 3.8), in which
they found that the Y*^{X} chromosome out-competed the normal Y as a
pairing partner. [Hunt & LeMaire, on the other hand, interpret the
improved fertility of XY*^{X} females as compared to XY^{POS} females as
being due to the absence of Y-specific DNA, implying that the
inappropriate expression of Y genes may be a cause of oocyte failure.]

4.7 $XY*^XY*^X$ and $XXY*^XY*^X$ females

Hunt and LeMaire (1992b) also reported MI data for females carrying two $Y*^X$ chromosomes. No breeding information is given in the paper, but it is worth discussing the MI data here, because it illustrates one of the pitfalls mentioned earlier concerning the use of MI data as a measure of sex chromosome pairing. In the $XY*^XY*^X$ female analysed, the majority of cells (62%) had all three sex chromosomes present as univalents, while 23% had an $XY*^X$ bivalent with a $Y*^X$ univalent and 15% had a $Y*^XY*^X$ bivalent and an X univalent. In the $XXY*^XY*^X$ female analysed, all cells had an XX bivalent, but 72% of these cells also had a $Y*^XY*^X$ bivalent. The authors conclude that the presence of a univalent X chromosome interferes with $Y*^X$ pairing. However, if we accept that there is strong selection during pachytene against cells with unsynapsed sex chromosomes, the interpretation of the MI data is very different. In the $XY*^XY*^X$ females it is likely that virtually all pachytene cells had at least one univalent sex chromosome (as is the case in XYY females) and will have been selected against during pachytene. At MI we are seeing those that survived the elimination process, perhaps because the univalent sex chromosomes achieved self-synapsis. In the $XXY*^XY*^X$ females it is likely that virtually all pachytene cells will have XX bivalents (as in XXY females), while the two $Y*^X$ chromosomes will be paired in some cells but not in others. In the absence of self-synapsis, the cells with two $Y*^X$ univalents should be strongly selected against, while those with $Y*^XY*^X$ bivalents should progress unscathed, resulting in a marked preponderance of cells with $Y*^XY*^X$ bivalents at MI. Clearly, pachytene SC data are needed before any firm conclusions can be drawn.

4.8 Unpaired sex chromosomes in females – conclusions

This review of the information available for a range of examples of females with unpaired sex chromosomes, provides considerable support for Burgoyne & Baker's (1984) suggestion that Miklos' concept of a 'meiotic quality control' is applicable to female meiosis. It also provides support for Speed's (1986) suggestion that non-homologous synapsis can satisfy pairing requirements. Nevertheless, the female meiotic data pose some interesting new questions. Why, for example, should X-Y pairing be so much more efficient during male than female meiosis (Hunt & LeMaire, 1992b). Of direct relevance to Miklos' model is the question as to whether the hypothetical 'pairing sites' are limited to the pairing and exchange ('pseudoautosomal') region of the X and Y, or might they be present over the length of the X (and Y?), with those outside the pseudoautosomal region becoming inoperative when the X (and Y?) is 'inactivated' during male (but not female) meiosis? With the increasing number of variant mouse sex chromosomes becoming available, and the prospect of being able to produce tailor-made deletions using telomere-mediated deletion techniques (Farr et al., 1991), we may soon be in a position to address these questions.

5 General conclusions

In the light of these examples of male and female mice with unpaired sex chromosomes, there can be little doubt that the presence of unpaired sex chromosomes is causally related to gametogenic failure. Although we have restricted our review to the mouse, we are certain this conclusion is equally applicable to man. For example, it comes as no surprise to us that two men with sex chromosome pairing failure due to loss of the X chromosomal 'pseudoautosomal' region, had total spermatogenic arrest after pachytene (Gabriel-Robez, et al., 1990). It is particularly important to keep in mind the consequences of sex chromosome pairing failure, when Y-chromosomal rearrangements are being used to map putative Y genes involved in spermatogenesis (see Vogt et al., this volume). For example, ring Y chromosomes will inevitably lead to spermatogenic failure, because they prevent X-Y pairing; only if the block is clearly premeiotic should one start to consider that the breakpoint involved in forming the ring may have disrupted a Y-chromosomal gene.

By restricting ourselves to sex chromosome pairing failure we have avoided the more complex questions raised by the association between balanced autosomal translocations and gametogenic impairment. Mittwoch & Mahadevaiah (1992) have carried out a thorough pachytene analysis of male and female mice with a range of autosomal rearrangements, and found that pachytene cells in which there were pairing anomalies associated with the rearranged chromosomes, were more likely to have asynapsis of other chromosome axes. They therefore concluded that 'pairing failure at meiosis is primarily a symptom, rather than a cause, of gametogenic arrest', although they suggested that 'once pairing failure is established, it could secondarily increase the probability of gametogenic failure.' Our analysis of the relationship between unpaired sex chromosomes and gametogenic failure leads us to a different emphasis. We accept that balanced autosomal rearrangements may, through position effects or through disruption of genes at translocation breakpoints, cause some impairment of the gametogenic process. This in turn may lead to increased levels of asynapsis. However, since our analysis of the consequences of sex chromosome univalence convinces us that chromosome asynapsis alone can cause gametogenic failure, we believe that chromosomal asynapsis is the major factor responsible for the increased post-synaptic germ cell loss in most carriers of balanced autosomal translocations.

6 References

Burgoyne, P.S. (1982) Genetic Homology and crossing over in the X and Y chromosomes of mammals. **Hum. Genet.**, 61, 85-90.
Burgoyne, P.S. and Baker, T.G. (1981) Oocyte depletion in XO mice and their XX sibs from 12 to 200 days *post partum*. **J. Reprod. Fert.**, 61, 207-212.
Burgoyne, P.S. and Baker, T.G. (1984) Meiotic pairing and gametogenic failure, in **Controlling Events in Meiosis** (eds C.W Evans and H.G. Dickinson), The Company of Biologists, Cambridge, pp. 349-362.
Burgoyne, P.S. and Baker, T.G. (1985) Perinatal oocyte loss in XO mice and its implications for the aetiology of gonadal dysgenesis in XO women. **J. Reprod. Fert.**, 75, 633-645.
Burgoyne, P.S. and Biddle, F.G. (1980) Spermatocyte loss in XYY mice:

addendum to Cytogenet. Cell Genet. *23*: 84-89. **Cytogenet. Cell Genet.**, 28, 143-144.

Burgoyne, P.S. and Biggers, J.D. (1976) The consequences of X-dosage deficiency in the germ line: impaired development *in vitro* of preimplanatation embryos from XO mice. **Devl Biol.**, 51, 109-117.

Burgoyne, P.S. et al. (1992a) Fertility in mice requires X-Y pairing and a Y-chromosomal 'spermiogenesis gene' mapping to the long arm. **Cell**, 71, 1-20.

Burgoyne, P.S. and Palmer, S.J. (1991) The genetics of XY sex reversal in the mouse and other mammals. **Semin. Dev. Biol.**, 2, 277-284.

Burgoyne, P.S., Sutcliffe, M.J. and Mahadevaiah, S.K. (1992b) The role of unpaired sex chromosomes in spermatogenic failure. **Andrologia**, 24, 17-20.

Cattanach, B.M. et al. (1982) Male, female and intersex development in mice of identical chromosome constitution. **Nature**, 300, 445-446.

Cattanach, B.M. and Pollard, C.E. (1969) An XYY sex-chromosome constitution in the mouse. **Cytogenetics**, 8, 80-86.

Cattanach, B.M., Pollard, C.E. and Hawkes, S.G. (1971) Sex-reversed mice: XX and XO males. **Cytogenet. Cell Genet.**, 10, 318-337.

Cattanach, B.M. et al. (1990) Illegitimate pairing of the X and Y chromosomes in *Sxr* mice. **Genet. Res.**, 56, 121-128.

Chandley, A.C. and Fletcher, J.M. (1980) Meiosis in *Sxr* male mice I. Does a Y-autosome rearrangement exist in sex-reversed (*Sxr*) mice? **Chromosoma**, 81, 9-17.

Das, R.K. and Kar, R.N. (1981) A 41,XYY male mouse. **Experientia**, 37, 821-822.

Eicher, E.M. (1982) Primary sex determining genes in mice, in **Prospects for Sexing Mammalian Sperm** (eds R.P. Amann and G.E. Seidel Jr), Colorado Associated University Press, Boulder, pp. 121 -135.

Eicher, E.M. et al. (1991) The mouse Y* chromosome involves a complex rearrangement, including interstitial positioning of the pseudo-autosomal region. **Cytogenet. Cell Genet.**, 57, 221-230.

Eicher, E.M. and Washburn, L.L. (1983) Inherited sex reversal in mice: identification of a new primary sex-determining gene. **J. Exp. Zool.**, 228, 297-304.

Eicher, E.M. and Washburn, L.L. (1986) Genetic control of primary sex determination in mice. **Ann. Rev. Genet.**, 20, 327-360.

Eicher, E.M. et al. (1982) *Mus poschiavinus* Y chromosome in the C57BL/6J murine genome causes sex reversal. **Science**, 217, 535-537.

Evans, E.P., Beechey, C.V. and Burtenshaw, M.D. (1978) Meiosis and fertility in XYY mice. **Cytogenet. Cell Genet.**, 20, 249-263.

Evans, E.P., Burtenshaw, M.D. and Brown, B.B. (1980) Meiosis in Sxr male mice: II. Further absence of cytological evidence for a Y -autosome rearrangement in sex-reversed (Sxr) mice. **Chromosoma**, 81, 19-26.

Evans, E.P., Burtenshaw, M.D. and Cattanach, B.M. (1982) Meiotic crossing over between the X and Y chromosomes of male mice carrying the sex-reversing (Sxr) factor. **Nature**, 300, 443-445.

Farr, C. et al. (1991) Functional reintroduction of human telomeres into mammalian cells. **Proc. Natl Acad. Sci. USA**, 88, 7006-7010.

Gabriel-Robez, O. et al. (1990) Deletion of the pseudoautosomal region and lack of sex-chromosome pairing at pachytene in two infertile men carrying an X;Y translocation. **Cytogenet. Cell Genet.**, 54, 38 -42.

Gubbay, J. et al. (1990) A member of a novel family of embryonically expressed genes mapping to the sex-determining region of the mouse Y chromosome. **Nature**, 346, 245-250.

Gubbay, J. et al. (1992) Inverted repeat structure of the *Sry* locus in mice. **Proc. Natl Acad. Sci. USA**, 89, in press.

Hale, D.W. et al. (1991) Synapsis and obligate recombination between the sex chromosomes of male laboratory mice carrying the Y* rearrangement. **Cytogenet. Cell Genet.**, 57, 231-239.

Hannappel, E. and Drews, U. (1979) Mosaic character of spermatogenesis in carriers of the sex reversed factor in the mouse. **Hormone Metab. Res.**, 11, 682-689.

Hannappel, E., Siegler, W. and Drews, U. (1980) Demonstration of 2n spermatids in carriers of the 'sex reversed' factor in the mouse by feulgen cytophotometry. **Histochemistry**, 69, 299-306.

Hunt, P.A. (1991) Survival of XO mouse fetuses: effect of parental origin of the X chromosome or uterine environment? **Development**, 111, 1137-1141.

Hunt, P.A. and Eicher, E.M. (1991) Fertile male mice with three sex chromosomes: evidence that infertility in XYY male mice is an effect of two Y chromosomes. **Chromosoma**, 100, 293-299.

Hunt, P.A. and LeMaire, R.M. (1992a) Differences in meiotic pairing behavior of the sex chromosomes in male and female mammals, in **Contemporary Approaches to the Study of Meiosis** (eds S. Heyner and F. Haseltine), AAAS, Washington, in press.

Hunt, P.A. and LeMaire, R. (1992b) Sex-chromosome pairing: evidence that the behavior of the pseudoautosomal region differs during male and female meiosis. **Am. J. Hum. Genet.**, 50, 1162-1170.

Kot, M.C. and Handel, M.A. (1990) Spermatogenesis in XO,Sxr mice: role of the Y chromosome. **J. Exp. Zool.**, 256, 92-105.

Kundu, S.C., Winking, H. and Gropp, A. (1983) Meiosis in XY sex -reversed mice. **XV International Congress of Genetics, New Delhi**. Abstract 271, Oxford and IBH Publishing Company, New Delhi.

Levy, E.R. and Burgoyne, P.S. (1986) Diploid spermatids: a manifestation of spermatogenic impairment in XO*Sxr* and T31H/+ male mice. **Cytogenet. Cell Genet.**, 42, 159-163.

Lovell-Badge, R. and Robertson, E. (1990) XY female mice resulting from a heritable mutation in the murine primary testis-determining gene, *Tdy*. **Development**, 109, 635-646.

Lyon, M.F., Cattanach, B.M. and Charlton, H.M. (1981) Genes affecting sex differentiation in mammals, in **Mechanisms of Sex Differentiation in Animals and Man** (eds C.R. Austin and R.G. Edwards),Academic Press, New York, pp. 329-386.

Lyon, M.F. and Hawker, S.G. (1973) Reproductive lifespan in irradiated and unirradiated XO mice. **Genet. Res.**, 21, 185-194.

Mahadevaiah, S.K., Lovell-Badge, R. and Burgoyne, P.S. (1993) *Tdy* -negative XY, XXY and XYY female mice: breeding data and synaptonemal complex analysis. **J. Reprod. Fert.**, in press.

Mahadevaiah, S., Setterfield, L.A. and Mittwoch, U. (1988) Univalent sex chromosomes in spermatocytes of *Sxr*-carrying mice. **Chromosoma**, 97, 145-153.

Matsuda, Y., Hirobe, T. and Chapman, V.M. (1991) Genetic basis of X-Y chromosome dissociation and male sterility in interspecific hybrids. **Proc. Natl Acad. Sci. USA**, 88, 4850-4854

McLaren, A. and Monk, M. (1982) Fertile females produced by inactivation of an X-chromosome of '*sex-reversed*' mice. **Nature**, 300, 446-448.

McLaren, A. et al. (1988) Location of the genes controlling H-Y expression and testis determination on the mouse Y chromosome. **Proc. Natl Acad. Sci. USA**, 85, 6442-6445.

Miklos, G.L.G. (1974) Sex-chromosome pairing and male fertility. **Cytogenet. Cell Genet.**, 13, 558-577.

Palmer, S.J. and Burgoyne, P.S. (1991) The *Mus musculus domesticus Tdy* allele acts later than the *Mus musculus musculus Tdy* allele: a basis for XY sex-reversal in C57BL/6-Y^{POS} mice. **Development**, 113, 709-714.

Palmer, S.J., Mahadevaiah, S.K. and Burgoyne, P.S. (1990) XYY spermatogenesis in XO/XY/XYY mosaic mice. **Cytogenet. Cell Genet.**, 54, 29-34.

Rathenberg, R. and Müller, D. (1973) X and Y chromosome pairing and disjunction in a male mouse with an XYY sex-chromosome constitution. **Cytogenet. Cell Genet.**, 12, 87-92.

Roberts, C. et al. (1988) Molecular and cytogenetic evidence for the location of *Tdy* and H-Y on the mouse Y chromosome short arm. **Proc. Natl Acad. Sci. USA**, 85, 6446-6449.

Speed, R.M. (1986) Oocyte development in XO fetuses of man and mouse: the possible role of heterologous X-chromosome pairing in germ cell survival. **Chromosoma**, 94, 115-124.

Sutcliffe, M.J., Darling, S.M. and Burgoyne, P.S. (1991) Spermatogenesis in XY, XYSxr^a and XOSxr^a mice: a quantitative analysis of spermatogenesis throughout puberty. **Mol. Reprod. Dev.**, 30, 81-89.

Taketo-Hosotani, T. et al. (1989) Development and fertility of ovaries in the B6.Y^{DOM} sex-reversed female mouse. **Development**, 107, 95-105.

Tease, C. (1990) Sex chromosome configurations in pachytene spermatocytes of an XYY mouse. **Genet. Res.**, 56, 129-133.

Tease, C. and Cattanach, B.M. (1989) Sex chromosome pairing patterns in male mice of novel *Sxr* genotypes. **Chromosoma**, 97, 390-395.

Tease, C. and Fisher, G. (1986) Further examination of the production -line hypothesis in mouse foetal oocytes. **Chromosoma**, 93, 447-452.

Tease, C. and Fisher, G. (1988) Chromosome pairing in foetal oocytes of mouse inversion heterozygotes, in **Kew Chromosome Conference III** (ed. P.E. Brandham), HMSO, London, pp. 293-297.

Vogt, P. et al. (1993) The 'AZF'-function of the human Y chromosome during spermatogenesis. *Ibid.*

Winsor, E.J.T., Ferguson-Smith, M.A. and Shire, J.G.M. (1978) Meiotic studies in mice carrying the sex reversal (*Sxr*) factor. **Cytogenet. Cell Genet.**, 21, 11-18.

7 Note added in proof

In section 2 'General Considerations', it was stated that "if there is
selection against cells with unsynapsed sex chromosomes prior to MI,
the frequency of univalence at MI will underestimate the primary
incidence of pairing failure." We have recently collected data on the
frequency of XYY cells throughout meiosis in XY/XYY mosaics, in order
to monitor the extent of the selection against XYY cells as meiosis
proceeds. The frequency of XYY cells was 68% in spermatogonia and 61%
at pachytene, but the frequency dropped to 35% by diakinesis
suggesting that many XYY cells were eliminated at the end of
pachytene. However, the frequency of XYY cells then increased
dramatically to 84% at MI, although the estimated frequency of XYY-
derived cells at MII was only 14%. We interpret this as evidence for
a 'piling up' of arrested XYY MI spermatocytes prior to their
elimination. This interpretation is supported by the fact that the
relative frequency of MI as compared to diakinesis spermatocytes is
elevated compared to normal XY males. Furthermore, a quantitative
cytological analysis of the spermatogenic failure in XSxr^{a}O males by
Sutcliffe et al. (1991) has documented a significant increase in the
incidence of meiotic metaphases compared with controls. If we accept
that in XYY males and XSxr^{a}O males it is the presence of univalent sex
chromosomes that is responsible for the meiotic failure, then the MI
frequencies wil *overestimate* rather than underestimate the incidence
of sex chromosome univalence (at least in the male).

These findings are important with respect to the interpretation of
all MI frequency data, but are particularly pertinent for XY* males
(section 3.5), and for XYY*^{X} males (section 3.8). In XY* males the
increase in univalence between pachytene and diakinesis/MI has been
taken as evidence for the separation of XY* bivalents. While this may
occur, there will also be an increase in univalent frequency due to
the piling up of arrested MI cells in which the X and Y* failed to
pair. In XYY*^{X} males, MI freqency data have been used to argue that
the Y*^{X} chromosome is more efficient in pairing than the normal Y.
While this may also be true, the difference in pairing frequencies
could be a distortion due to the 'piling up' of cells with one
configuration (in which the Y*^{X} is paired) in preference to other
configurations. Clearly, it is important to collect more meiotic data
for these genotypes, with the relative frequencies of the various
configurations being assessed at pachytene, diakinesis and MI.

20 Mammalian sex determination: what we have learnt from a chromosomal design fault

P. KOOPMAN

University of Queensland, Australia

1 Introduction

The nature of differences between males and females, and how these differences arise, have always held a deep fascination for artists, philosophers and scientists. From the study of genetically accessible organisms such as *Drosophila melanogaster* and *Caenorhabditis elegans*, it is commonly accepted that sex determination in mammals involves pathways of gene regulation, and that the choice of pathway is dependent on a switch mechanism. Understanding mammalian sex determination at the molecular level relies on isolating and studying all the genes in these pathways.

The peculiar biology and genetics of sex determination have resulted in one of these genes being singled out for special attention by molecular geneticists. The pioneering experiments of Jost demonstrated that the presence of testes is required for male development of rabbit fetuses; if testes are removed, females develop (Jost 1947). This observation suggested that in eutherian mammals, testis development is the key event that determines the sexual fate of the embryo. Cytogenetic analyses showed that sex in mammals is determined by the presence or absence of the Y chromosome (Ford et al. 1959; Jacobs and Strong 1959; Welshons and Russell 1959). Whilst this fact is now taken as axiomatic, it came as a surprise when first discovered as it differed from the prevailing paradigms of *Drosophila* and *C.elegans* where the ratio of X chromosomes to autosomes provides the sex-determining switch. Combining these observations led to the theory that the Y chromosome contains a dominant locus that triggers maleness in mammals by directing testis formation from the bipotential embryonic gonad. This locus was said to encode 'testis-determining factor' and was dubbed *TDF* in humans and *Tdy* in mice.

This paper describes the strategy used to clone a candidate for *TDF/Tdy*, and the evidence that this candidate really is the genetic switch that determines the choice of the male pathway in mammalian embryos.

Chromosomes Today   Volume 11. Edited by A.T. Sumner and A.C. Chandley. Published in 1993 by Chapman & Hall, London. ISBN 0 412 47670 3

2 Cloning a candidate for *TDF*

2.1 Normal and Abnormal X-Y interchange
The human X and Y chromosomes share a 2.6 Mb region of
homology at the ends of their short arms. This region is
referred to as the X-Y pairing region or the
pseudoautosomal region, and is the site of recombination
between the X and the Y during male meiosis (Chandley 1988)
(Figure 1a). The location of *TDF* amongst Y-unique sequence
close to the pseudoautosomal boundary was inferred from the
study of XX males. Most XX males possess Y-derived DNA
sequences, transferred to the paternal X chromosome by
aberrant X-Y interchange proximal to the pseudoautosomal
boundary (Figure 1b). Study of the Y-unique DNA content of
these XX males places *TDF* on the distal part of the Y short
arm, close to the pseudoautosomal boundary. (Affara et al.
1986; Muller et al. 1986; Vergnaud et al. 1986).

2.2 Cloning of *ZFY* and its exclusion as a candidate for *TDF*

In 1987 an XX male was described with only 280 kb of Y-

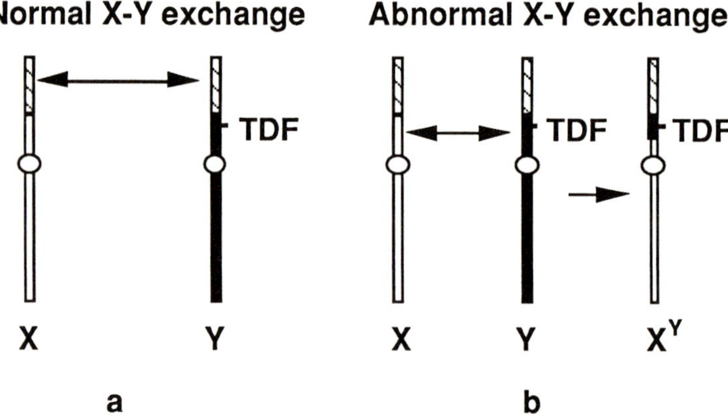

Fig 1   Normal and abnormal X-Y interchange in humans.

Normally, crossing over between the X and Y chromosomes
(indicated by a double headed arrow) occurs within the
shared "pseudoautosomal" region (a, cross-hatched).
Occasionally, crossing over occurs proximal to this region
(b), transferring some Y-unique material to the X
chromosome. As these abnormal X chromosomes can confer
maleness on XX children, *TDF* must map close to the
pseudoautosomal region.

unique DNA adjacent to the pseudoautosomal boundary (Page
et al. 1987). Of this only 140kb overlapped with a
deletion found in an XY female WHT 1013 (Figure 2a). The
zinc finger gene *ZFY* was found in this interval. Although
the position and evolutionary conservation of *ZFY* was
compatible with a sex-determining role, six lines of
evidence indicated that *ZFY* was not *TDF*:

1. A virtually identical sequence, *ZFX*, was found on the
X chromosome. It was proposed that *ZFX* and *ZFY* might be
functionally identical, and that X-inactivation could
provide a gene dosage mechanism of sex determination.
However, *ZFX* was found to escape X-inactivation,
eliminating this hypothesis (Schneider-Gädicke et al.
1988).

2. The marsupial homologue of *ZFY* does not map to the Y-
chromosome (Sinclair et al. 1988)

3. Human *ZFY* was found to be expressed ubiquitously
(Schneider-Gädicke et al. 1988), contrary to
expectations of *TDF*.

4. The two mouse homologues of *ZFY*, *Zfy-1* and *Zfy-2*,
were found to have normal structure and expression in a
line of mice known to be mutant in *Tdy* (Gubbay et al.
1990a).

5. *Zfy-2* is not expressed in developing mouse testes.
Zfy-1 is expressed in normal testes, but not in W^e/W^e
testes lacking germ cells. As testes can form in the
absence of *Zfy-1* and *Zfy-2* expression, neither can be
Tdy (Koopman et al. 1989).

6. Goodfellow and colleagues identified four
masculinized XX individuals ("XX males") possessing Y-
unique sequences not including *ZFY* (Palmer et al. 1989).

2.3 Cloning of *SRY*

The four XX males analysed by Goodfellow and colleagues
were all positive for Y-specific markers located in the
35kb immediately adjacent to the pseudoautosomal boundary
(Figure 2b). This 35kb region was searched for conserved
Y-linked sequences and an open reading frame corresponding
to the gene *SRY* (Sex-determining region Y gene) was found
(Sinclair et al. 1990). An homologous gene, *Sry* was found
on the mouse Y (Gubbay et al. 1990b). The evidence that
these represent the *TDF* and *Tdy* loci respectively is
summarized in the following sections. The discrepancy
between the newly-defined sex-determining region of the
human Y and that defined by Page and colleagues (1987) was
resolved when a second deletion involving the *SRY* region

was found in the XY female WHT 1013 (Figure 2c.) (Page et
al. 1990).

3   **Properties of *SRY/Sry***

3.1 Position on the Y chromosome
In humans, *SRY* maps to the 35kb region adjacent to the
pseudoautosomal boundary; this is the smallest part of the
human Y known to confer maleness, and no other genes have
been reported mapping to this region. In mice, the minimum
translocation of the Y known to confer maleness is defined
by the Sxr' region (Figure 3). Southern blotting shows
that *Sry* is present in XXSxr' DNA, confirming that *Sry* maps
to the Sxr region as expected of *Tdy*.

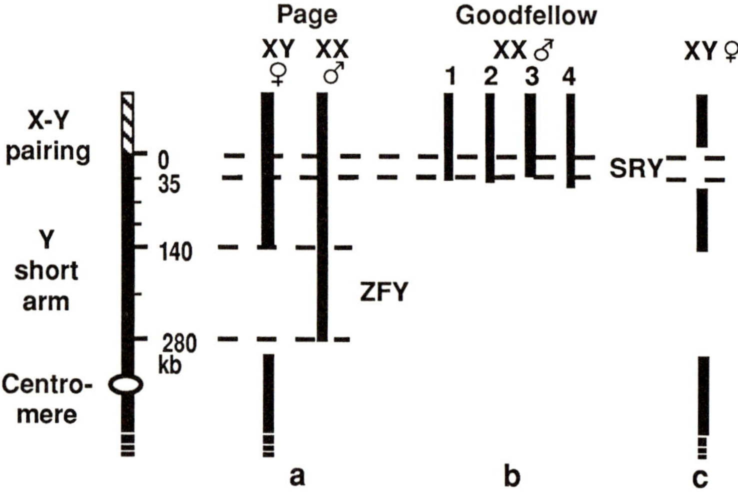

Fig. 2   Positional cloning of *ZFY* and *SRY*

Deletion analysis of one XY female (WHT 1013) and one Y-
positive XX male identified a 140kb region of the Y short
arm in which *TDF* was thought to reside (a). *ZFY* was cloned
from a cosmid walk in that region (Page et al. 1987).
Subsequently, four masculinized XX patients ("XX males")
all lacking *ZFY* and all possessing about 35kb of Y-unique
sequence, were identified by Goodfellow and colleagues
(Palmer et al. 1989; Sinclair et al. 1990). *SRY* was found
in this 35kb region (b). The apparent dilemma between the
Page and Goodfellow mapping of *TDF* was resolved by the
discovery of a second deletion in patient WHT 1013 in the
region of *SRY* (c) (Page et al. 1990).

3.2 Evolutionary Conservation

Noah's ark blotting confirmed that *SRY/Sry* is present on the Y chromosomes of many mammalian species (Sinclair et al. 1990). These observations have recently been extended to include two marsupial species, *Sminthopsis macroura* and *Macropus eugenii* (Foster et al. 1992). It appears that *SRY/Sry* took on its testis-determining function prior to the divergence of marsupials and eutherian mammals 80-180 million years ago.

Tiersch and colleagues (1991) tested the phylogenetic conservation of *SRY/Sry* using Southern blots of DNA from various non-mammalian species. These included birds (which have a ZZ/ZW sex determining system where females are heterogametic), reptilian species having temperature-dependent, genetic or ZZ/ZW sex determining mechanisms, and lower vertebrates. In all cases the *SRY* probe detected sequences common to both males and females. It is not yet clear whether these sequences represent functional homologues of *SRY/Sry* itself or of the conserved family of genes related to *Sry* in the mouse genome (Gubbay et al. 1990b).

3.3 DNA binding

Sequence analysis of genomic clones containing human and mouse *SRY* open reading frames showed that they contain a conserved, 237bp region with homology to a motif that has come to be known as an HMG box. HMG boxes are found in a growing number of eukaryotic genes encoding amongst other things the eponymous high mobility group (HMG) proteins and a number of transcription factors.

The HMG box is known to confer DNA-binding ability on several of these transcription factors. For example, the T-cell transcription factor TCF-1 binds to the sequence AACAAAG (van de Wetering et al. 1991). As SRY is more closely related to this transcription factor than to any other HMG box protein, Harley and colleagues (Harley et al. 1992) tested the ability of *in vitro*-produced SRY protein to bind to this sequence. Binding was confirmed by gel retardation assays, and mutations in positions 2,4,5 and 6 of the heptanucleotide sequence abolished binding, indicating that binding is sequence-specific. In addition, mutant SRY protein that was engineered to contain the amino acid substitutions found in five XY females bound little or no DNA, implying that DNA binding by SRY is required for sex determination. Similar results have been reported for *SRY* containing a frameshift mutation found in a human XY female (Nasrin et al. 1991).

SRY has recently been shown to bind to AACAAT, the target sequence of the yeast HMG box protein Rox-1, with 6-fold higher affinity than to the TCF-1 binding site (Giese et al. 1992). SRY binding to the AACAAT motif induces a bend in the DNA helix; Giese and colleagues propose that this bending juxtaposes normally distant regulatory

sequences, facilitating the assembly of higher-order
nucleoprotein structures involved in gene regulation.

3.4 Mutations in *SRY/Sry* corresponding with sex reversal

One would expect to find mutations in *SRY/Sry* underlying at
least some cases of XY sex reversal if this gene is to
qualify as a candidate for *TDF/Tdy,*. Indeed, *de novo*
mutations in the HMG box region in two XY women were found
soon after the discovery of *SRY* (Berta et al. 1990; Jäger
et al. 1990). Recently, three more *de novo* mutations have
been described (Hawkins et al. 1992; McElreavey et al.
1992), confirming beyond reasonable doubt that *SRY* is *TDF*.
Of these five *de novo* mutations, three clearly inactivate

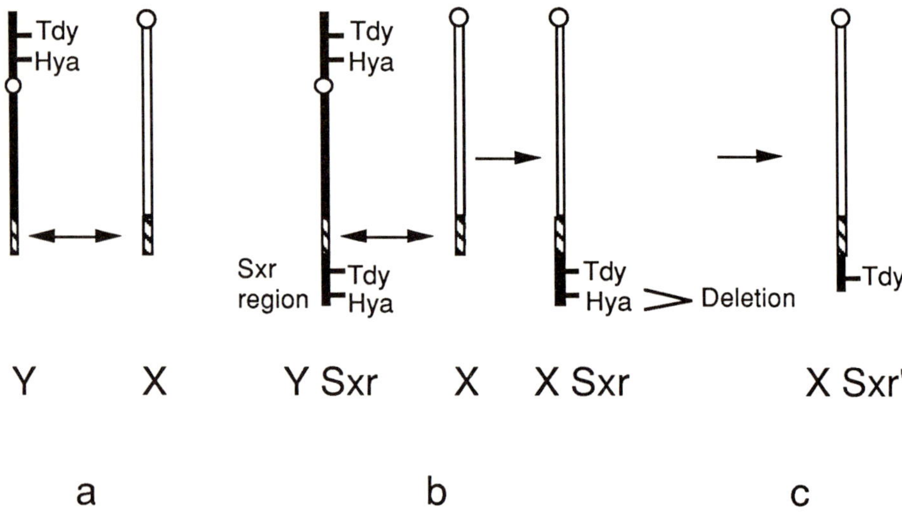

Fig. 3   Generation of the Sxr' mutation in mice

In mice the pseudoautosomal region (cross-hatched) maps to
the end of the long arm on both the X and Y chromosomes
(a). Duplication and translocation of some Y short arm
material onto the end of the Y long arm (b) places this
material distal to the pseudoautosomal region, where it can
be transferred onto an X chromosome during normal meiotic
exchange. This unusual X chromosome (XSxr) can cause sex
reversal and confer H-Y antigen expression, implying that
the genes *Tdy* and *Hya* are in the translocated fragment (the
Sxr region). Amongst *Sxr* mice, a variant (Sxr') was
observed which gives rise to XX males which are H-Y
negative. This is thought to be due to a deletion of part
of the Sxr region from an XSxr chromosome (c). The
resulting Sxr' region represents the smallest translocation
of Y chromosome material known to confer maleness on an XX
background.

the *SRY* protein (one frameshift and two premature stop
codons) whilst the other two change the protein more subtly
(one Gly ➡ Arg and one Met ➡ Ile substitution). No
mutations causing XY sex reversal have been described
outside the HMG box of *SRY*, raising the possibilty that
this is the only domain critical to the male determining
function of *SRY*.

Two further mutations in *SRY* have been found in XY
females which are shared by their XY fathers (Berta et al.
1990; J.R. Hawkins and P. Goodfellow, unpublished data).
It is thought that these variants cause conditional sex
reversal depending on other genetic or environmental
factors. The observation that one of these mutations
reduces but does not abolish DNA binding activity may
explain its partial penetrance (Harley et al. 1992).

Whilst a proportion of XY females carry mutations in the
HMG box of *SRY*, the great majority clearly do not. These
individuals are likely to have mutations in other parts of
the *SRY* coding sequence, in regulatory regions of the *SRY*
gene, or in other genes in the male-determining pathway.

As more mutations in the *SRY* HMG box are found to be
associated with XY sex reversal, a clearer picture will
emerge of the amino acid positions which are critical to
the male determining ability of *SRY*. It is noteworthy that
in sequencing of *SRY* in over 100 normal XY males, not one
polymorphism in the HMG box region has been detected.

Lovell-Badge and Robertson (1990) have described a line
of mice in which XY females are generated due to a mutation
which must involve *Tdy* itself. Southern blotting has
revealed that *Sry* is deleted in these XY female mice
(Gubbay et al. 1990b). This deletion encompasses only
about 11kb of DNA, and involves no genes other than *Sry*
(Gubbay et al. 1992).

3.5 Expression of *Sry*

In the mouse, the primordial gonad or genital ridge
develops as a swelling on the mesonephros at about 10.5 dpc
(days *post coitum*) and appears identical in XX and XY
embryos up until 11.5 dpc. By 12.5 dpc, the male genital
ridge is characterised by the presence of testis cords, due
to the differentiation and alignment of Sertoli cells. It
is expected, therefore, that *Tdy* would be expressed in the
genital ridge prior to this overt differentiation. Studies
using both PCR and *in situ* hybridisation reveal the
presence of *Sry* transcripts only between 10.5 and 12.5 dpc,
specifically in cells of the genital ridge. Transcripts
have not been detected elsewhere in the embryo (Koopman et
al. 1990).

Further information on the type of cell expressing *Sry*
in the genital ridge was obtained through the study of
mouse embryos homozygous for the W[e] mutation. Urogenital
ridges from 11.5 dpc embryos homozygous for W[e] lack germ

271

cells, yet showed similar levels of *Sry* expression to
control non-mutant embryos by PCR analysis (Koopman et al.
1990). *Sry* is therefore expressed by somatic cells in the
developing testis, in accordance with the predictions made
about *Tdy*.

4   Sex reversal in mice transgenic for *Sry*

To directly address the function of *Sry*, transgenic mice
were constructed by injecting this gene on a 14kb genomic
DNA fragment, into fertilised mouse embryos. In one third
of the transgenic XX embryos, testicular development was
found at 14.5 dpc (Koopman et al. 1991). Amongst
transgenic animals allowed to go to term, 3 were found to
be chromosomally female (XX). One of these animals,
m33.13, was sex-reversed with normal external and internal
male genitalia. This mouse was a normal male in all repects
but for his unusually small testes. Histological
examination of sections of the testes revealed normal
tubules, Leydig cells, peritubular myoid cells and Sertoli
cells, however, spermatogenesis was absent and the germ
cells were degenerating (Koopman et al. 1991). This result
was predicted as a similar phenotype is found in all male
mice with two X chromosomes.

Detailed sequence analysis of the 14kb fragment failed
to identify the presence of any genes other than *Sry*; we
conclude that *Sry* is the only gene on the Y chromosome
required for initiating testis development and is therefore
equivalent to *Tdy*.

Not suprisingly, some of the XX mice transgenic for L741
were not sex reversed. Transgenes are susceptible to
position effects which may suppress their expression
depending on their site of integration into the genome.
Founder mouse m32.10 had multiple copies of the *Sry*
transgene and was a fertile female, allowing a transgenic
line to be bred and expression of *Sry* to be examined
amongst the offspring. Two findings have resulted from the
analysis of this line (N. Vivian, P. Koopman and R. Lovell-
Badge, manuscript in preparation). First, *Sry* was found to
be expressed in embryos of this line, ruling out a lack of
expression due to position effects as an explanation for
the lack of sex reversal. Second, some of the XX progeny
of line 32.10 develop as males. These observations imply
that a threshold of *Sry* activity is required for its
function in male development, and that *Sry* expression in
line 32.10 is close to this threshold; those mice
expressing *Sry* below the threshold level develop as females
whilst the few whose expression exceeds the threshold level
develop as males. An accurate assay for *Sry* expression is
required to test this hypothesis and to define the
threshold level.

5   Conclusions

In a relatively short time we have learnt much about the genetic signal that determines the choice of developmental pathways in the embryo - the choice between male and female. The gene *SRY/Sry* has be identified and shown beyond reasonable doubt to be the single, Y-borne switch gene on which this choice hinges. It seems that this one gene represents a common sex-determining switch that unites all mammals and distinguishes them from other classes of animals. *SRY/Sry* achieves its important function in development by being activated specifically in the genital ridge and specifically at the time the developmental "decision' is being made whether to develop as a testis or to continue on the default course of ovarian development. *SRY/Sry* is known to bind to specific DNA sequences and this activity appears to be necessary for male sex determination. By analogy with other HMG proteins, it is likely that *SRY/Sry* engages the testis-determining pathway by transcriptional activation of downstream genes.

Whilst some may consider the placement of the primary testis-determining locus a design-fault of the human Y chromosome, it is this single factor that has most facilitated the cloning of *SRY*. The identification and study of *SRY/Sry* has provided a point of entry into dissection of the genetic cascade leading to male development.

Although much is known about *SRY/Sry* and its function in testis determination, much remains to be discovered. The precise temporal and spatial regulation of *Sry* must be due to genes "upstream" in the pathway. *SRY/Sry* appears to control the transcription of "downstream" genes, and these too remain to be identified. Identification of these other genes and an understanding of their interaction is required before the process of sex determination can be fully described at the molecular level.

6   References

Affara, N.A.et al. (1986) Variable transfer of Y-specific sequences in XX males. **Nucl. Acids Res.**, 14, 5375-5387.

Berta, P.et al. (1990) Genetic evidence equating *SRY* and the male sex determining gene. **Nature**, 348, 448-450.

Chandley, A.C. (1988) Meiosis in man. **Trends. Genet.**, 4, 79-84.

Ford, C.E.et al. (1959) A sex chromosome anomaly in a case of gonadal dysgenesis (Turner's Syndrome). **Lancet**, i, 711-713.

Foster, J.W.et al. (1992) The Human Sex Determining Gene

SRY Detects Homologous Sequences on the Marsupial Y Chromosome. Manuscript submitted.

Giese, K. Cox, J. and Grosschedl, R. (1992) The HMG Domain of Lymphoid Enhancer Factor 1 Bends DNA and Facilitates Assembly of Functional Nucleoprotein Structures. **Cell**, 69, 185-195.

Gubbay, J.et al.(1990a) Normal structure and expression of *Zfy* genes in XY female mice mutant in *Tdy*. **Development**, 109, 647-653.

Gubbay, J. et al. (1990b) A gene mapping to the sex-determining region of the mouse Y chromosome is a member of a novel family of embryonically expressed genes. **Nature**, 346, 245-250.

Gubbay, J. et al. (1992) Inverted repeat structure if the Sry locus in mice. **Proc. Natl Acad Sci. USA**, in press.

Harley, V.R.et al. (1992) DNA Binding Activity of Recombinant SRY from Normal Males and XY Females. **Science**, 255, 453-456.

Hawkins, J.R.et al. (1992) Mutational analysis of SRY: nonsense and missense mutations in XY sex reversal. . **Hum. Genet.**, 88, 471-474.

Jacobs, P.A. and Strong, J.A. (1959) A case human intersexuality having possible XXY sex determining mechanism. **Nature**, 183, 302-303.

Jäger, R.J. et al. (1990) A human XY female with a frame shift mutation in the candidate testis-determining gene *SRY*. **Nature**, 348, 452-454.

Jost, A. (1947) Recherches sur la differentiation sexuelle de l'embryon de lapin. **Arch. Anat. Microsc. Morphol. Experim.**, 36, 271-315.

Koopman, P. et al.(1989) *Zfy* gene expression patterns are not compatible with a primary role in mouse sex determination. **Nature**, 342, 940-942.

Koopman, P.et al. (1991) Male development of chromosomally female mice transgenic for *Sry*. **Nature**, 351, 117-121.

Koopman, P. et al. (1990) Expression of a candidate sex-determining gene during mouse testis differentiation. **Nature**, 348, 450-452.

Lovell-Badge, R. and Robertson, E. (1990). XY female mice resulting from a heritable mutation in the primary testis-determining gene, *Tdy*. **Development**, 109, 635-646.

McElreavey, K.D.et al. (1992) XY Sex Reversal Associated with a Nonsense Mutation in SRY. **Genomics**, 13, 838-840.

Muller, U. et al. (1986) Deletion mapping of the testis determining locus with DNA probes in 46, XX males and in 46, XY and 46, X, dic(Y) females. **Nucl. Acids Res.**, 14, 6489-6505.

Nasrin, N. et al. (1991) DNA-binding properties of the product of the testis-determining gene and a related protein. **NATURE**, 354, 317-320.

Page, D.C. et al. (1990) Additional deletion in sex-determining region of human Y chromosome resolves paradox of X,t(Y,22) female. **Nature**, 346, 279-281.

Page, D.C. et al. (1987) The sex-determining region of the human Y chromosome encodes a finger protein. **Cell**, 51, 1091-1104.

Palmer, M.S.et al. (1989) Genetic evidence that *ZFY* is not the testis-determining factor. **Nature**, 342, 937-939.

Schneider-Gädicke, A.et al. (1988) ZFX has a gene structure similar to ZFY, the putative human sex determinant and escapes X inactivation. **Cell**, 57, 1247-1258.

Sinclair, A. et al. (1988) Sequences homologous to *ZFY* , a candidate human sex-determining gene, are autosomal in marsupials. **Nature**, 36, 780-783.

Sinclair, A.H. et al. (1990) A gene from the human sex-determining region encodes a protein with homology to a conserved DNA-binding motif. **Nature**, 346, 240-244.

Tiersch, T.R. Mitchell, M.J. and Wachtel, S.S. (1991) Studies on the phylogenetic conservation of the SRY gene. **Hum. Genet.**, 87, 571-573.

van de Wetering, et al.(1991) Identification and cloning of TCF-1, a T lymphocyte-specific transcription factor containing a sequence-specific HMG box. **EMBO J.**, 10, 123-132.

Vergnaud, G. et al. (1986) A deletion map of the human Y chromosome based on DNA hybridization. **Am. J. Hum. Genet**, 38, 109-124.

Welshons, W.J. and Russell, L.B. (1959) The Y chromosome as the bearer of male determining factors in the mouse. **Proc.Natl.Acad.Sci. USA**, 45, 560-566.

21 Evolution of the mammalian XY pairing segment

W. SCHEMPP and R. TODER
University of Freiburg, Germany

The highly heteromorphic mammalian X and Y chromosomes are thought to have evolved from a homologous chromosome pair (Ohno 1967). Consequently, homoeologous segments may still exist on the highly differentiated mammalian sex chromosomes representing a relic of such an ancestral pair of autosomes. This evolutionary process would provide the basis for the bipartite structure of the eutherian Y chromosome: the still homologous pairing segment is shared and recombines between the X and Y chromosome, and thus could ensure the proper segregation of the sex chromosomes during male meiosis, while the Y-specific region, including the testis-determining factor gene(s) must avoid recombination, otherwise the chromosomal basis of sex determination would break down.

Molecular evidence for the existence of strictly homologous sequences on the heteromorphic sex chromosomes recombining frequently in male meiosis has been found in the human (Cooke et al., 1985; Simmler et al., 1985), the mouse (Keitges et al., 1985; Soriano et al., 1987), and the chimpanzee (Weber et al., 1988). Furthermore, by pairwise comparison of X and Y sequences at the boundary between the homologous and the sex-specific parts, Ellis et al., (1990) have demonstrated that homology between the sex chromosomes was maintained by recombination in the sequences telomeric to an Alu insertion site in the great apes and the Old World monkey species macaque and gelada baboon. As these sex-chromosomal sequences show a seemingly autosomal inheritance behaviour, they have been termed pseudoautosomal (Burgoyne, 1982). While the pseudoautosomal pairing segments of the human and the chimpanzee can be regarded as strictly homoeologous, that of the mouse differs with respect to chromosomal banding appearance (Somssich et al., 1981) and genetic content (Harbers et al., 1990), and therefore cannot be homoeologized in a direct linear manner.

Our own chromosomal replication and DNA hybridization studies have revealed that, in a representative sample of primate species (reviewed in Schempp et al., 1989), and in a variety of non-primate eutherian mammals (Schempp and Toder, 1992), a homoeologous early replicating segment in a distal position of the X and Y chromosomes coincides with the conservation and the specific localization of the human derived pseudoautosomal probe 113F, which belongs to a group of repeated elements that are now called STIR (subtelomeric interspersed repeat) elements (Rouyer et al., 1990).

Chromosomes Today Volume 11. Edited by A.T. Sumner and A.C. Chandley. Published in 1993 by Chapman & Hall, London. ISBN 0 412 47670 3

Thus, in spite of structural rearrangements, the highly heteromorphic X and Y chromosomes of a wide range of eutherian mammals share a homoeologous early replicating segment in a telomeric or subtelomeric position, a location that is compatible with meiotic crossing over in the male sex.

It may be concluded that this X-Y homologous segment is necessary for the regular meiotic disjunction of the highly heteromorphic sex chromosomes.

The question arises, does this X-Y homologous segment in eutherian mammals represent a relic of an ancestral homology of the sex chromosome pair, or does it represents an acquisition de novo?

Comparative mapping data on the conservation and localization of human X- and Y-specific gene sequences in species of the mammalian orders Marsupialia and Monotremata have led to the assumption that in eutherian mammals, the fusion of two autosomal blocks with an ancestral mammalian sex chromosome pair has occured. An almost complete reduction of the ancestral homologous segment of the Y chromosome then followed, resulting in the present day eutherian X and Y chromosomes. While from the human X chromosome only those genes that are located on the long arm, are X-linked in marsupials and monotremes, those of the short arm of the human X chromosome are autosomal in marsupial and monotreme species (reviewed in Graves and Watson, 1991).

Therefore it may be concluded that the homoeologous early replicating segments (pseudoautosomal region) still existing on the highly differentiated sex chromosomes of a wide range of eutherian mammals are rather a relic of this de novo autosomal acquisition than a relic of the putative ancestral chromosome pair.

Following this argument, evolution of the eutherian Y chromosome has started with a considerably reduced element containing genes for male differentiation and presumably a segment homologous to the X chromosome allowing for recombination. Interestingly, and as already mentioned, there is a considerable difference in structure and organization of the pseudoautosomal and the sex determining regions when comparing the human and the mouse Y chromosomes. While in the human the pseudoautosomal and the sex determining regions are located in close proximity to each other on the terminal short arm of the Y chromosome (Vergnaud et al., 1986; Sinclair et al., 1990), in the mouse these segments are located at opposite ends of the Y chromosome, the sex determining region being on the short arm (Gubbay et al., 1990) and the pseudoautosomal region representing the distal long arm segment (Keitges et al., 1985; Harbers et al., 1986). Recent molecular analysis of the steroid sulfatase (STS) locus may explain some of the human-mouse differences. In humans the X-linked STS gene is located about 5 Mb from the pseudoautosomal region and escapes X-inactivation, while a nonfunctional pseudogene of STS is located on the long arm of the Y chromosome (Yen et al., 1988). Interestingly in the mouse, Sts is pseudoautosomal (Keitges et al., 1985). Comparative studies in primates strongly suggest that the ancestral STS gene was pseudoautosomal and that a pericentric inversion of the Y chromosome in the early primate lineage has disrupted the former pseudoautosomal location of STS (Yen et al.,1988; Schempp and Toder, 1992). These STS data have led to the assumption that this pericentric inversion of the Y chromosome in the early primate lineage has disrupted the originally more extended pseudoautosomal region, transferring its proximal part, including STS, to the long arm. Simultaneously the sex determining region from the long arm has come into close proximity with the pseudoautosomal region on the distal short arm of the Y chromosome (Ellis and Goodfellow, 1989). On the X, STS escapes inactivation, and on the Y it is a pseudogene.

An analogous situation to STS was demonstrated recently for the Kallmann (KALIG-1/ADML) locus. In humans the X-linked KALIG-1/ADML gene is located some 1.5 Mb proximal to the STS gene and likewise escapes inactivation, while a homologue on the long arm of the human Y chromosome presumably is not expressed (Franco et al., 1991; Legouis et al., 1991). Again comparative studies in primates indicate that the ancestral KALIG-1/ADML gene was pseudoautosomal and has become sex-specific consecutively to a pericentric inversion in the Y chromosome (Franco et al., 1991). A schematic representation of this pericentric inversion of the Y chromosome in the early primate lineage is given in Fig. 1.

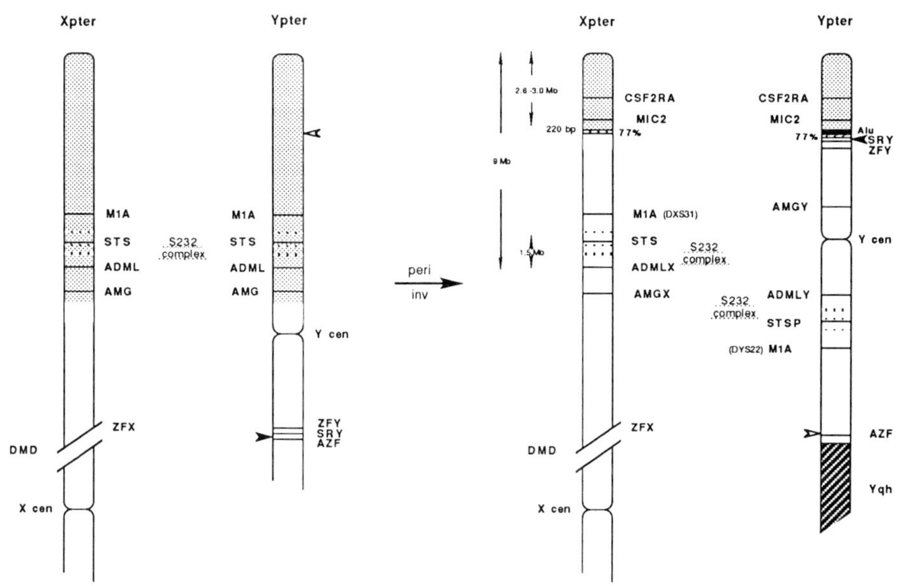

Fig. 1. Schematic representation of a pericentric inversion in the Y chromosome that is postulated to have occured during early primate evolution disrupting an originally contiguous large pseudoautosomal region including STS and ADML and the ancestral male-specific linkage group which contains genes for sex determination and male sexual differentiation.

In order to trace this postulated rearrangement of the Y chromosomes in primates, as landmarks the pseudoautosomal and the Y-specific euchromatic region , including the sex determining region, may be used.

Our first interest was concentrated to the question, whether the close proximity of the testis-determining region (SRY) to the Y pseudoautosomal region also holds true for the Y chromosomes of the great apes.

In contrast to the X chromosomes, the Y chromosomes have undergone extensive re-arrangements during evolution of the great apes (Weber et al., 1986). Nevertheless, as is already described for the ZFY gene (Müller and Schempp, 1991), SRY maps closely to the pseudoautosomal segment in a telomeric or subtelomeric position of the Y chromosomes of the great apes. While, on the other hand, a low copy repeat (FR-35-II) of the proximal short arm of the human Y chromosome is obviously included in the Y chromosomal rearrangements during hominoid evolution (Toder et al., in preparation).

Thus, despite cytogenetically visible structural alterations within the euchromatic parts of the Y chromosomes of the human and the great apes, a segment of the Y chromosomes of these species, including the pseudoautosomal and the testis-determining regions, seems to be more strongly conserved than the rest of the Y chromosomes.

It may be speculated, that the close proximity of the testis-determining region to the pseudoautosomal boundary on the Y chromosomes of the human and the great apes prevents the extension of the pseudoautosomal region into the sex chromosome-specific sequences. Indeed, the translocation of TDF to the X chromosome by non-homologous recombination in human male meiosis results in XX males and/or XY females who are both sterile; the transmission of such a new recombinant sex chromosome is thus excluded.

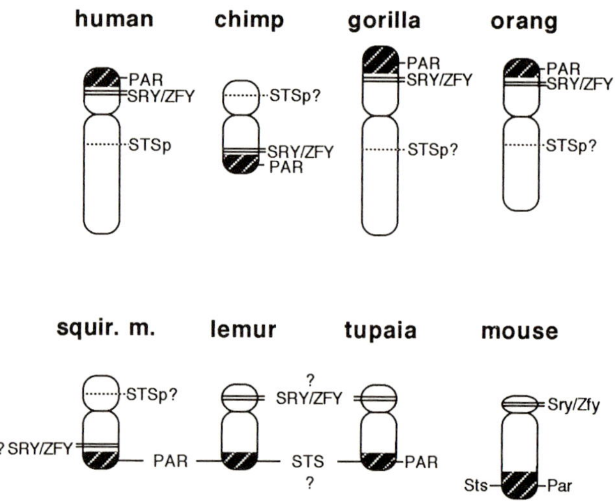

Fig. 2. Comparative mapping data of the Y chromosomes of different primate species (squir.m. means squirrel monkey) together with that of the tree shrew (tupaia) and of the mouse. For explanation see text.

In Fig. 2 our comparative in situ project is summarized schematically. Both, the human and the mouse Y chromosomal situations are well characterized. While in the human, the sex determining region (SRY/ZFY) is in close proximity to the pseudoautosomal region (PAR), and a pseudogene for STS (STSp) is on the Y long arm, the mouse Y chromosome seems to represent the ancestral situation: the mouse Sts gene is located pseudoautosomally on the long arm and the sex determining region is located on the short arm. Our current studies are directed towards the Y chromosomal situation of the prosimian species (lemur), and the tupaias as forerunner primate species. Here the question arises whether their Y chromosomes are comparable to that of the mouse Y chromosome, reflecting the ancestral situation of a more extended pseudoautosomal region, as this is shown in Fig. 1.

ACKNOWLEDGEMENT

This work was supported by grants from the Deutsche Forschungsgemeinschaft.

References

Burgoyne, P.S. (1982) Genetic homology and crossing over in the X and Y chromosomes of mammals. Hum Genet 61: 85-90

Cooke, H.J., Brown, W.A.R. and Rappold, G.A. (1985) Hypervariable telomeric sequences from the human sex chromosomes are pseudoautosomal. Nature 317:687-692

Ellis, N. and Goodfellow, P.N. (1989) The mammalian pseudoautosomal region. TIG 5: 406-410

Ellis, N. et al. (1990) Evolution of the pseudoautosomal boundary in old world monkeys and great apes. Cell 63:977-986

Franco, B. et al. (1991) A gene deleted in Kallmann's syndrome shares homology with neural cell adhesion and axonal path-finding molecules. Nature 353: 529-536

Graves JAM, Watson JM (1991) Mammalian sex chromosomes: evolution of organization and function. Chromosoma 101:63-68

Gubbay, J. et al. (1990) A gene mapping to the sex-determining region of the mouse Y chromosome is a member of a novel family of embrionically expressed genes. Nature 346: 245-250

Harbers, K. et al. (1986) High frequency of unequal recombination in pseudoauto-somal region shown by proviral insertion in transgenic mouse. Nature 324: 682-685

Harbers, K. et al. (1990) Structure and chromosomal mapping of a highly poly-morphic repetitive DNA sequence from the pseudoautosomal region of the mouse sex chromosomes. Cytogenet. Cell Genet. 53: 129-133

Keitges, E. et al. (1985) X-linkage of steroide sufatase in the mouse is evidence for a functional Y-linked allele. Nature 315: 226-227

Legouis, R. et al. (1991) The candidate gene for the X-linked Kallmann syndrome encodes a protein related to adhesion molecules. Cell 67: 423-435

Müller, G. and Schempp, W. (1991) Comparative mapping of ZFY in the hominoid apes. Hum Genet 88:59-63

Ohno, S. (1967) Sex chromosomes and sex-linked genes. Springer, Berlin Heidelberg New York

Rouyer, F. et al. (1990) An interspersed repeated sequence specific for human sub-telomeric regions. EMBO J. 9: 505-514

Schempp, W. and Toder, R. (1992) Molecular cytogenetic studies on the evolution of sex chromosomes in primates. In: J.A.M. Graves and K. Reed (eds) Proc of the 1992 Boden Res Conf: Mammalian sex chromosomes and sex-determining genes: their differentiation, autonomy and interactions in gonad differentiation and function. Harwood Academic Publishers (in press)

Schempp, W., Weber, B. and Müller, G. (1989) Mammalian sex-chromosome evo-lution: a conserved homoeologous segment on the X and Y chromosomes in primates. Cytogenet. Cell Genet. 50: 201-205

Simmler, M.-C. et al. (1985) Pseudoautosomal DNA sequences in the pairing region of the human sex chromosomes. Nature 317: 692-697

Sinclair, A.H. et al. (1990) A gene from the human sex-determining region encodes a protein with homology to a conserved DNA-binding motif. Nature 346: 240-244

Somssich, I., Hameister, H. and Winking, H. (1981) The pattern of early replicating bands in the chromosomes of the mouse. Cytogenet. Cell Genet. 30: 222-231

Soriano, P. et al. (1987) High rate of recombination and double crossovers in the mouse pseudoautosomal region during male meiosis Proc Natl Acad Sci, USA 84: 7218-7220

Vergnaud, G. et al. (1986) A deletion map of the human Y chromosome based on DNA hybridization. Am J Hum Genet 38: 109-124

Weber, B., Schempp, W. and Wiesner, H. (1986) An evolutionarily conserved early replicating segment on the sex chromosomes of man and the great apes. Cytogenet Cell Genet 43: 72-78

Weber, B., Weissenbach, J. and Schempp, W. (1988) X-Y crossing over in the chimpanzee. Hum Genet 80: 301-303

Yen, P.H. et al. (1988) The human X-linked steroid sulfatase gene and a Y-encoded pseudogene: evidence for an inversion of the Y chromosome during primate evolution. Cell 55: 1123-1135

Meiosis and aneuploidy

22 The questionable role of the synaptonemal complex in meiotic chromosome pairing and recombination

J. LOIDL

University of Vienna, Austria

Introduction

The synaptonemal complex (SC) is a mainly proteinaceous structure that is present between closely paired chromosomes at meiotic prophase. It occurs almost universally in higher eukaryotes, its structure is highly conserved and its presence is virtually always linked to the ability to perform crossing-over. Therefore it has been attributed a fundamental function in meiotic chromosome pairing and recombination. However, over the last years, evidence has accumulated that the recognition of homologous chromosomes is independent of the SC. Moreover, it has been argued that the initiation of meiotic crossing-over does not require an SC, either.

Recently, studies in yeast have contributed much to our present understanding of meiotic pairing and recombination. Yeast is a particularly favourable system to study meiosis as many sporulation deficient mutants are known which help to elucidate the genetic pathways involved in recombination and pairing. The disadvantage of yeast as a cytologically not very amenable system has been overcome by the introduction of spreading techniques for yeast SCs (Dresser and Giroux 1988, Loidl et al. 1991).

Here, I want to summarize recent cytological findings obtained from a range of organisms, including budding and fission yeasts, on SC formation and behaviour and I will address the question of its functions.

Chromosomes Today   Volume 11. Edited by A.T. Sumner and A.C. Chandley. Published in 1993 by Chapman & Hall, London. ISBN 0 412 47670 3

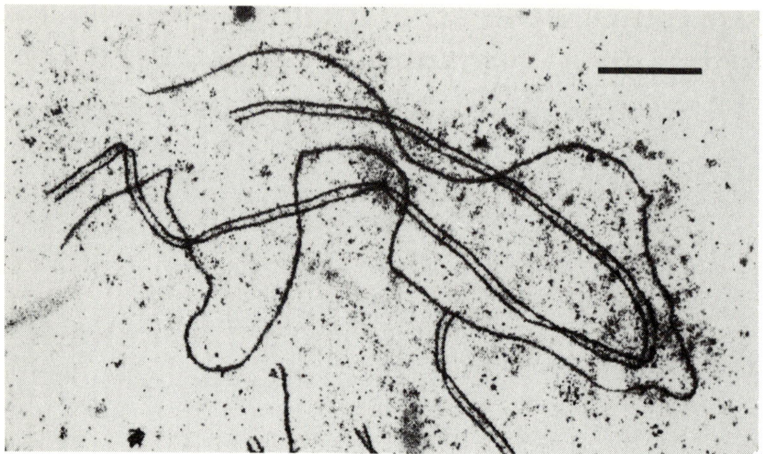

Fig. 1. Meiotic pairing of three homologous chromosomes in a polyploid *Achillea*. Electron micrograph of a section of a spread pachytene nucleus. Two chromosomes are synapsed, the third is merely aligned. Note that the aligned axis exceeds the others in length. Bar represents 2μm.

Presynaptic alignment (PSA)

In most eukaryotes, the synapsis of homologous chromosomes is believed to be preceded by their alignment at some distance. Under favourable conditions this presynaptic alignment (PSA) can be observed as the association of axial elements. Often, however, these are not yet developed at leptotene to zygotene in some chromosome regions whilst in others synapsis already starts by the formation of transverse filaments of the SC (e.g., Loidl and Schweizer 1992). Therefore, PSA is not always obvious but there is circumstantial evidence for it from the distant alignment of superfluous axial elements which do not participate in two-by-two synapsis during pachytene in polyploids (see, e.g., Loidl and Jones 1986, Loidl et al. 1990) (Fig. 1). In the budding yeast, *Saccharomyces cerevisiae* we have overcome the problem of poor visibility of the frail and fragmentary axial elements during early meiotic prophase by the delineation of the chromosomes themselves by in situ hybridization. By using this approach we could visualize PSA directly as the association of homologous DNA sequences (H. Scherthan, J. Loidl,

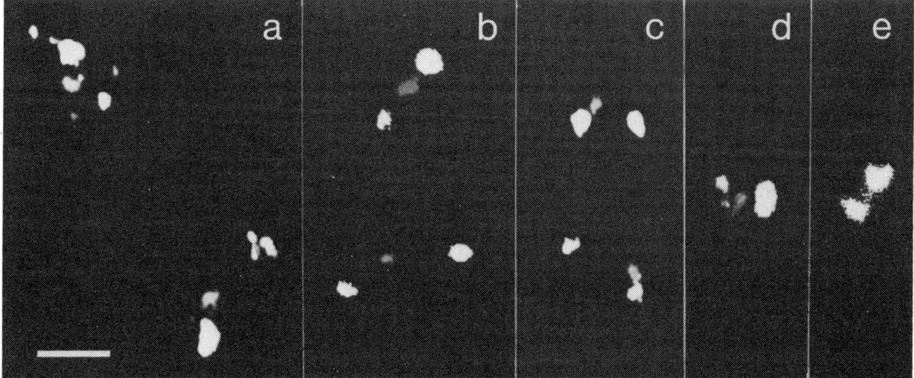

Fig. 2. Visualization of chromosomes no. V of *Saccharo-myces cerevisiae* by in situ hybridization with composite probes of chromosome-specific cloned DNA. Three subregions on each chromosome are highlighted after fluorescence detection of labelled hybrid DNAs by laser scanning micro-scopy. Condensation and pairing behaviour of the homologues during meiotic prophase is shown.(**a**) Chromosomes at interphase or early meiotic prophase. Fluorescence signals are dispersed indicating that chromosomes are largely decondensed; they appear to occupy irregularly shaped domains. (**b**) The linear arrangement of three distinct fluorescent dots on each chromosome indicates that conden-sation has commenced. (**c**) Condensed homologues are aligned roughly in parallel at a distance too large to be bridged by the SC. (**d**) Condensation has increased further; two of the three labelled subregions are paired, only the region to the left still appears as two separate signals. This stage may be interpreted as zygotene. (**e**) The pairs of hybridization signals are completely fused, indicating the intimate pairing of homologues. This stage corresponds to pachytene. Chromosomes are about 50% longer than the regions marked by the hybridization signals. (Pictures by H. Scherthan and J. Loidl, for details see Scherthan et al. - submitted). Bar represents 2μm.

T. Schuster and D. Schweizer - submitted) (Fig. 2).

This demonstration of PSA can be taken as evidence that meiotic search for and recognition of homology take place without the involvement of the SC (see Loidl 1990). It has been suggested that homologue search employs some process of DNA strand invasion, heteroduplex formation and base matching (e.g., Smithies and Powers 1986, Carpenter 1987). These events may be accompanied by non-reciprocal recombination events that can be recorded as gene conversions (e.g., Haber et al. 1991).

Possible functions of the axial elements
From ultrastructural pictures it is obvious that the chromatin loops emanating from pachytene bivalents are attached to the axial elements (Weith 1985). From this it seems that the axial elements provide the physical link between the chromatin and the transverse filaments at pachytene and thus are simply a structural element of the pachytene SC. I would like to speculate, however, that axial elements serve a more sophisticated function in the homology search process. PSA, reflecting homologue recognition, has so far been found to occur only between chromosome regions that have formed axial element segments already (e.g., Loidl and Schweizer 1992). Also the roughly simultaneous occurrence of double strand breaks (which have been interpreted as being part of a homology search mechanism) and the advent of first axial element precursors in *S. cerevisiae* (Padmore et al. 1991) suggest a functional relationship. In the fission yeast, *Schizosaccharomyces pombe*, all functions of meiosis, including crossing-over, can be exerted in the absence of a complete tripartite SC, but it does have axial element fragments. (J. Bähler, J. Loidl, T. Wyler and J. Kohli, in preparation) (Fig. 3). These observations suggest that the minimum structural requirement for meiotic recombinogenic pairing are axial elements rather than mature SCs.

We must take into consideration that a major problem with homology search is the large number of possible combinations of pairs of DNA segments that have to be compared by a homology search mechanism. The homology search process can be imagined as requiring the shuffling of chromosomes (which are gigantic on a molecular scale) within the nucleus and repeated DNA unwinding and heteroduplex formation. It would make the process a lot easier if (i) only a fraction of DNA sequences were actively or passively involved in homology testing events (i.e., if the number of potential testing sites were kept low) and if (ii) PSA is mediated by some early recognition events that would promote subsequent homologous interacti-

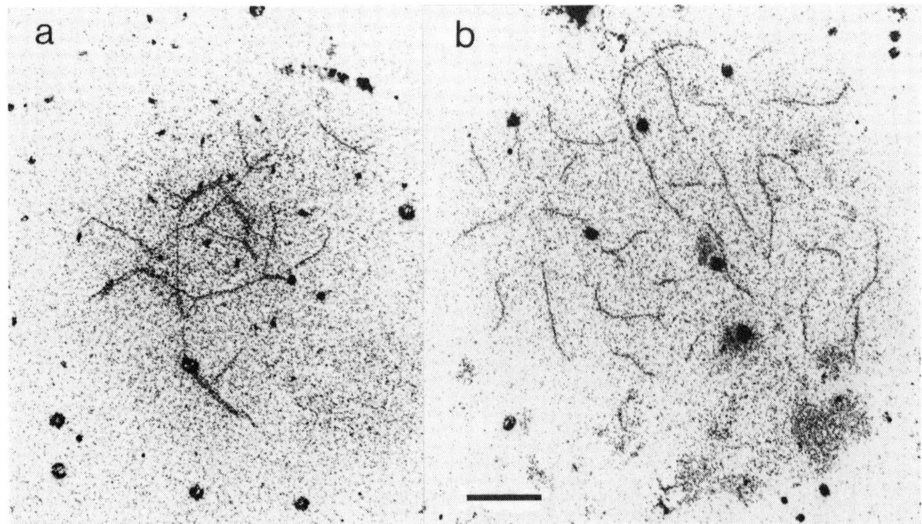

Fig. 3. Electron micrographs of meiotic prophase nuclei of
Schizosaccharomyces pombe. (**a**) Axial element-like structures associate transiently in a seemingly irregular
manner without developing transverse filaments and central
element. It is uncertain whether this stage corresponds to
the period of homologue recognition and/or reciprocal
recombination. (**b**) At a later stage, axial fragments are
separate. (Pictures by J. Bähler and J. Loidl) Bar
represents 1μm.

ons of the remaining recognition sites. I propose that the
axial elements could serve in organizing chromosome structure and positioning which allows the preselection of
homology testing sites and the mutual promotion of homologous contact sites.

Axial elements could provide the scaffold along which
the chromatin is lined up as loops in the appropriate
sequence to expose corresponding DNA segments for mutual
recognition (Fig. 4). This is a modification of the preselection theory by Stern et al. (1975). The original theory
says that less than 1% of total chromosomal DNA is
directly associated with SC proteins at meiosis (whereas
the rest is in the loops) and is thereby preselected as
the potential substrate for crossing-over. Similarly, only
a small portion of chromosomal DNA could be caught by the
proteins of the axial elements and serve a function in

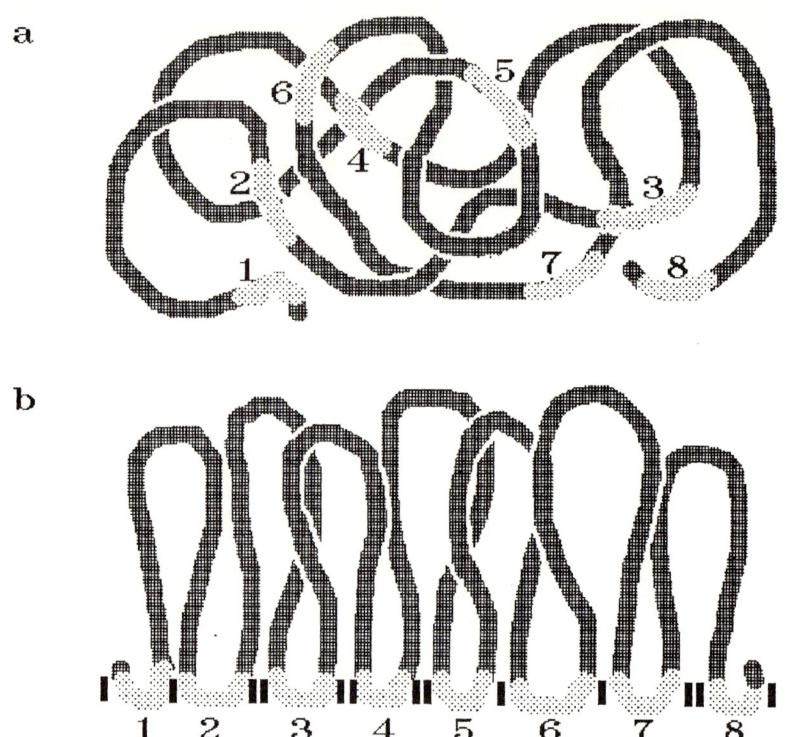

Fig. 4. Model for the organization of chromatin during homologue recognition. (**a**) In the absence of a longitudinal scaffold the chromatin of each chromosome occupies a roughly spherical domain in the nucleus. This corresponds to the situation at interphase and early meiotic prophase as depictured in Fig. 2a. The potential homology recognition sites are in no particular sequence and therefore unlikely to match with the homologous chromosome. (**b**) The presence of a scaffold as constituted by the axial element could bring corresponding regions on homologous chromosomes into register. In addition, only a subset of potential homology recognition sites could become exposed to mutual testing at the pairing surface. The portion of chromatin that contains DNA segments involved in homologue search might persist in a decondensed state.

homology testing (see also Pearlman et al. 1992) (Fig.4).

In addition, the DNA-fraction associated with the axial elements could be delayed in prophase condensation as it otherwise might constitute an obstacle to the molecular homology testing process.

We may assume that the homology search process that leads to PSA employs a large number of molecular homology testing events between all potential recognition sites on chromosomes. The minimum requirement for bivalent formation by the whole chromosome complement is one recognition site per chromosome. However, particularly in plants with large chromosomes, there seem to exist several individual homology recognition sites per chromosome. The number of homology testing events could be reduced if there existed some presynaptic chromosome disposition or other restrictions to randomness in homologue search. Presynaptic alignment of axial elements could contribute to the reduction of testing events by coordinating the action of neighbouring recognition sites. As soon as PSA is mediated between two chromosomes by some early recognition sites, it will, for sterical reasons, promote subsequent homologous interactions of the remaining recognition sites on the same homologous pair. This would prevent them from extending search to non-homologous chromosomes and thus keep the number of unsuccessful homology testing events low. Also, if many recognition sites accumulated between homologous chromosomes, they would take precedence over occasional ectopic interactions that might occur during the search process. This could lead to their resolution and ensure that crossovers predominantly involve allelic sites.

If one follows the above argument that the local formation of axial elements is sufficient for homologue search and recognition, then it remains to be explained why the majority of eukaryotes develop a complete, tripartite and continuous SC.

What comes first, synapsis or crossing-over?
The answer to this question is relevant to considerations of possible SC functions. For many years it was widely agreed that the function of the SC is to provide the intimate connection of homologous chromosome regions that is required for crossing-over. The concept was that primary homologous recognition processes (which may be accompanied by non-reciprocal recombination events) would

bring about presynaptic alignment which in turn would
promote SC formation between homologous chromosomes.
Within the framework of the SC another round of recognition and recombination events would cause crossing-over.
(For this "classical" view of the origin of crossovers see
Loidl 1990, 1991.) This hypothesis gains support from a
large number of cases where initial homoeologous multi-
valent synapsis at zygotene is corrected to homologous
bivalent SCs at pachytene. Consequently, chiasmate connec-
tions are found only between homologous chromosomes at
metaphase I (for lit. see Jenkins and Rees 1991). The most
convincing explanation for the feasibility of pairing
correction is that crossovers do not precede SC formation
at zygotene and form only following the establishment of
homologous bonds at pachytene. The concept of a two-step
(chromosome- and sequence-) homology recognition has the
advantage that it can explain why crossing-over usually
does not involve homologous sequences on non-homologous
chromosomes.

On the other hand, recent studies in *S. cerevisiae*
have suggested that crossovers could represent a subset of
the initial sites of gene conversion. Allelic and ectopic
gene conversions are believed to accompany molecular
events that are involved in homology testing (e.g., Roeder
1990, Haber et al. 1991). According to this view the sites
of primary homologous contact and crossovers (and perhaps
also SC initiation sites) would be identical (Padmore et
al. 1991). The SC would only succeed the recombination
event. One argument in favour of the direct origin of
crossing-overs from presynaptic recombination processes is
that all recombination deficient yeast mutants characterized so far reduce ectopic recombination to the same extent
as allelic recombination. Were pre-synaptic recombination
(in the course of homologue search) and crossing-over two
different events, there should exist a class of mutants
that is proficient in ectopic recombination but deficient
in the exchange events that occur after chromosome pairing
(Bhargava et al. 1992).

Circumstantial evidence for crossovers without SCs
comes also from the incidence of crossovers between
dispersed related sequences which can be recorded as chro-

mosomal translocations (Jinks-Robertson and Petes 1986, Lichten et al. 1987). These crossovers seem not to be accompanied by synapsis of the regions involved. However, a more extensive cytological examination will be necessary. Moreover, in *S. cerevisiae* cultures that had been returned from early meiotic prophase to vegetative growth, some recombination was monitored. This, too, nourished the suspicion that some crossing-over could occur prior to SC formation (Olson and Zimmermann 1978). Finally, a yeast mutant (*zip1*) was found which develops no tripartite SC and yet has a crossing-over frequency of about 50% as compared to the wild-type (G.S. Roeder, personal communication).

Another proposal for the SC-crossing-over relationship was made by Carpenter (1987). She argued that local formation of (homologous and non-homologous) SC might be required for homology testing and therefore also for ectopic recombination events. However, as mentioned above, cytological evidence for non-homologous SC tracts during zygotene is scarce.

Possible functions of the SC
The hypothesis that a complete, tripartite SC is required for the formation of crossovers is challenged by its lack in the fungus *Aspergillus nidulans* and in *S. pombe*. Moreover, evidence for crossing-over (or, at least, crossing-over initiation) preceding SC formation, that has come from studies in *S. cerevisiae* (see above) have cast doubt on the dogma that the SC is an inevitable precondition for crossing-over.

In accordance with the view that the initiation of crossing-over precedes SC formation, Padmore et al. (1991) suggested that the SC might be required to convert Holliday-intermediates that form in the course of homologue recognition, into crossovers. However, this function would not call for a continuous end-to-end SC which is found in most organisms, nor makes the concept allowance for those exceptional cases, where crossing-over can occur in the absence of an SC.

The observation that *Aspergillus nidulans* and *Schizo-*

saccharomyces pombe, which both lack regular SCs, are devoid of crossing-over interference led to the proposal that the SC might serve as the carrier of a signal for crossing-over interference (Egel-Mitani et al. 1982, and lit. cit. therein). This concept is elaborated by a model of King and Mortimer (1990) according to which early nodules, which are believed to indicate the sites of primary homologue recognition, tend to convert into late nodules which correspond with crossovers. By doing so, they initiate polymerization of an SC component which grows along all the bivalent and repels early nodules. Thus only one to a few nodules per bivalent would remain where crossing-over takes place. In its simplest form (i.e., assuming that SC formation by zipper-like growth of the central element itself is the interference signal), the model by King and Mortimer (1990) does not conform with observations in higher plants. There, a large number of sites where the central element is initiated (see Gillies et al. 1990), meets a usually much lower number of chiasmata. Therefore, a new, hence unobserved, component of the SC would have to be invoked as the signal-trans-mitting structure. In *Allium fistulosum* it is particularly evident, that the pericentric regions where chiasmata are localized are not the first regions to be synapsed but rather come late (Albini and Jones 1987). Thus, although the model is attractive by suggesting a structural basis to crossing-over interference, it seems likely that the SC, and in particular the central element, is not the structure in question.

Maguire (1982, 1984) proposed that components of the SC could be important for chiasma maintenance and sister chromatid cohesion. Chiasma maintenance means that chias-mata can be stable until the onset of anaphase I only if distal to them the sister chromatids hold together. Other-wise bivalents would fall apart precociously and mis-segregate at the first meiotic division (Maguire 1984). In regions proximal to chiasmata, sister chromatids have to stay together even until anaphase II in order to ensure a regular second meiotic division. Maguire et al. (1991) speculated that sister chromatid cohesiveness might be somehow provided for by prior involvement in completed synapsis. This concept is not supported by the limited SC formation in the turbellarian *Mesostoma ehrenbergii*, which, in regions other than a small distal SC segment in only three of its five bivalents, does not even form axial elements (Oakley and Jones 1982, Croft and Jones 1989). Clearly, *Mesostoma* does not rely on the SC for performing orderly meiotic chromosome and chromatid separation.

In cases where SC formation is only local, crossing-over is also limited to these regions, as in the orthopteran *Chloealtis conspersa* (Moens and Short 1983) and male *M. ehrenbergii* (Croft and Jones 1989). However, in organisms where synapsis is complete, recombination events are not restricted to the sites where pairing is initiated. *A. fistulosum* is a good example of this. This observation suggests that it could be one function of the SC to confer proximity to chromosome regions that are not involved in primary homologue recognition (see Maguire 1977). This function would not conflict with evidence from cases where SC formation is limited or missing altogether. Whereas in organisms without SCs, crossing-over could occur only at sites where chromosomes have made homologous contact, a continuous SC would make any chromosomal region capable of homologous recombination. Thus, the organism would be provided with a wider range of possibilities for chiasma distribution. This concept would lend support to the "classical" hypothesis that - at least some - crossovers are initiated only after SC formation.

However, it may be over-simplified to attribute to the SC only one function, however convincing this may be. Like other complex biological structures it may have acquired (and lost) a variety of functions in the course of its evolution. The search has to be continued.

Acknowledgements. I am grateful to Dr. G.H. Jones and Dr. M.J. Moses for their valuable critical comments on the manuscript. This work was supported in part by the Austrian Fonds zur Förderung der wissenschaftlichen Forschung (Grant no. P7843-B).

References

Albini, S.M. and Jones, G.H. (1987) Synaptonemal complex spreading in *Allium cepa* and *A. fistulosum*. I. The initiation and sequence of pairing. **Chromosoma** 95, 325-338.

Bhargava, J., Engebrecht, J. and Roeder, G.S. (1992) The *rec102* mutant of yeast is defective in meiotic recombination and chromosome synapsis. **Genetics** 130, 59-69.

Carpenter, A.T.C. (1987) Gene conversion, recombination nodules, and the initiation of meiotic synapsis. **BioEssays** 6, 232-236.

Croft, J.A. and Jones, G.H. (1989) Meiosis in *Mesostoma ehrenbergii ehrenbergii*. IV. Recombination nodules in spermatocytes and a test of the correspondence of late recombination nodules and chiasmata. **Genetics** 121, 255-262.

Dresser, M.D. and Giroux, C.N. (1988) Meiotic chromosome behavior in spread preparations of yeast. **J. Cell Biol.** 106, 567-573.

Egel-Mitani, M., Olson, L.W. and Egel, R. (1982) Meiosis in *Aspergillus nidulans*: another example for lacking synaptonemal complexes in the absence of crossover interference. **Hereditas** 97, 179-187.

Gillies, C.B., Dollin, A.E. and Dai, K. (1990) Chromosomal and genetic factors influencing synaptonemal complex formation. **Chromosomes Today** 10, 297-310.

Haber, J.E. et al. (1991) The frequency of meiotic recombination in yeast is independent of the number and position of homologous donor sequences: implications for chromosome pairing. **Proc. Natl. Acad. Sci. USA** 88, 1120-1124.

Jenkins, G. and Rees, H. (1991) Strategies of bivalent formation in allopolyploid plants. **Proc. R. Soc. Lond. B.** 243, 209-214.

Jinks-Robertson, S. and Petes, T.D. (1986) Chromosomal translocations generated by high-frequency meiotic recombination between repeated yeast genes. **Genetics** 114, 731-752.

King, J.S. and Mortimer, R.K. (1990) A polymerization

model of chiasma interference and corresponding computer simulation. **Genetics** 126, 1127-1138.

Lichten, M., Borts, R.H. and Haber, J.E. (1987) Meiotic gene conversion and crossing over between dispersed homologous sequences occurs frequently in Saccharomyces cerevisiae. **Genetics** 115, 233-246.

Loidl, J. (1990) The initiation of meiotic chromosome pairing: the cytological view. **Genome** 33, 759-778.

Loidl, J. (1991) Coming to grips with a complex matter. A multidisciplinary approach to the synaptonemal complex. **Chromosoma** 100, 289-292.

Loidl, J., Ehrendorfer, F. and Schweizer, D. (1990) EM analysis of meiotic chromosome pairing in a pentaploid *Achillea* hybrid. **Heredity** 65, 11-20.

Loidl, J. and Jones, G.H. (1986) Synaptonemal complex spreading in *Allium*. I. Triploid *A. sphaerocephalon*. **Chromosoma** 93, 420-428.

Loidl, J., Nairz, K. and Klein, F. (1991) Meiotic chromosome synapsis in a haploid yeast. **Chromosoma** 100, 221-228.

Loidl, J. and Schweizer, D. (1992) Synaptonemal complexes of *Xenopus laevis*. **J. Heredity**, in press.

Maguire, M.P. (1977) Homologous chromosome pairing. **Phil. Trans. R. Soc. Lond. B.** 277, 245-258.

Maguire, M.P. (1982) Evidence for a role of the synaptonemal complex in provision for normal chromosome disjunction at meiosis II in maize. **Chromosoma** 84, 675-686.

Maguire, M.P. (1984) The mechanism of meiotic homologue pairing. **J. Theor. Biol.** 106, 605-615.

Maguire, M.P., Paredes, A.M. and Riess, R.W. (1991) The desynaptic mutant of maize as a combined defect of synaptonemal complex and chiasma maintenance. **Genome** 34, 879-887.

Moens, P.B. and Short, S. (1983) Synaptonemal complexes of bivalents with localized chiasmata in *Chloealtis conspersa* (Orthoptera). In **Kew Chromos. Conf. II.** (eds P.E. Brandham and M.D. Bennett), George Allen & Unwin, London, pp. 99-106.

Oakley, H.A. and Jones, G.H. (1982) Meiosis in *Mesostoma ehrenbergii ehrenbergii* (Turbellaria, Rhabdocoela). I.

Chromosome pairing, synaptonemal complexes and chiasma localisation in spermatogenesis. **Chromosoma** 85, 311-322.

Olson, L.W. and Zimmermann, F.K. (1978) Meiotic recombination and synaptonemal complexes in *Saccharomyces cerevisiae*. **Molec. Gen. Genet.** 166, 151-159.

Padmore, R., Cao, L. and Kleckner, N. (1991) Temporal comparison of recombination and synaptonemal complex formation during meiosis in *S. cerevisiae*. **Cell** 66, 1239-1256.

Pearlman, R.E., Tsao, N. and Moens, P.B. (1992) Synaptonemal complexes from DNase-treated rat pachytene chromosomes contain $(GT)_n$ and LINE/SINE sequences. **Genetics** 130, 865-872.

Roeder, G.S. (1990) Chromosome synapsis and genetic recombination: their roles in meiotic chromosome segregation. **Trends Genet.** 6, 385-389.

Smithies, O. and Powers, P.A. (1986) Gene conversions and their relation to homologous chromosome pairing. **Phil. Trans. R. Soc. Lond. B.** 312, 291-302.

Stern, H., Westergaard, M. and von Wettstein, D. (1975) Presynaptic events in meiocytes of *Lilium longiflorum* and their relation to crossing-over: a preselection hypothesis. **Proc. Natl. Acad. Sci. USA** 72, 961-965.

Weith, A. (1985) The fine structure of euchromatin and centromeric heterochromatin in *Tenebrio molitor* chromosomes. **Chromosoma** 91, 287-296.

23 Meiotic nodules in vascular plants

S.M. STACK, J.D. SHERMAN, L.K. ANDERSON and
L.S. HERICKHOFF
Colorado State University, USA

1 Introduction

Nodules have been observed in association with the synaptonemal
complex (SC) in a wide variety of eukaryotic organisms (see Carpenter 1979 and subsequently many other reports). However, this
discussion is restricted to nodules in vascular plants, including tomato (*Lycopersicon esculentum*, 2n = 24) on which we have
done most of our work. Tomato is well suited for studying nodules in whole mount spreads of SCs because we can identify each
of tomato's twelve SCs (Stack 1982; Sherman and Stack 1992) and
determine whether tomato's SCs are in early, middle, or late
pachytene (Stack and Anderson 1986a,b).
We identify nodules in plants by the following combination
of characteristics:

1) Nodules are recognized only when they are in association
with axial cores and central elements of SCs from leptotene
through early diplotene.
2) Nodules are ellipsoidal structures that are 30 to 150 nm
in their longest dimension.
3) Nodules have fuzzy borders, unlike stain precipitates of
similar size and shape.
4) Nodules can be stained with uranyl acetate - lead citrate
(UP), phosphotungstic acid (PTA), or silver (Ag40).

Because of the way nodules are defined, we cannot say with
confidence that nodules occur exclusively in association with
axial cores and SCs because there are no other criteria for
identifying nodules, e.g., antibodies, that would permit their
identification out of the context of axial cores and SCs. For
the same reason there is no way to know whether all nodules are
related.

2 Leptotene

Nodules are observed first in leptotene where they are associated with axial cores. Sometimes nodules can be seen before
there is any clear indication of presynaptic alignment (Fig. 1).
When presynaptic alignment is evident, nodules are often found
between associated axial cores (Fig. 2). This was first reported by Albini and Jones (1987) in *Allium cepa*, and we have made
similar observations in a number of other species including tomato, *Lilium longiflorum*, *Tradescantia edwardsiana*, and *Psilotum
nudum* (a primitive vascular plant) (unpublished observations and
Anderson and Stack 1988). In spreads of SCs from *P. nudum*, nodules were often observed in association with fibers connecting
axial cores (see Figs. 3-6 in Anderson and Stack 1988). Such
connecting fibers are sometimes seen in other plant species as

Chromosomes Today   Volume 11. Edited by A.T. Sumner and A.C. Chandley. Published in
1993 by Chapman & Hall, London. ISBN 0 412 47670 3

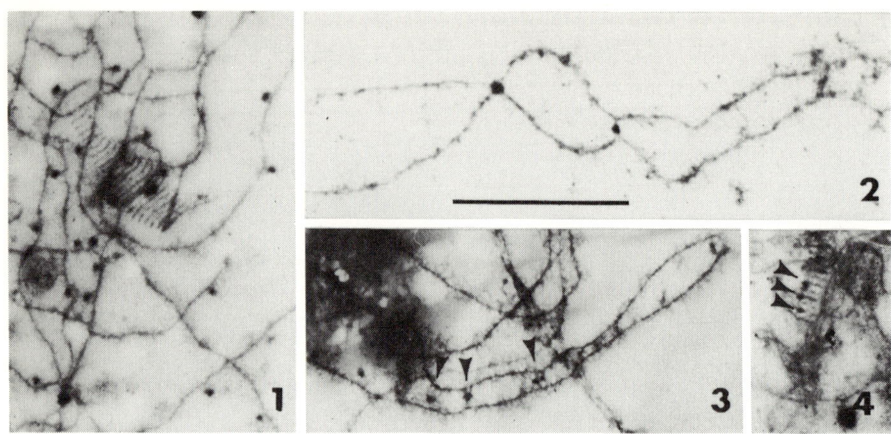

Fig. 1. UP-stained nodules and axial cores during leptotene in tomato. A polycomplex is visible in the center. Figure 1 is from Stack and Anderson (1986a) with permission from the American Journal of Botany. Fig. 2. Two *Lilium longiflorum* axial cores with nodules at association sites. Fig. 3. Tomato axial cores that are presynaptically aligned with nodules and fibers between cores (arrowheads). Fig. 4. Tomato leptotene polycomplex with nodules (e.g., arrowheads). The bar represents 2 μm.

well (Fig. 3). These fibers link axial cores and might be involved in drawing them together during synapsis.

Sometimes nodules can be found on polycomplexes during leptotene (Fig. 4). If polycomplexes represent polymerized excess SC components that are not associated with chromatin, then nodules may be attracted to SC components regardless of the presence of chromatin.

3 Zygotene

In early zygotene short segments of SC form in distal euchromatin first (Fig. 5). Many, but not all, of these segments have one or more nodules on the central element. This suggests that synapsis is preferentially initiated where nodules are located. However, since some short segments of SC have no nodules, nodules may not be absolutely required at synaptic initiation sites. Alternatively, nodules may be required, but sometimes they are lost from a synapsed segment.

As zygotene progresses, additional nodules are added to extended SCs (Fig. 6). These nodules are often close together, sometimes even touching, and do not show interference. Nodules can also occur on opposite sides of the central element, i.e., above and below the frontal plane of the SC (Stack and Anderson 1986a,b).

During zygotene, nodules vary in size (Figs. 5-7). It is possible that nodules are assembled in place, and size differences reflect different stages in assembly. Alternatively, there may be several distinct types of nodules.

At least in tomato, lateral elements of SCs in heterochromatin are thinner and lighter staining than lateral elements in

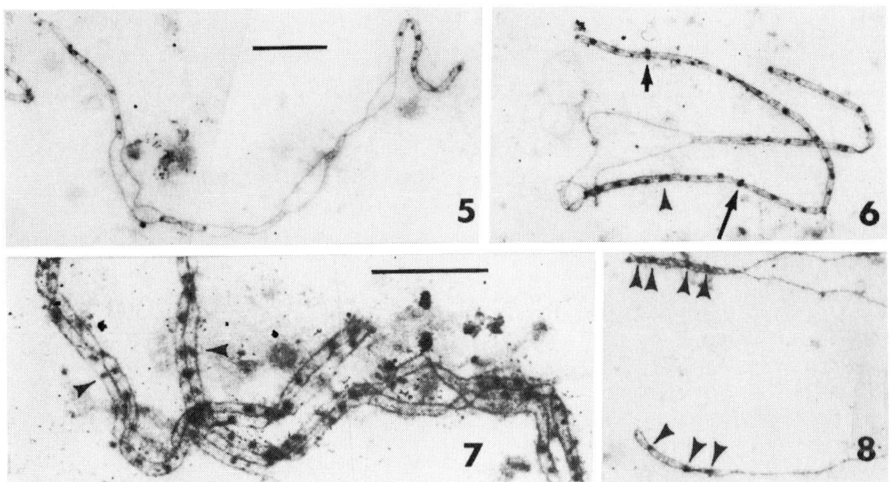

Fig. 5-8. Zygotene SCs. Fig. 5. UP-stained tomato bivalent with five synapsed segments. Nodules occur primarily on the distal, darker segments of SC that run through euchromatin. Nodules vary in size. The lighter staining SC and axial cores run through pericentric heterochromatin where there are no nodules. Fig. 6. UP-stained tomato bivalent with numerous nodules on long synapsed distal segments. Nodules can occur close together, even touching along the long axis of the SC (arrowhead), side-by-side across the long axis of the SC (short arrow), and above and below the frontal plane of the central element (long arrow at twist in the SC). Synapsis has just begun in the proximal pericentric heterochromatin where the SC and axial cores are stained more lightly and there are no nodules. Figure 6 is from Stack and Anderson (1986b) with permission from Chromosoma. Fig. 7. UP-stained tetraploid potato (*Solanum tuberosum*) SCs. There is extensive triple (three parallel lateral elements) and quadruple (four parallel lateral elements) synapsis. Many nodules occur on central elements. Nodules can occur at the same level on two adjacent central elements in triple synapsis (arrowheads). Figure 7 is from Sherman et al. (1989) with permission from Genome. Fig. 8. Part of a silver-stained (Ag40) trivalent from tomato trisomic for chromosome 10. Above are two ends synapsed homologously while the third chromosome below has partially synapsed as a non-homologous foldback. Nodules are visible on both synapsed segments (arrowheads). The bar in Figure 5 represents 2 μm for Figures 5, 6, and 8. The bar in Figure 7 also represents 2 μm.

euchromatin (Figs. 5,6). The large blocks of pericentric heterochromatin synapse last, and few nodules are found associated with these segments of SCs. Possibly tightly packed heterochromatin inhibits association of nodules or components of nodules with SCs (Stack and Anderson 1986a,b).

In trisomic and polyploid tomatoes and potatoes, triple and quadruple synapsis is common (Fig. 7). Here three or four lateral elements are linked in parallel by two or three central elements, respectively. Nodules can associate with adjacent central elements, sometimes at the same level (Sherman et al. 1989).

Foldback synapsis in tomato is almost certainly nonhomologous because molecular analysis indicates little duplication in the tomato genome (Tanksley et al. 1988). Some zygotene foldbacks do not contain nodules, again indicating that nodules may not be required for synapsis. On the other hand, some foldbacks have nodules, so zygotene nodules can occur in nonhomologously synapsed regions (Fig. 8).

4 Early pachytene

During early pachytene in tomato, the number of nodules per bivalent ranges from one to over twenty (Figs. 9, 10). Such variation is typical of early pachytene in other plants as well. Just as in zygotene, nodules are much more common in euchromatin than in heterochromatin.

In general, when there are more than three nodules on an SC, the nodules tend to vary in size like those observed in zygotene. When there are three or less nodules on an SC, the nodules tend to be large and more uniform in size like those observed later in pachytene (Figs. 9, 10). Because the latter nodules agree closely with the number and size of nodules in middle to late pachytene (see below), there appears to be a precipitous loss of nodules in early pachytene with the result that by middle pachytene, each SC has one to three nodules. Based on behavior and size, there may be two types of nodules on SCs in early pachytene, small nodules that are lost and large nodules that are retained.

Nodules usually have been observed only after PTA- or UP-staining, although Loidl (1987) reported sometimes observing silver-stained nodules in *A. ursinum*. Recently we developed a reliable method for silver staining nodules (Sherman et al. 1992) in a variety of plants. In the course of this study, we found that tomato nodules are differentially sensitive to temperature during silver staining (Sherman et al. 1992). If silver staining is performed at 40°C (Ag40), nodules are observed from zygotene through early diplotene at the same frequencies as after UP staining. If silver staining is performed at 50°C (Ag50), middle to late pachytene nodules occur at the same frequency as after UP staining. However, zygotene and early pachytene nodules are less common, occurring at the same frequency as nodules in middle to late pachytene. If staining is performed at 60°C (Ag60), all nodules are lost. Since most silver staining protocols utilize incubation at 60°C, this may explain why nodules are not usually observed on silver-stained SCs.

These observations again indicate that there may be two types of nodules in zygotene and early pachytene, those that are sensitive to silver-staining at 50°C and those that are not. Sensitive nodules and insensitive nodules may correspond to nodules that subsequently are normally lost and normally retained, respectively. Staining early pachytene nodules at 50°C may just hasten loss of nodules that would be lost normally after a few more hours of development, i.e., by middle pachytene (Sherman et al. 1992). From these observations it is not clear whether there are two types of nodules from the beginning or one type that gives rise to two types.

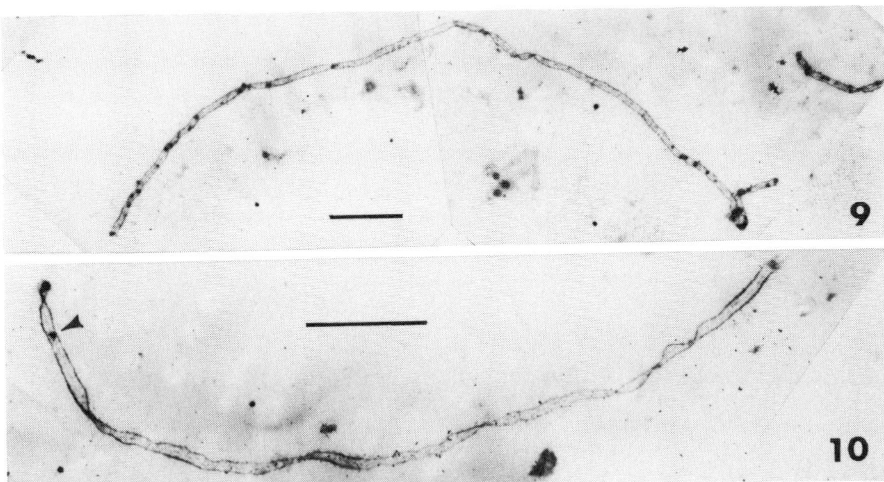

Fig. 9. UP-stained early pachytene tomato SC with numerous nodules in the distal euchromatin and few if any in the proximal pericentric heterochromatin. Fig. 10. UP-stained early pachytene tomato SC with one nodule (arrowhead). Both figures are from Stack and Anderson (1986a) with permission from the American Journal of Botany. Bars represent 2 µm.

5 Middle and late pachytene

During middle and late pachytene in tomato there are only one to three nodules per SC, which are usually located in distal euchromatin and rarely in proximal heterochromatin (Figs. 11-14) (Stack and Anderson 1986a,b; Sherman et al. 1992). Middle and late pachytene nodules are larger and more uniform in size than those observed in zygotene and early pachytene. These nodules are similar to the largest nodules on early pachytene SCs, suggesting that they have been retained from early pachytene. Although we have observed many examples of nonhomologous synapsis in tomato during middle to late pachytene, we have never observed it in association with nodules (e.g., Fig. 13).

Middle and late pachytene nodules are often referred to as "recombination nodules" ("RNs") or "late nodules" in contrast to the "early nodules" of zygotene and early pachytene. RNs are thought to lie at sites of crossovers and future chiasmata (Carpenter 1979). A comparison of chiasmata and RNs in reference to position and frequency strongly suggests a 1:1 relationship. For instance, in *A. fistulosum*, both chiasmata and RNs are localized near centromeres (Levan 1933; Albini and Jones 1984). In *L. longiflorum*, the number of chiasmata at diakinesis closely matches the number of middle to late pachytene RNs (Stack et al.1989). In normal diploid tomato, univalents have not been observed at diakinesis, indicating every homologous pair is held together by at least one chiasma (Rick, personal communication). This is in line with our observation that we invariably see at least one RN per SC in normal diploid tomatoes that have been properly spread and stained (Sherman et al. 1992). Similarly,

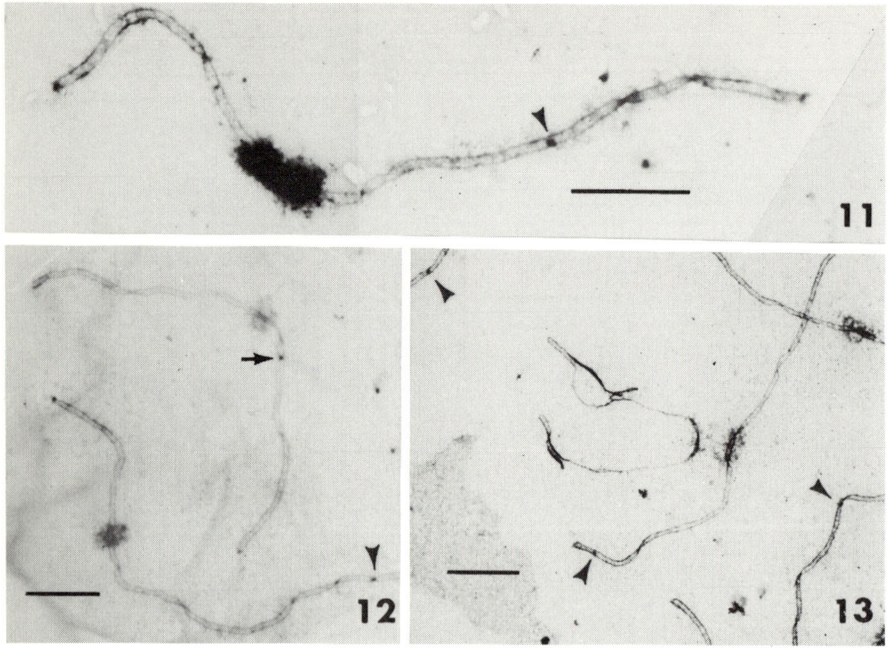

Figs. 11-13. Late pachytene SCs of tomato. Kinetochores are
prominent fuzzy structures about one micrometer in diameter.
Fig. 11. UP-stained SC with one RN on the SC in euchromatin (arrowhead). There seem to be two kinetochores in tandem on this
SC. Fig. 12. Two UP-stained SCs with one RN each. The RN in
the lower SC is in distal euchromatin (arrowhead). The RN in
the upper SC is in pericentric heterochromatin (arrow). Fig.
13. Silver-stained SCs from a tomato that is trisomic for
chromosome 5. Each end of the univalent is partially folded
back and nonhomologously synapsed on itself. There are no RNs
in these foldbacks, but RNs are visible in surrounding SCs (arrowheads). Bars represent 2 μm.

in an analysis of heterozygotes for five different translocations in tomato, we found that distributions of RNs at middle to
late pachytene correspond to distributions of chiasmata at diakinesis (Herickhoff et al. unpublished observations). In the
same study, we also found that RNs show interference within synapsed arms of quadrivalents.

For tomato the frequency of RNs is close to the frequency of
crossing over indicated by the size of the classical gene linkage map. Specifically there is an average of twenty-one RNs
per set of SCs and a total gene map length of 1063 cM (÷ 50 =
21.3 crossovers; Stack and Anderson 1986a; Tanksley and
Mutschler 1990). However, if the comparison is made for individual chromosomes there are inconsistencies. Some SCs have
more RNs and others have fewer RNs than predicted by the gene
linkage map (Sherman et al. unpublished observations). In addition, with a length of 1276 cM, the restriction fragment length
polymorphism (RFLP) linkage map for tomato is longer than the

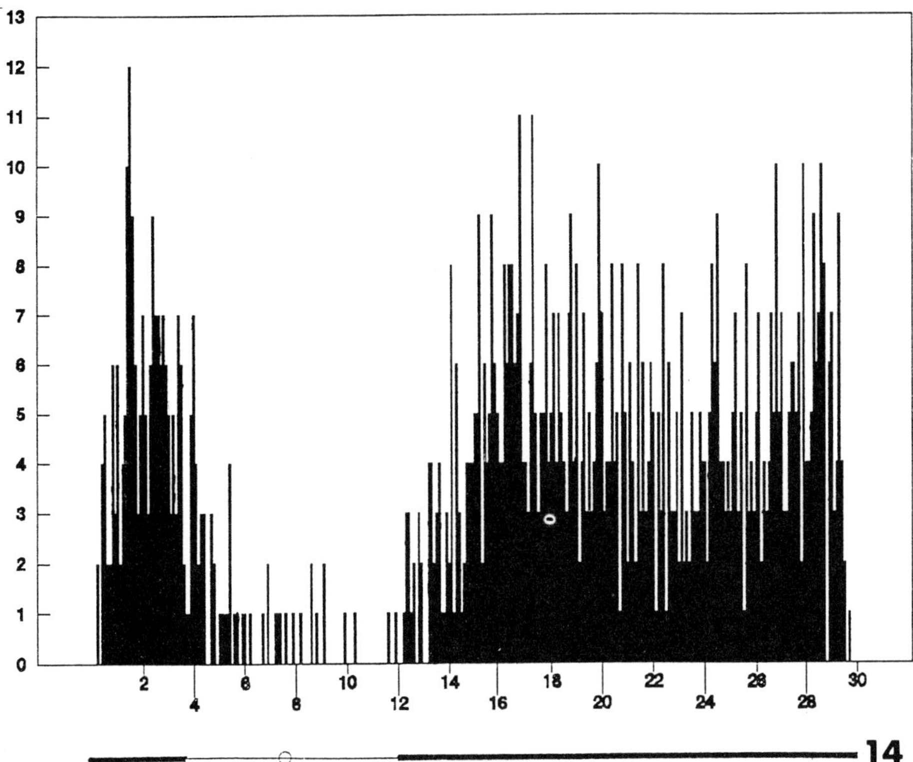

Fig. 14. Histogram of RN distribution on tomato SC 1 (n = 252 SCs). Numbers on the ordinate indicate RNs. The abscissa represents the average length of SC 1 (30 μm) divided into 0.1 μm segments. Numbers indicate micrometers. Below the histogram is a diagram of SC 1 drawn to scale. Distal euchromatin is indicated by thick lines. Proximal (pericentric) heterochromatin is indicated by thin lines on either side of an open circle at the kinetochore. While one would conclude from the diagram that a few RNs were observed in kinetochores, this appearance must be due to small misalignments because no RNs were observed within kinetochores.

gene map (Tanksley et al. 1992), indicating an average of (1276 ÷ 50 =) 25.5 crossover events and an expectation of 25.5 RNs (in contrast to the average of twenty-one observed). While there are several possible explanations for these apparent inconsistencies, we suggest that two factors are primarily responsible: Lack of saturation of the gene linkage map and different amounts of crossing over in male versus female gametes.

Because of the requirement for terminal markers on every chromosome, it is probable that the gene linkage map is not saturated and is thus too short. Since the RN map and the gene linkage map are currently about the same size, saturation should result in a gene linkage map that is longer than the RN map and closer in length to the RFLP map. Recently, Tanksley et al.

(1992) reported that the RFLP linkage map is near saturation with 1030 RFLP markers, an average separation between markers of only 1.2 cM, and a total length of 1276 cM. This RFLP map is significantly longer than the (21 x 50 =) 1050 cM RN map. This might be explained by either natural or artifactual loss of RNs, but we think the close agreement between RN and chiasma frequencies makes this unlikely. Is there anything about the way RN maps or linkage maps are prepared that could explain these differences in length?

The tomato RN map is based on observations of primary microsporocytes only, whereas linkage maps are usually based on recombination in both primary microsporocytes and primary megasporocytes. In this regard, de Vicente and Tanksley (1991) used RFLP markers to compare rates of recombination in primary microsporocytes and primary megasporocytes from a *L. esculentum* X *L. pennellii* hybrid. They found 18% more recombination in female gametes than in male gametes. If there is also more crossing over in female gametes of *L. esculentum*, then linkage maps relying on data from both male and female gametes should be longer than the RN map. Other differences may include (1) environmental and varietal effects on crossover rates (e.g., Gavrilenko 1984), differences in crossover rates between specific markers in primary microsporocytes and primary megasporocytes (de Vicente and Tanksley 1991; Thomas and Rothstein 1991), and additional recombination events, i.e., gene conversions, that might contribute to linkage maps but not RN maps (Stack and Anderson 1986b; Carpenter 1987). Also, RFLP maps are based on crossover rates in *L. esculentum* X *L. pennellii* hybrids which do not have exactly the same crossover rate and pattern as diploid tomato (Rick 1969). All of these factors considered, the resemblance between tomato linkage maps and the RN map is remarkable and indicative of a close relationship between RNs and crossover events.

Because we can identify each of tomato's twelve SCs, it has been possible to determine the frequency and distribution of RNs along each SC. From an ongoing analysis, we have prepared histograms to show the distribution of RNs on approximately 250 of each of tomato's twelve SCs (e.g., SC number 1 in Fig. 14). From these histograms a number of conclusions can be drawn:

1) Only a few RNs occur on SC in heterochromatin. However, these few indicate that there is a low level of crossing over in heterochromatin.
2) We have observed no RNs in SC associated with kinetochores, indicating no crossing over in kinetochores.
3) There are no RNs near telomeres, indicating no crossing over within about 0.3 μm of the ends of SCs.
4) RNs are not equally distributed along SC in euchromatin. There are fewer RNs near the euchromatin-heterochromatin border and near telomeres.

6 Early diplotene

During early diplotene SCs begin to break down. This occurs by desynapsis and/or fragmentation of SCs. In tomato desynapsis precedes fragmentation. Desynapsis is retarded at telomeres and RNs (Figs. 15,16)(Stack and Anderson 1986a). In *L. longiflorum*

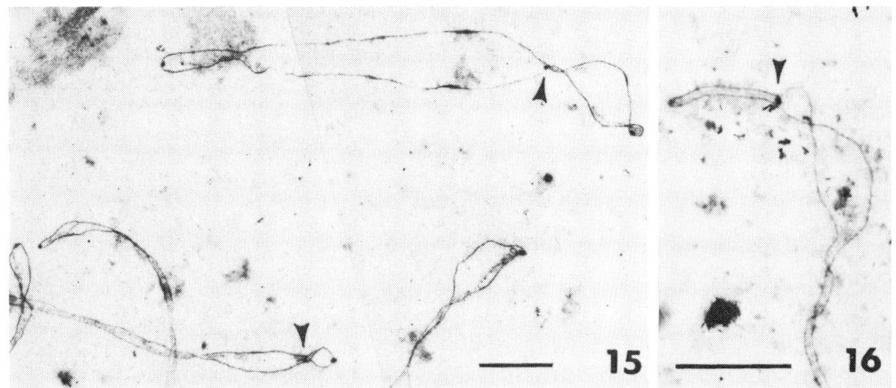

Fig. 15. Silver-stained early diplotene tomato SCs. The SCs have partly desynapsed. Nodules are visible in two areas that have not yet desynapsed (arrowheads) The ends of SCs have darkly stained telomeres. Fig. 16. UP-stained segment of a dip- lotene tomato SC where desynapsis has progressed to a nodule (arrow). Note the surrounding stain precipitates with sharp borders and irregular shapes. Bars represent 2 μm.

fragmentation occurs before desynapsis (Stack et al. 1989), and RNs are found on some but not all fragments. With elimination of SCs in each species, RNs are no longer identifiable.

7 Questions

The preceding observations raise a number of questions. Are nodules confined to prophase I of meiosis? Are nodules assem- bled and disassembled only in association with axial cores and SCs? Are RNs (late nodules) derived from early nodules, or are there two types of nodules initially? Do nodules have more than one function, for example, synapsing homologues, searching for homology, converting genes, crossing over, and forming chiasma- ta? All of these functions are much to ask of one structure, but the functions are related, and nodules are large and poten- tially complex. In comparison, a ribosome with a diameter of 20 nm contains approximately eighty-two proteins as well as four large RNA molecules (Alberts et al. 1989). On this basis, a nod- ule with a diameter of 100 nm has 125X the volume of a ribosome, and proportionally could contain (82 X 125 =) 10,250 proteins (and many RNA molecules?). A structure as large and potentially complex as a nodule could perform a number of functions.

8 Acknowledgements

We thank Dan Peterson for carefully reading the manuscript. This research was supported in part by grants DCB-8718170 and DCB-8918613 from the NSF and grant 1978-670 from the Colorado Agricultural Experiment Station.

9. References

Alberts, B. et al. (1989) **Molecular Biology of the Cell.** 2nd ed. Garland Publishing, Inc., NY, p. 211.

Albini, S., and Jones, G. (1984) Synaptonemal complex-associated centromeres and recombination nodules in plant meiocytes prepared by an improved surface-spreading technique. **Exp. Cell Res.**, 155, 588-592.

Albini, S.M., and Jones, G.H. (1987) Synaptonemal complex spreading in *Allium cepa* and *A. fistulosum*. I. The initiation and sequence of pairing. **Chromosoma**, 95, 324-338.

Anderson, L.K. and Stack, S.M. (1988) Nodules associated with axial cores and synaptonemal complexes during zygotene in *Psilotum nudum*. **Chromosoma**, 97, 96-100.

Carpenter, A.T.C. (1979) Synaptonemal complex and recombination nodules in wild type *Drosophila melanogaster* females. **Genetics**, 92, 511-541.

Carpenter, A.T.C. (1987) Gene conversion, recombination, and the initiation of meiotic synapsis. **Bioessays**, 6, 232-236.

Gavrilenko, T.A. (1984) Effect of temperature on crossing over in tomatoes. **Tsitol. Genet.**, 18, 347-352.

Levan, A. (1933) Cytological studies in Allium. IV. Allium fistu losum. **Svensk Bot. Tidskr.**, 27, 211-232.

Loidl, J. (1987) Synaptonemal complex spreading in *Allium ursinum*: pericentric asynapsis and axial thickenings. **J. Cell Sci.**, 87, 439-448.

Rick, C.M. (1969) Controlled introgression of chromosomes of *Lycopersicon pennellii* into *Lycopersicon esculentum*: Segregation and recombination. Genetics, 62,753-768.

Sherman, J.D. and Stack, S.M. (1992) Two-dimensional spreads of synaptonemal complexes from solanaceous plants. V. tomato (*Lycopersicon esculentum*) karyotype and idiogram. **Genome**, 35, 354-359.

Sherman, J.D., Stack, S.M., and Anderson, L.K. (1989) Two dimensional spreads of synaptonemal complexes from solanaceous plants. IV. Synaptic irregularities. **Genome**, 32, 743--753.

Sherman, J.D., Herickhoff, L.A. and Stack, S.M. (1992) Silver staining two types of meiotic nodules. **Genome**, in press.

Stack, S. (1982) Two-dimensional spreads of synaptonemal complexes from solanaceous plants. I. The technique. **Stain Technol.**, 57, 265-272.

Stack, S., and Anderson, L. (1986a) Two-dimensional spreads of synaptonemal complexes from solanaceous plants. II. Synapsis in *Lycopersicon esculentum* (tomato). **Am. J. Bot.**, 73, 264-281.

Stack, S., and Anderson, L. (1986b) Two-dimensional spreads of synaptonemal complexes from solanaceous plants. III. Recombination nodules and crossing over in *Lycopersicon esculentum* (tomato). **Chromosoma**, 94, 253-258.

Stack, S.M., Anderson, L.K., and Sherman, J.D. (1989) Chiasmata and recombination nodules in *Lilium longiflorum*. **Genome**, 32, 486-498.

Tanksley, S.D. et al. (1988) Conservation of gene repertoire but not gene order in pepper and tomato. **Proc. Nat. Acad. Sci. USA**, 85, 6419-6423.

Tanksley, S.D. et al. (1992) High density molecular linkage maps of tomato and potato: Biological inferences and practical applications. **Genetics**, in press.
Tanksley, S.D. and Mutschler, M.A. (1990) Linkage map of the tomato (*Lycopersicon esculentum*) (2 N = 24), in **Genetic maps. Book 6. Plants** (ed S.J. O'Brien), Cold Spring Harbor Laboratory Press, Cold Spring Harbor, pp. 6.3-6.15.
Thomas, B.J. and Rothstein, R. (1991) Sex, maps, and imprinting. **Cell**, 64, 1-3.
de Vicente, M.C., and Tanksley, S.D. (1991) Genome-wide reduction in recombination of backcross progeny derived from male versus female gametes in an interspecific cross of tomato. **Theor. Appl. Genet.**, 83, 173-178.

24 The origin of non-disjunction in humans

T. HASSOLD and S. SHERMAN

Emory University School of Medicine, USA

1 Introduction

Trisomy is the most commonly identified chromosome abnormality in humans, occurring in at least 4% of all clinically recognized pregnancies (Hassold and Jacobs, 1984). Some trisomies are compatible with livebirth as approximately 0.3% of all newborn infants have an additional sex chromosome or chromosome 13, 18, or 21. However, most trisomic fetuses do not survive to term. They are associated either with clinically recognized spontaneous abortions, where they account for approximately 25% of all such fetuses, or they terminate as sub-clinical spontaneous abortions. Estimates of trisomy frequency for this latter type of pregnancy outcome are not available but evidence from cytogenetic studies of sperm and oocytes suggest that as many as 10% - 15% of all human conceptuses may be trisomic (e.g., Pellestor, 1991).

Despite the high frequency of trisomy and its obvious clinical importance, we still know relatively little about its origin. In large part, this is due to the difficulties inherent in obtaining the relevant meiotic material, i.e., the testicular and ovarian tissues in which the non-disjunctional events occur. Thus, most studies of human non-disjunction have focused on determining the parent and meiotic stage of origin of the additional chromosome. Studies beginning in the 1960s made use of blood group polymorphisms to evaluate the origin of sex chromosome trisomy and during the 1970s and 1980s chromosome heteromorphisms were used to determine the origin of autosomal trisomies, especially trisomy 21. However, this approach also has had limitations owing to the relatively low level of informativeness of blood group and chromosome polymorphisms.

The recent identification of DNA polymorphisms has eliminated this latter problem. Highly polymorphic markers are now available for all human chromosomes and, for many, centromere-specific polymorphisms are available as well. Thus, it is possible to determine the parental origin of any trisomy and both the parent and meiotic stage of origin of several trisomies. Furthermore, the use of DNA polymorphisms now makes it possible to ask new questions regarding the origin of human trisomies. One of the most important of these is whether or not abnormal genetic recombination is associated with non-disjunction in humans. In yeast, meiotic mutants affecting recombination increase the likelihood of non-disjunction (e.g., Surosky and Tye, 1988), and in female Drosophila, X-chromosome non-disjunction appears to occur more frequently in bivalents

Chromosomes Today Volume 11. Edited by A.T. Sumner and A.C. Chandley. Published in 1993 by Chapman & Hall, London. ISBN 0 412 47670 3

that have undergone zero or two exchanges than those with one exchange (Merriam and Frost, 1964). This suggests that, for a given bivalent, there may be an optimal number of exchanges for normal disjunction. The application of DNA polymorphism analysis now makes it possible to determine whether this relationship also holds for human non-disjunction.

In this review, we summarize our molecular studies of the parent and meiotic stage of origin of non-disjunction in over 500 trisomic conceptuses and describe our preliminary studies of recombination in trisomies 16, 21, and the sex chromosome trisomies.

2 Methodology

2.1 Study population

Trisomic conceptuses were ascertained from one of three sources. First, all cases of trisomies 2, 4, 7, 10, 11, 14, 15, 16, and 22, six cases of trisomy 13 and seven cases of trisomy 21 were identified as part of cytogenetic studies of spontaneous abortions conducted at hospitals in Honolulu, Hawaii (Hassold et al., 1980) or Atlanta, Georgia (Hassold, unpublished observations). Second, ten cases of trisomy 21 were ascertained as therapeutic abortions, most of which were detected on prenatal diagnosis for advanced maternal age. Finally, all remaining cases of trisomy 13 and trisomy 21, and all cases of sex chromosome trisomy involved liveborn individuals; most of these were ascertained because of phenotypic abnormalities, although some individuals with sex chromosome trisomy were identified as part of cytogenetic studies of unselected newborn infants (e.g., see Robinson et al., 1990).

For trisomies 13, 21, and the sex chromosome trisomies, there was no significant effect of ascertainment on the parent or meiotic stage of origin of trisomy. Thus, for these trisomies, we have pooled the results from the different ascertainment categories.

2.2 DNA studies

For determinations of parental origin of trisomy, we used a total of 109 DNA probes or primer sets from chromosome 2, 4, 7, 10, 11, 13, 14, 15, 16, 21, 22, and the X chromosome. Most of these detected restriction fragment length polymorphisms (RFLPs) or variation in the number of tandem repeats (VNTRs) in Southern blotting experiments, although for chromosome 21, PCR-based assays were used to detect variation in microsatellites.

Determinations of meiotic stage of origin of trisomy were made by first identifying centromeric markers for which the parent of origin was heterozygous, and then observing whether heterozygosity had been maintained in the trisomic offspring (consistent with meiosis I non-disjunction) or had been reduced to homozygosity (consistent with meiosis II non-disjunction). For chromosomes 13, 16, and the X chromosome, these determinations were based on probes detecting variation in alpha satellite centromeric sequences. For chromosome 21, for which a highly polymorphic centromeric marker has not yet been described, any of five pericentromeric markers (D21S215, D21S120, D21S13, D21S16, and D21S192) were used to infer the meiotic stage of non-disjunction.

Studies of recombination and construction of centromere-based genetic maps for chromosome 21, the X chromosome, and the XY pseudoautosomal region were carried out as previously described (Hassold et al., 1991; Sherman, et al., 1991). For these analyses, the evaluation of recombination depended on identifying the parent of origin of trisomy, and on identifying DNA markers for which he/she was heterozygous. The markers were then evaluated in the trisomic offspring to determine if heterozygosity was maintained (non-reduction) or reduced to homozygosity (reduction). Recombination was considered to have occurred between two adjacent loci if, in the trisomic individual, one locus was reduced and the other was non-reduced. Multiple pairwise comparisons of markers were then used to construct a genetic map of a chromosome region or of the entire chromosome (Hassold et al., 1991; Sherman et al., 1991).

3 Results

3.1 The parent and meiotic stage of origin of trisomy

To date, we have initiated molecular studies of the origin of trisomy in over 600 trisomic fetuses or liveborns; results on 506 cases in which we have been able to specify the parental origin of the additional chromosome are provided in Table 1. Of the 506 trisomies, 421 (83.2%) were maternally derived, with maternal meiosis I errors being approximately three times as common as maternal meiosis II errors. Thus, it is clear that non-disjunction at maternal meiosis I is the most frequent cause of trisomy in humans.

However, our results also provide evidence for variation in non-disjunctional mechanisms among the different chromosomes. For example, there was significant variation in the proportion of paternally-derived trisomies, with the most obvious difference being between the 47,XXY condition, for which almost 50% of cases were paternal in origin, and trisomy 16, for which we have yet to identify a paternal trisomy. In general, there seems to be an association between chromosome size and likelihood of paternal non-disjunction, as paternally-derived cases accounted for 23.8% (10/42) of trisomies involving autosomes 2-15, but only 4.9% (14/284) of trisomies involving chromosomes 16, 21, or 22.

Among maternally derived trisomies, the proportion of meiosis I to meiosis II cases also varied among chromosomes. That is, for trisomy 21 and the sex chromosome trisomies, approximately one-third of maternal trisomies were apparently due to meiosis II non-disjunction. In contrast, virtually all cases of trisomy 16 were attributable to non-disjunction at maternal meiosis I, suggesting the existence of a mechanism of non-disjunction which is restricted to, or more likely to involve, chromosome 16.

3.2 Recombination and non-disjunction

We have now initiated studies evaluating the possible association of aberrant recombination and non-disjunction for five different conditions, namely sex chromosome trisomy and trisomy 21 of paternal origin, and sex chromosome trisomy, trisomy 21, and trisomy 16 of maternal origin. We have constructed trisomy-based genetic maps for three of these conditions and compared them to conventional linkage maps from normal meioses, and these are shown in Fig. 1.

Table 1. Molecular studies of parent and meiotic stage of origin in human trisomies (data from Jacobs et al., 1989; Hassold et al., 1991; Sherman et al., 1991; and unpublished observations).

		PATERNAL			MATERNAL			
Trisomy	No. of Cases	Meiosis I	Meiosis II	Meiosis I or II	Meiosis I	Meiosis II	Meiosis I or II	% Paternal
2	4	---	---	2	---	---	2	50
4	4	---	---	1	---	---	3	25
7	3	---	---	---	---	---	3	0
10	4	---	---	---	---	---	4	0
11	1	---	---	---	---	---	1	0
13	7	---	---	2	1	---	4	29
14	8	---	---	2	---	---	6	25
15	11	---	---	3	---	---	8	27
16	62	---	---	---	51	1	10	0
21	211	2	8	4	97	35	65	7
22	11	---	---	---	---	---	11	0
XXY	133	58	---	---	40	13	22	44
XXX	47	---	3	---	24	10	10	6

3.2.1 47,XXY of paternal origin

To date, we have studied 41 paternally derived 47,XXYs and their parents, testing six polymorphic loci spanning the pseudoautosomal region. These results have been described previously in Hassold et al. (1991). For our analyses, the 41 individuals were scored as being non-reduced or reduced to homozygosity at each locus, with all individuals being scored as non-reduced at the centromere (sex/cen) since they had received both their father's X and Y chromosomes. Six of the 41 individuals were reduced at one or more of the pseudoautosomal loci, consistent with crossing-over in this region. However, in the remaining cases, including 33 which were informative for the telemeric DXYS14 and DXYS20 loci, all loci were heterozygous; i.e., there was no evidence for crossing-over. Since normal male meioses are associated with a single obligatory exchange in this region (Page et al., 1987), these observations indicate an association between failure of pseudoautosomal region recombination and XY chromosome non-disjunction.

To further test these observations, we constructed a genetic map of the pseudoautosomal region based on the meioses involved in XY chromosome non-disjunction, and compared it to a map based on normal meiotic events (Fig. 1B). Under the assumption of complete interference, the trisomy-based map was highly significantly shorter than the normal map (13 cM vs. 44 cM), confirming the association of reduced recombination and XY chromosome non-disjunction.

Thus, our results provide evidence that most 47,XXYs of paternal origin result from meioses in which the X and Y chromosomes failed to recombine. However, we cannot yet conclude that this is the cause of non-disjunction; i.e., it could be that the primary event is an abnormality in formation or maintenance of pairing, with the failure to recombine simply being a by-product of this abnormal process. Nevertheless, our results make it clear that the non-disjunctional event is typically due to errors in prophase I, and not to processes acting at metaphase/anaphase I which disrupt chromosome segregation. Thus, future studies of paternal sex-chromosome non-disjunction can focus on why the X and Y chromosomes fail to recombine.

3.2.2 Trisomy 21 of paternal origin

Cytological studies of male meioses indicate that, as for the XY bivalent, the two chromosomes 21 are typically held together by a single chiasma (Laurie and Hulten, 1985). Therefore, we were interested in knowing whether or not paternal non-disjunction for chromosome 21 was associated with a failure of recombination. To date, the small number of paternally derived cases (n = 14) precludes the construction of a trisomy-based male genetic map of chromosome 21. Nevertheless, our preliminary results already indicate that failure to recombine is relatively unimportant in the genesis of paternally derived trisomy 21. That is, the majority of cases apparently involved errors at meiosis II and, in those cases involving meiosis I errors, cross-overs usually were detected. Thus, it seems likely that the factors affecting paternal chromosome 21 non-disjunction are different than those associated with XY non-disjunction.

3.2.3 Sex chromosome trisomy of maternal origin

In 1990, Morton et al. presented a preliminary trisomy-derived genetic map of the X chromosome, based on analyses of 65 47,XXYs and 47,XXXs of maternal origin. Figure 1A provides a schematic representation of the map and, for comparison, a normal female map of the X chromosome based on the same genetic markers. The trisomy-related map was significantly shorter than the conventional map (106 cM vs 173.5 cM), suggesting an association between reduced recombination and maternal X chromosome non-disjunction. Furthermore, among 34 cases of presumptive meiosis I origin in which at least five non-centromeric loci were evaluated, 6 (18%) showed no evidence of crossing-over between the non-disjoined X chromosomes. Thus, some cases of maternally-derived sex chromosome trisomy apparently are associated with failure of recombination of the X chromosome bivalent.

However, there was also evidence for an effect of increased recombination on X chromosome non-disjunction as, in the pericentromeric region, the trisomy-based map was longer than the normal map, although not significantly so. Further analyses of the pericentromeric region will be important in determining whether or not "chromosome entanglement" (Bridges et al., 1916) due to pericentromeric recombination is a significant contributor to X chromosome non-disjunction in humans.

3.2.4 Trisomy 21 of maternal origin

Previous studies carried out in our (Sherman et al., 1991) and other laboratories (Warren et al., 1987) indicated an association between reduced recombination and trisomy 21 of maternal origin. However, neither of these studies separated cases by meiotic stage of origin to insure that the effect was restricted to trisomies of meiosis I origin. Thus, we have now used pericentromeric markers on chromosome 21 to infer the meiotic stage of origin, and have identified 97 trisomies of meiosis I and 35 trisomies of meiosis II origin. Using these cases, we constructed two different genetic maps and compared them to a normal female map (Fig. 1C).

The meiosis I trisomy map was significantly shorter than the normal map (24 cM vs 98 cM), confirming the earlier reports of reduced recombination in maternal trisomy 21. In studies comparing the distribution of cross-overs between the meiosis I trisomy map and the normal map, we were unable to identify any obvious differences; i.e., there were no "cold spots" of recombination associated with the meiosis I trisomy map. Instead, the reduction in recombination was constant along the entirety of 21q, indicating that the recombination deficiency is attributable to trisomy-generating meioses in which the two chromosome 21s failed to pair and/or to recombine.

Surprisingly, the meiosis II trisomy map was also significantly shorter than the normal map (56 cM vs 98 cM). The reason for this is not yet clear, but it may be due to occasional misclassifications of the stage of origin of trisomy. For example, our assay does not discriminate between meiosis II and mitotic non-disjunction. If a significant proportion of cases scored as meiosis II trisomy are, in fact, mitotic in origin, this will act to reduce the length of the meiosis II map. This is because mitotic non-disjunction results in two identical chromosomes which, in our assay, will appear as non-recombinants.

3.2.5 Trisomy 16 of maternal origin

To date, we have identified 62 cases of maternally derived trisomy 16

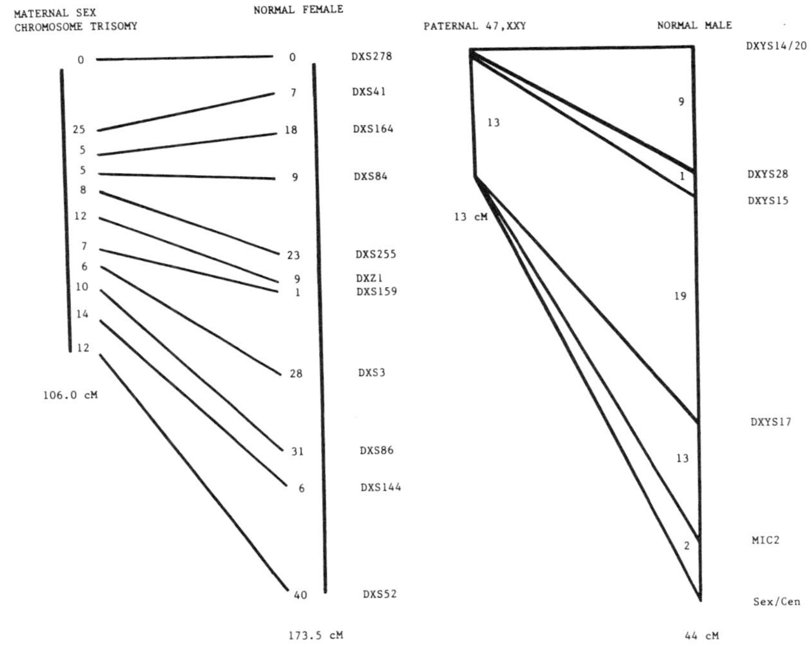

Fig. 1. Trisomy-based genetic linkage maps (and, for comparison, normal maps) for (A) the X chromosome (B) the pseudoautosomal region and (C) chromosome 21.

Maternal
Meiosis I
Maternal
Meiosis II
D21S215/S120/
S16/S13
6
4
2
D21S110
6
19
25
1
2
5
2
24 cM
D21S1/S11
2
1
18
13
SOD1/D21S28
D21S58
12
16
D21S17
7
D21S55
ETS2
5
D21S15
7
D21S19
7
D21S112
56 cM
23
COL6A1
99 cM

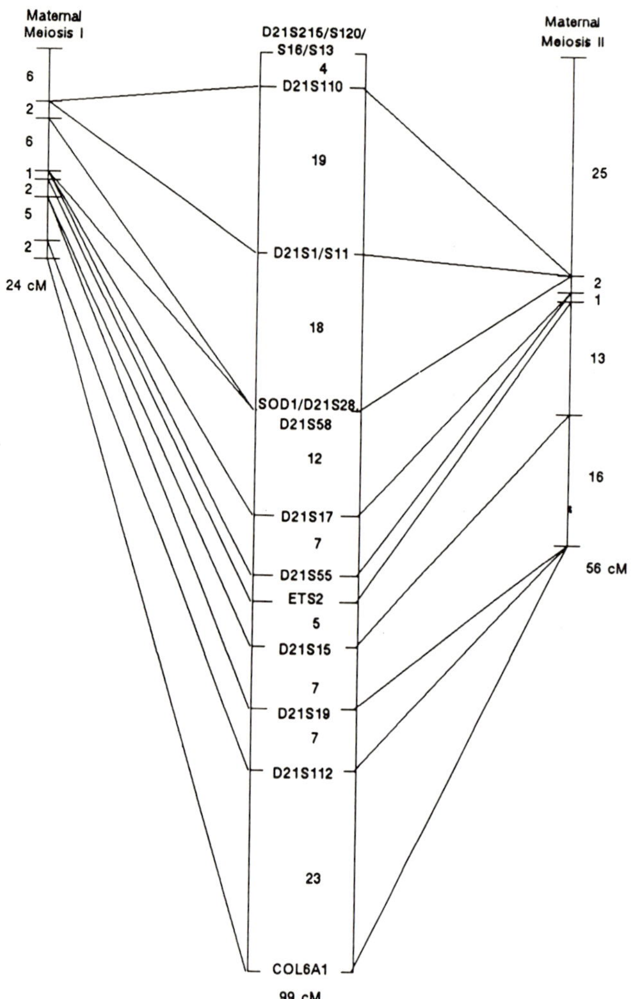

1C

of which 51 were due to meiosis I errors. Using the meiosis I cases, we constructed a preliminary genetic map, which included loci from the subtelomeric region of 16p, the centromere, and distal 16q, and thus should span most of the genetic length of the chromosome. The overall length of the map was 120 cM, which is somewhat shorter than the 198 cM value reported for the normal female map (Kozman et al., 1992). However, our map contained one gap on 16p where we were unable to link adjacent loci. Additionally, several of our markers are not yet represented on the normal female map, so that we have not been able to directly compare the two types of maps. Thus, we do not yet know if the observed reduction in map length is a significant one.

However, a second line of evidence suggests that even if recombination deficiency is associated with trisomy 16, total absence of recombination is relatively unimportant in the genesis of this condition. That is, we have detected cross-overs in over 90% of the meiosis I cases for which a sufficient number of loci were tested; in contrast, the vast majority of paternal 47,XXY and maternal trisomy 21 cases involve meioses in which recombination between the non-disjoined chromosomes did not occur. Thus, it seems likely that the mechanisms of non-disjunction leading to maternal trisomy 16 differ from those associated with maternal trisomy 21 or paternal 47,XXY.

3.2.6 Recombination and non-disjunction: summary

Our preliminary studies suggest that most cases of paternal 47,XXY and a large proportion of cases of maternal trisomy 21 result from meioses in which the non-disjoined chromosomes failed to recombine. This situation also applies to a smaller proportion of cases of maternal sex chromosome trisomy, although for this condition increased pericentromeric recombination may also be important. Studies of paternal trisomy 21 and maternal trisomy 16 have failed to identify an association between recombination abnormalities and non-disjunction.

Acknowledgements
We gratefully acknowledge our collaborators at the John F. Kennedy Institute (Glostrop, Denmark), the Wessex Regional Genetics Laboratory (Salisbury, England), McMaster University (Hamilton, Canada), and the University of Michigan (Ann Arbor, USA), and our own laboratory staff at Emory University. This work was supported by NIH contract HD92907 and NIH grant HD21341.

4. References

Bridges, C.B. (1916) Nondisjunction as proof of the chromosome theory of heredity. **Genetics,** 1, 1-52, 107-163.

Hassold, T. et al. (1980) A cytogenetic study of 1000 spontaneous abortions. **Ann. Hum. Genet.,** 44, 151-178.

Hassold T. and Jacobs P.A. (1984) Trisomy in man. **Annu. Rev. Genet.,** 18, 69-97.

Hassold, T.J. et al. (1991) XY chromosome nondisjunction in man is associated with diminished recombination in the pseudoautosomal region. **Am. J. Hum. Genet.,** 49, 253-260.

Jacobs, P.A. et al. (1989) The origin of sex chromosome aneuploidy, in

Molecular and Cytogenetic Studies of Non-disjunction. (eds. Hassold, T.J. and Epstein, C.J.) Alan R. Liss, New York, NY, pp. 135-151.

Kozman, H. et al. (1992) Combination of the genetic and cytogenetic physical maps of chromosome 16. **Cytogenet. Cell Genet.**, 60, 172.

Laurie, D.A. and Hulten, M.A. (1985) Further studies on bivalent chiasma frequency in human males with normal karyotypes. **Ann. Hum. Genet.**, 49, 189-201.

Merriam, J.R. and Frost, J.N. (1964) Exchange and nondisjunction of the X-chromosomes in female Drosophila melanogaster. **Genetics**, 49, 109-122.

Morton, N.E. et al. (1990) A centromere map of the X chromosome from trisomies of maternal origin. **Ann. Hum. Genet.**, 54, 39-47.

Page, D.C. et al. (1987) Linkage, physical mapping, and DNA sequence analysis of pseudoautosomal loci on the human X and Y chromosomes. **Genomics**, 1, 243-256.

Pellestor, F. (1991) Frequency and distribution of aneuploidy in human female gametes. **Hum. Genet.**, 86, 283-288.

Robinson, A. et al. (1990) Sex chromosome aneuploidy: The Denver prospective study, in **Birth Defects: Original Article Series,** Vol. 26, No. 4, (eds. Evans, J.A. Hamerton, J.L. Robinson, A.) Wiley-Liss, New York, pp. 59-115.

Sherman, S.L. et al. (1991) Trisomy 21: Association between reduced recombination and non-disjunction. **Am. J. Hum. Genet.**, 49, 608-620.

Surosky, R.T. and Tye, B-K. (1988) Meiotic disjunction of homologs in Saccharomyces cerevisiae is directed by pairing and recombination of the chromosome arms but not by pairing of the centromeres. **Genetics**, 119, 273-287.

Warren, A.C. et al. (1987) Evidence for reduced recombination on the nondisjoined chromosomes 21 in Down syndrome. **Science**, 237, 652-654.

25 Studies on maternal age-related aneuploidy in mammalian oocytes and cell cycle control

U. EICHENLAUB-RITTER

University of Bielefeld, Germany

Introduction

In the human, about 60% of all oocytes ovulated in women over 40 years of age may be chromosomally unbalanced (Hassold & Jacobs, 1984). Although rises in aneuploidy with increased maternal age are much lower in other mammals, the CBA/Ca mouse has been established as an appropriate model system to study the mechanisms responsible for errors in chromosome distribution (e.g. Brook et al, 1984; Eichenlaub-Ritter et al., 1988). We recently obtained evidence that predisposition to nondisjunction may be correlated to a shortened cell cycle in oocytes from aged females (Eichenlaub-Ritter & Boll, 1989). At the molecular level, these alterations in maturation could be based on disturbances in protein phosphorylation since a drop in cAMP-levels, and the transient deactivation/reactivation of cAMP-dependent protein kinase (PKA; reviewed by Schultz, 1988), and the Ca^{2+}-phospholipid-dependent protein kinase (PKC; Bornslaeger et al., 1986) have been implicated in comitment to resume meiosis in mammalian oocytes. Also, alterations in the phosphorylation status and activity of the catalytic subunit of maturation promoting factor (MPF), $p34^{cdc2}$ (Choi et al., 1991), and the c-*mos*-kinase (Zhao et al., 1990) could lead to disturbances in cell cycle progression. We therefore tried to affect the pattern of protein phosphorylation in germinal vesicle (GV)-stage oocytes of young mice by transient exposure to drugs which influence, directly or indirectly, the activity of PKA, like isobutyl-1-methylxanthine (IBMX) and forskolin (Boernslaeger et al., 1985), or that of PKC, like phorbol esters (Urner & Schoerderet-Slatkine, 1984). In the latter series of experiments, consequences of treatment with a tumor-promoting phorbol, phorbol 12,13-dibutyrate (PdBu) on maturation and correct chromosome distribution was compared to that of a nonpromoter, phorbol 12,13-diacetate (PdAc). In a third group a simultaneous exposure to PdBu and retinoic acid (RA) was employed since retinoids can antagonize the activity of tumor promoters (Urner & Schoerderet-Slatkine, 1984). The effect of the drug-treatments on protein phosphorylation patterns in oocytes, and on disturbances in cell cycle progression, spindle formation and chromosome distribution after recovery were analysed (Gülle & Eichenlaub-Ritter, 1990).

Besides alterations in protein phosphorylation, disturbances in gene expression may contribute to a loss of cell cycle control, and, possibly, fidelity of chromo-

Chromosomes Today Volume 11. Edited by A.T. Sumner and A.C. Chandley. Published in 1993 by Chapman & Hall, London. ISBN 0 412 47670 3

some distribution. Thus, overexpression of the human
CDC25B gene which is known to control the G2- to M-phase
transition in mitosis may induce loss of cell cycle control and transformation in virally transfected cells
(Nagata et al., 1991). In oocytes, duration of M-phase
and the critical transition from meta- to anaphase
appears to involve the activity of *the* c-*mos* protein. Ab-
lation of c-*mos* message by injection of anti-sense oligonucleotides into mouse oocytes can result in formation of
aberrant spindles, or a total block in first metaphase
and inhibition of polar body formation (PbF) (Zhao et
al., 1990). We therefore compared the levels of c-*mos*
mRNA in mouse oocytes being at high to those at low risk
for aneuploidy using PCR techniques (Sobek & Eichenlaub-
Ritter, 1991). In these experiments actin was chosen as
an intrinsic marker of transcription because it appears
to be constitutively expressed in GV-stage oocytes
(Bachvarova et al., 1989).

Materials and Methods

Fully grown, GV-intact oocytes were isolated from ovaries
of young (2-4) or aged (>9month) CBA/Ca mice at diestrous
of the natural cycle, and spontaneously matured *in vitro*
in modified M2-medium (Eichenlaub-Ritter et al., 1988).
GVBD and PbF were determined at hourly intervals under a
dissection microscope. At defined times after isolation
from ovaries, oocytes were processed for indirect anti-
tubulin immunofluorescence (Eichenlaub-Ritter et al.,
1986). Chromosomal constitution of *in vitro* matured oocytes was assessed after spreading and C-banding (Eichenlaub-Ritter & Boll, 1989). In nocodazole experiments,
oocytes were exposed for 1h to the drug, after 7 or 8h of
maturation, and recovered in M2 medium (Eichenlaub-Ritter
& Boll, 1989). The effect of transient exposure to drugs
interfering with the activity of kinases was studied in
oocytes placed directly, after isolation from follicles,
into medium containing IBMX (100µM, 5h), forskolin (50µM,
3h), PdBu (2,5ng/ml, 3h), PdAc (200ng/ml,3h), or PdBu
(2,5ng/ml) plus RA (100µM, 3h). Concentration of drugs
was chosen so that all oocytes remained blocked in the
GV-stage for the duration of the treatment, but could resume maturation after release from the drug. Protein
phosphorylation patterns were determined in GV-intact
oocytes from hormonally untreated CBA mice subjected to
one of the drugs, and cultured for 4h in M2 medium containing ^{32}P-orthophosphate (Endo et al., 1986). After
cell lysis aliquots of labelled proteins were separated
by 2D polyacrylamide gel electrophoresis, followed by autoradiography (Endo et al., 1986).
 To compare the relative concentration of c-*mos* mRNA in
oocytes being at high risk for aneuploidy, after a 3h

block in GV-stage by PdBu, to that in cells being at low risk for aneuploidy, after 3h of IBMX-exposure, oocytes were rapidly frozen after addition of known concentrations of rabbit globin mRNA. Since globin is not transcribed in oocytes, it can be used as an extrinsic marker to assess the quantity of RNA preparations (Manejwala et al., 1991). Total RNA was extracted from typically 450 oocytes for each group (Chirgwin et al., 1979). After reverse transciption (RT) aliquots were incubated with i) c-*mos* specific pairs of primers, ii) globin specific pairs of primers, or iii) actin specific primers, and subjected to 30 cycles of PCR. Aliquots of each sample were separated on agarose gels, stained with ethidium bromide and recorded on polaroid film (linear sensitivity range). The identity of the respective bands of reaction product was verified by Southern blotting with nested primers or specific probes (Sobek & Eichenlaub-Ritter, 1991). Quantification was done by densitometry. Linear increases in globin mRNA could be shown to give linear increases in signals obtained from cDNA bands after RT-PCR (Manejwala et al., 1991). After correcting for differences in mRNA content which were determined by comparison of globin signals in controls and the two treatment groups, the relative abundance of c-*mos* message was estimated by comparing signals of c-*mos* cDNA relative to actin cDNA in the two treatment groups (Eichenlaub-Ritter & Sobek-Klocke, in preparation).

Results

Cell cycle and aneuploidy in nocodazole-exposed oocytes from young and aged CBA mice
Oocytes from young (2-4month) or aged (>9month) diestrous mice undergo maturation with different timing. GVBD appears slightly delayed in the oocytes from the aged females whereas the first polar body is emitted much earlier than in those cells from young mice (Eichenlaub-Ritter & Boll, 1989). Accordingly, in groups of oocytes processed for indirect anti-tubulin immunofluorescence after defined times of *in vitro* maturation the percentage of cells in more advanced stages of meiosis is always greater in those oocytes obtained from the aged mice compared to young mice (Fig.1B). At the same time, aneuploidy levels (conservatively calculated as twice the percentage of hyperploids) are significantly elevated in the *in vitro* matured cells from aged females (Fig.1A).

When oocytes from both age-groups are placed into nocodazole-containing medium after 7 or 8h of maturation, all spindle microtubules become disassembled (Eichenlaub-Ritter & Boll, 1989). After release they are able to resume meiosis, although with altered kinetics. For instance, after 11h of culture which includes the 1h noco-

Fig.1A: Aneuploidy (2x hyperploids) of *in vitro* matured oocytes of young (y) and aged (a) mice of controls or oocytes recovered from a 1h block in nocodazole, applied either 7h or 8h after resumption of meiosis. Differences in aneuploidy are significant (p<0.05) between age-groups in controls but not in nocodazole-exposed oocytes.
Fig.1B: Percentage of cells in characteristic meiotic stages after defined times of *in vitro* maturation. Control: No treatment; Nocodazole: 1h drug-exposure after 8h of culture and recovery.

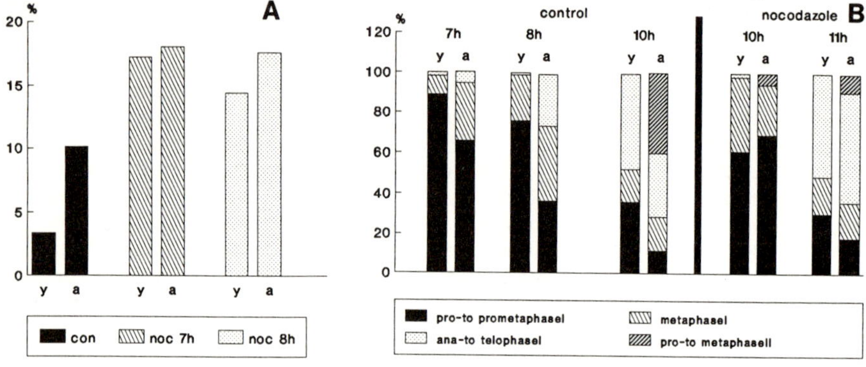

dazole block, far fewer cells from aged mice are in second meiosis (pro- to metaphase II) compared to the 10h control (Fig.1B). The time available for spindle formation and chromosome alignment, before anaphase is triggered, becomes reduced to about 1 to 2h in the nocodazole-exposed cells, and concomitantly aneuploidy levels are shown to rise in both age-groups (Fig.1B). However, not only does the nocodazole-block greatly reduce the age-specific differences in the cell cycle but there is also no more significant difference in nondisjunction observed between oocytes from young or aged mice (Fig.1A). This strengthens the hypothesis that there are correlations between cell cycle alterations and predisposition to aneuploidy in mammalian oocytes (Eichenlaub-Ritter & Boll, 1989).

Cell cycle and aneuploidy in oocytes after exposure to IBMX, forskolin or phorbols

Incubation of oocytes in IBMX, inhibiting phosphodiesterase, or forskolin, activating adenylate cyclase, both induce arrest of oocytes in the GV-stage (Bornslaeger & Schultz, 1985). After release from forskolin-treatment (n=137), GVBD may be triggered earlier than in untreated controls (n=132); in IBMX-exposed cells (n=56) it becomes slightly delayed (Fig.2A). Whereas forskolin-exposed oocytes (n=132) also proceed earlier into first anaphase compared to controls (n=82), time of polar body formation (PbF) is more variable in IBMX-treated oocytes (n=111), and, on the average, progression into the second meiotic division is delayed (Fig.2B). Spindle formation is not visibly affected by the drug-treatment in either group of oocytes.

Kinetics of oocyte maturation become also greatly altered when oocytes are transiently exposed to PdBu, PdAc or PdBu plus RA. In all experimental groups the time needed to undergo GVBD (determined in n=96, 108, and 78 cells, respectively) and to recover from phorbol-treatment is delayed compared to controls (Fig.2A). The period of recovery is shorter in oocytes treated with the nonpromoting phorbol PdAc or a combination of PdBu and RA compared to PdBu alone (Fig.2A). Accordingly, formation of the first polar body (PbF) is also delayed in all three experimental groups with respect to untreated controls (n for PdBu, PdAc and PdBu/RA was 149, 58, and 61, respectively; Fig.2B). However, in spite of this delay, the absolute time available for spindle formation, between GVBD and first anaphase or PbF, respectively, is much shorter in the PdBu and PdBu/RA group compared to control oocytes. To quantify this difference, the time at which 50% of all cells had undergone GVBD and PbF was

Fig.2: Kinetics of oocyte maturation. For each experimental group the time at which 50% of oocytes had undergone GVBD (A) or emitted a polar body (PbF) (B) was determined. The interval between these two events (PbF minus GVBD) (C) depicts the time available for spindle formation and chromosome alignment prior to anaphase I. Relative differences to the control are shown in (D).

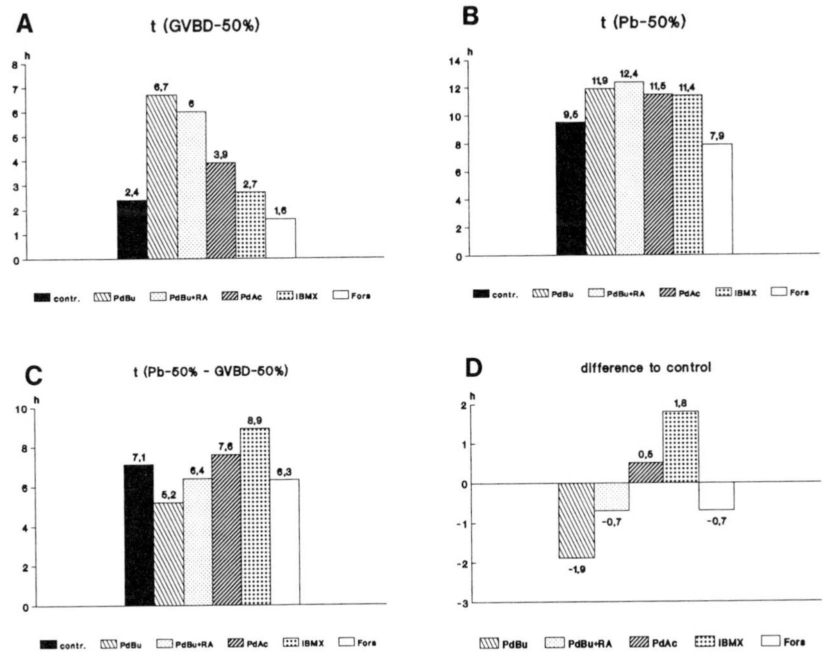

determined, the interval between these two times calculated (Fig.2C) and compared to controls (Fig.2D). From
these calculations it is estimated that the average time
available for spindle formation and chromosome alignment
is most dramatically reduced in oocytes transiently exposed to the tumor promoter PdBu. Addition of RA during
exposure to PdBu leads to a more normal cell cycle
(Fig.2D). Whereas prometa- to metaphase I is also reduced
in forskolin-exposed cells, it appears to be longer in
oocytes treated with PdAc or IBMX (Fig.2D).

Spindle structure and formation is not visibly disturbed in oocytes released from phorbol. However, lagging
of chromosomes in late telophase I (Fig.3) occured more
often in PdBu-recovering oocytes (11.6%) processed for
indirect anti-tubulin immunofluorescence after 9-10h of
maturation (n=119) as compared to controls (5.6%) which
were fixed for immunofluorescence after the same time of
culture (n=78). Although the total number of cells from
both groups which were in telophase I at the time of fixation has been low, this observation indicates that
PdBu-pretreatment may affect the synchrony between chromosome separation, anaphase trigger and cytokinesis.

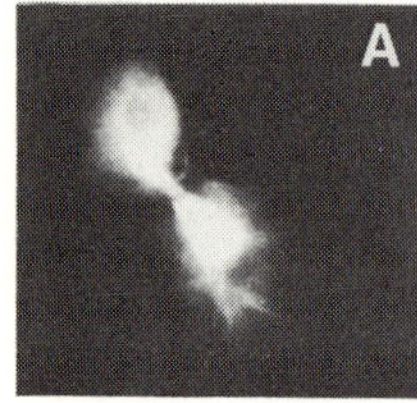
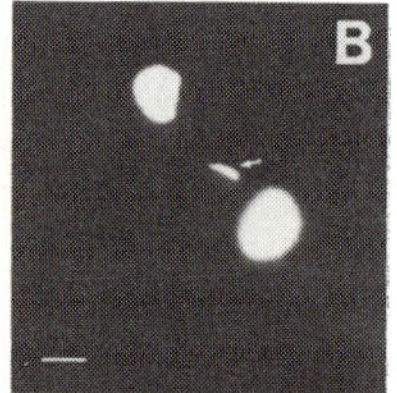

Fig.3: PdBu-recovered oocytes
with lagging of chromosomes
in late telophase I. A: Indirect anti-tubulin immunofluorescence; B: Dapi-stained
chromosomes in the same cell;
Bar: 5µm.

Besides causing disturbances in the cell cycle, PdBu
also induces significant rises in hyperploidy in oocytes

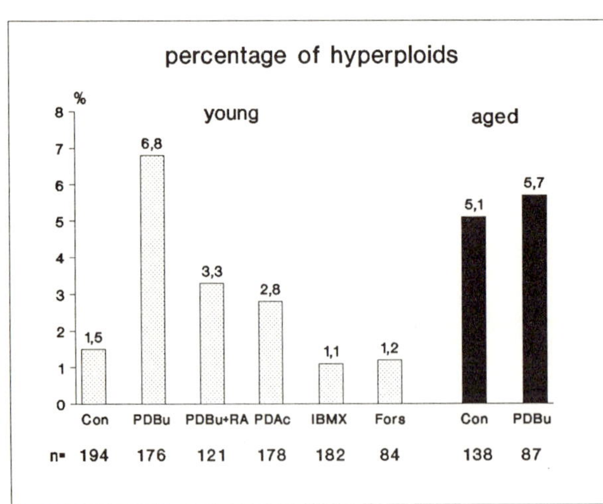

Fig.4: Hyperploidy
levels in control
oocytes (con) from
young (dotted bar)
or aged (solid bar)
mice, or cells from
PdBu, PdAc,
PdBu/RA, IBMX, or
forskolin (fors)-
exposed cells.
Significant diffe-
rences (p<0.01) are
only found between
control and PdBu-
treated oocytes,
and between con-
trols from young
and aged females
(see Fig.1).

from young mice compared to those of aged individuals (Fig.4). Addition of RA during PdBu-exposure does not release oocytes from the block in GV-stage. However, aneuploidy levels after recovery and maturation are no longer significantly different from control levels although the absolute percentage of hyperploids is slightly elevated (Fig.4). PdAc-exposure leads to moderate but not significant increases in aneuploidy, while forskolin, in spite of shortening prometaphase I, does not appear to affect chromosome distribution (Fig.4). Similarly, chromosomal constitution in IBMX-treated oocytes appears normal. Interestingly, aneuploidy in oocytes from aged females which have been exposed to PdBu is in the same range as that in PdBu-exposed oocytes from young mice (Fig.4), and does not seem higher than the previously determined rates of aneuploidy of *in vitro* matured control oocytes from aged females (Eichenlaub-Ritter & Boll, 1989).

Protein phosphorylation in drug-exposed mouse oocytes

When protein phosphorylation patterns were compared between oocytes from young and aged hormonally untreated CBA mice, which were temporarily blocked in GV-stage by IBMX, PdBu or PdAc, some characteristic differences were observed. Consistently a protein of M_r 20k is less phosphorylated or less abundant in oocytes, obtained from either age-group and treated with PdBu or PdAc, compared to those oocytes exposed to forskolin. This difference in protein phosphorylation may not be related to aneuploidy but rather reflect differences in substrates of PKA and PKC since the PdAc-exposed cells have reduced levels of the 20kDa phosphoprotein but are still at low risk for aneuploidy.

Fig.5: Differences in protein phosphorylation pattern of PdAc-exposed oocytes from young (y) and aged (a) mice. Two spots in the 34kDa range (arrows) are less prominent in the sample from aged females.

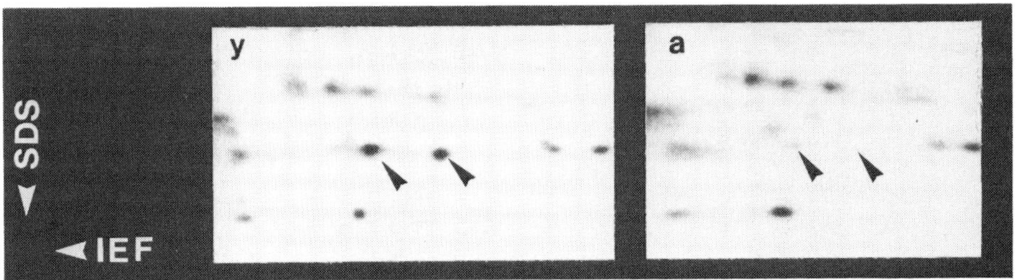

By contrast, two spots corresponding to proteins of M_r 34k are less prominent in 2D gels obtained from PdAc-exposed oocytes of aged females, compared to the same or the PdBu-exposed treatment group from young mice (Fig.5). These phosphoproteins correspond in molecular weight to

the catalytic subunit of MPF, p34^{cdc2}, which, inciden-
tally, is also visible as three bands, two phosphorylated
and one unphosphorylated form, on 1D-SDS gels of mouse
oocyte proteins (Choi et al., 1991). The alteration in
phosphorylation of the p34 protein(s) may therefore indicate differences in the phosphorylation of this protein
and can be related to the altered cell cycle kinetics and
predisposition to aneuploidy with advanced age. Apart
from these two differences in phosphorylation patterns,
we did not find any additional ones within the pI range
analysed (pI 5 to 7). Changes in protein phosphorylation
which might account for the altered cell cycle and aneuploidy in PdBu-exposed oocytes from young mice have so
far not been detected.

Comparison of relative amount of c-<u>mos</u> mRNA in oocytes at high and low risk for aneuploidy

We first compared the relative abundance of reaction product of *c-mos* cDNA to that of actin cDNA after simultaneously performed RT-PCR from the same number of oocytes
exposed to either, IBMX or PdBu. In two out of three
individual sets of experiments, there was more c-*mos*
reaction product than amplified actin cDNA in both treatment groups. This corresponds to calculations which indicate that *c-mos* as well as actin are abundant transcripts
in mouse oocytes (Bachvarova et al., 1989; Rosenberg,
pers. comm.). When the relative yields of c-*mos* reaction
product to actin product was compared between risk
groups, after correcting for differences in the initial
mRNA concentration as evident from the globin signal,
there appears to be always more c-*mos* as actin present in
the PdBu-treated oocytes compared to the IBMX-exposed

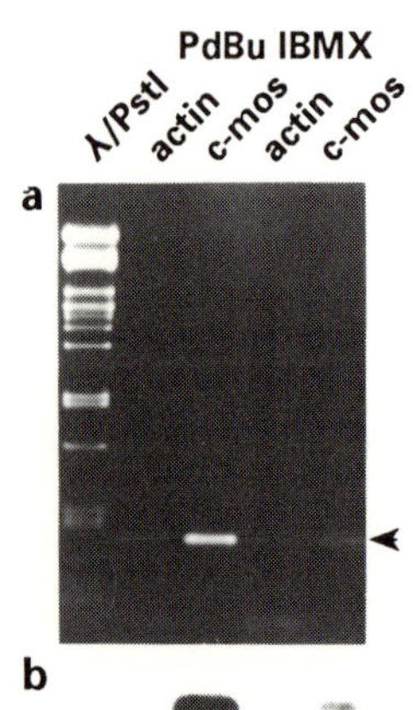

Fig.6: Presence of more amplification
product of c-*mos* relative to actin in
oocytes being at high risk for aneuploidy
(PdBu) compared to those not predisposed
to aneuploidy (IBMX). a: 1.5% agarose gel
of amplification products (400bp,
arrowhead) after RT-PCR, and /PstI
marker (left lane). B: Southern blot with
^{32}P-labelled, c-*mos* specific nested
primer.

cells. A most prominent example of differences between the relative abundance of *c-mos* to actin reaction product is shown in Fig.6.

However, the absolute differences between *c-mos* and actin levels are variable although they are always in the same range in a given experiment. This may be due to minor differences in the activity of the enzyme and efficiency of reverse transcription of the target mRNA's in individual sets of experiments. Still, the consistent prevalence of the level of *c-mos* to actin reaction product in oocytes being at high risk for aneuploidy compared to those being at low risk for nondisjunction argues for the presence of more *c-mos* mRNA in oocytes of PdBu as compared to IBMX treated oocytes. Further studies have to show whether higher concentrations of *c-mos* message in the oocytes may also induce rises in the concentration and activity of the *c-mos* protein kinase.

Discussion

Initially, when studying the cellular basis for the loss of fidelity of chromosome segregation in mammalian oocytes with advanced age of the female, we were unable to detect any obvious disturbances in meiosis I spindles or chromosome behavior in oocytes from aged mice (Eichenlaub-Ritter et al., 1986). However, our previous observations already implied that the cells from aged individuals may have altered cell cycles during *in vitro* maturation, possibly contributing to the high risk for aneuploidy. Therefore, if alterations in hormonal environment or cell-cell interactions in the ovary are responsible for the differences in the constitution of oocytes, they appear to affect the state of the oocyte already during arrest in the GV-stage, prior to its resumption of maturation.

The nocodazole experiments support the notion that correlations between cell cycle progression and aneuploidy exist because a reduction in the age-specific differences in maturation (Fig.1B) also abolishes the significant differences in aneuploidy levels between agegroups. Moreover, the similar absolute rates of numerical chromosome aberrations in nocodazole-treated oocytes from young and aged females imply that oocytes from older females may be no more sensitive to disturbances in their milieu than those from young mice. The overall increase in aneuploidy after nocodazole was suggested to result from the considerable shortening of prometaphase I in drug-exposed oocytes (Eichenlaub-Ritter & Boll, 1989).

In this study we can show that drugs like forskolin and IBMX which are able to block mouse oocytes in dictyate stage, presumably by affecting cAMP levels and activity of PKA (Schultz, 1988), do not influence the con-

stitution of oocytes from young mice in such a way that
they behave as being derived from aged females. Thus, in
spite of the reduction in first prometaphase (GVBD to
PbF) by forskolin, or the delayed GVBD and PbF in IBMX-
exposed oocytes, there is no increase in aneuploidy.
However, one has to consider that these alterations in
cell cycle progression are uniform in either delaying or
advancing both transition points, GVBD and PbF/anaphaseI.
Such a uniform pertubation in cell cycle progression may
therefore not result in a loss of fidelity of chromosome
separation. Generally, there appear to be a number of
feedback controls and checkpoints which ensure that
during normal division chromosomes will not disjoin before a metaphase spindle is present and chromosomes are
assembled at the spindle equator (Hoyt et al., 1991;
Eichenlaub-Ritter et al., in preparation), and neither
IBMX nor forskolin seem to affect these controls.

By contrast, coordination between spindle formation,
chromosome alignment and separation of homologues seems
to be disturbed in oocytes from young mice recovering
from a PdBu-block. This is already indicated by the frequent lagging of chromosomes during late telophase I
(Fig.3). Concomitantly, characteristic, phenotypically
similar alterations in the cell cycle as observed in aged
females are induced in the oocytes from young mice after
drug-exposure. Unlike IBMX-treatment these disturbances
are non-uniform in that GVBD is delayed (Fig.2) but PbF
and first anaphase are triggered prematurely, relative to
GVBD (Fig.2). At the same time aneuploidy becomes significantly increased to levels comparable to those observed
in untreated oocytes from aged females (Eichenlaub-Ritter
& Boll, 1989). This further supports the notion that
correlations between loss of cell cycle control and loss
of fidelity of chromosome segregation exist. One of the
effects of tumor promoters in general may be based on
their interference with cell cycle control which predis-
poses a cell to aneuploidy, and which in turn can then be
a further factor in multistep tumorigenesis of somatic
cells. In the oocyte system we have succeeded in showing
evidence for a primary, phorbol-induced cause of chromosomal imbalance, the influence of the tumor promoter on
cell cycle progression which may lead to asynchrony between disjoining of chromosomes and anaphase transition.

In contrast to results obtained with oocytes from
young mice, it was unexpected to find no increase in the
hyperploidy levels of PdBu-treated oocytes from aged mice
compared to untreated controls (Eichenlaub-Ritter & Boll,
1989). This result implies that the tumor promoter
affects the constitution and cell cycle regulation of the
"young" oocyte in a pathway converging with that disturbed already in "aged" cells. Therefore, PdBu-exposed
oocytes from young mice may serve as a model system for
further studies on mechanisms underlying age-related non-

disjunction. Since we do not find significant elevations in aneuploidy of oocytes exposed to the non-tumor promoter PdAc, our observation moreover suggest that there may be similarities in the etiology of age-related aneuploidy and the mechanisms leading to tumor promotion.

Cell cycle progression in all eukaryotes appears to be regulated by highly conserved mechanisms. Intrinsic and centrally to the G2- to M-phase transition and to anaphase trigger are the dephosphorylation/ phosphorylation and the re-/ deactivation cycles of a kinase, $p34^{cdc2}$, which is the catalytic subunit of MPF (for review see Nurse, 1990). It appears intriguing that two spots on 2D gels from oocyte proteins, corresponding to proteins with a relative molecular weight of 34k are less abundant in the oocytes from aged females compared to those from young mice. If these spots, in fact, correspond to the differentially phosphorylated forms of the $p34^{cdc2}$ kinase, this might explain the altered cell cycle. Thus $p34^{cdc2}$ needs to associate with a cyclin of the B-type prior to activation of MPF, which appears dependent on its phosphorylation status (Gould et al., 1991). Reduction in p34 protein content, or decreased phosphorylation could therefore delay pre-MPF formation, and GVBD. A reduction in $p34^{cdc2}$-protein concentration may on the other hand cause the premature anaphase transition. We are currently using immunoprecipitation with a $p34^{cdc2}$-specific antibody to confirm the identity of the p34 proteins. So far the observations of others (Choi et al., 1991) who also used cdc2-specific antibodies in Western blots, have already shown that $p34^{cdc2}$ exists in three electrophoretically distinct bands on 1D-SDS gels of proteins from mouse oocytes. Two of these which are phosphorylated may be identical to the two spots we are able to identify in our 2D gels.

In contrast to the oocytes from aged mice we were unable to find evidence for disturbances in protein phosphorylation patterns of PdBu-treated oocytes from young mice. Similar to its activity in inhibition of maturation proximally to the block in GV-stage by IBMX (Bornslaeger et al., 1986), the tumor promoter may therefore also effect oocytes from young mice in a step further downstream from pre-MPF formation on the signalling pathway leading to resumption of maturation. If phosphorylations induced by PdBu are for instance only transient they may not be detected after 3h of drug-treatment in our essay. However, tumor promoting phorbols have been shown to be able to transiently influence phosphorylation of transcription factors and thereby greatly affect gene expression at the transcriptional level (e.g. Angel et al., 1987). Out of many possible candidate genes the expression of which could be altered by the phorbol we observed only one, the c-*mos* proto-oncogene. Its product is known to bind to tubulin (Zhou et al., 1991), it may in-

fluence spindle formation, and as cytostatic factor may
regulate cell cycle progression in oocytes (Sagata et
al., 1989). Although it was not possible to directly
assess the number of transcripts in IBMX- and PdBu-exposed oocytes, the prevalence of c-*mos* amplification
product over that of actin implies that, in fact, there
may be more c-*mos* message in cells being at high risk for
aneuploidy (PdBu) compared to those at low risk (IBMX).
These preliminary results still need confirmation by
methods other than quantitative PCR which may be more reliable. Also, further studies have to show if the presence of more message will concomitantly lead to more
abundance of protein. However, our observations give at
least a first indication of a relationship between altered protein phosphorylation, altered gene expression,
disturbances in cell cycle control and predisposition to
aneuploidy in oocytes with advanced age of the female.

In particular, the ability of RA to reduce the alteration in maturation as well as the levels in aneuploidy
are intriguing. Retinoids are able to bind to receptors
belonging to the steroid hormone family of transcription
factors (Evans, 1988). Transactivation or repression of
these transcription factors with other, phorbol-inducible
transcription factors have been described (e.g. Jonat et
al., 1990). Besides, RA or carotinoids appear to be natural components of follicular fluid, and may be even indi-
cators of the quality of mammalian oocytes (Schweiger &
Zucker, 1988). Further studies are needed to establish
the importance of RA, RA receptors and their influence on
oocyte maturation in mammals. We hope that our observations will open up new perspectives to develop methods to
predict the quality of oocytes, or even positively influence oocyte maturation and constitution.

To conclude, our data imply that there are correlations between age-related aneuploidy, and disturbances in
cell cycle control during maturation which may influence
the synchrony in disjoining of homologues during
anaphase. Recent evidence which indicates that the
recombination frequencies of trisomics which are derived
by nondisjunction during first maternal meiosis in the
human are reduced compared to normally segregating
chromosomes (Sherman et al., 1991) does not contradict
our observations. One can speculate that a reduction in
the overall rate of chiasmata which need to become
resolved at the beginning of anaphase I may well bear on
the timing of anaphase trigger and could thus be
responsible for the difference in cell cycle kinetics as
well as nondisjunction rates between oocytes from young
and aged mice. Like in the PdBu-exposed oocytes, reduced
recombination could potentially induce premature anaphase
trigger, and loss of fidelity of chromosome separation.
Further studies have to show if it will be possible to
influence the oocyte's programme in a similar way as

reported for our mouse model to allow for better control
of chromosome segregation and reduce the risk for
aneuploid gametes.

References

Angel,P. et al. (1987) Phorbol ester inducible genes con-
    tain a common *cis* element recognized by a TPA-modu-
    lated transacting factor. Cell 49,729-739.
Bachvarova, R. et al. (1989) Amounts and modulation of
    actin mRNA in mouse oocytes and embryos. Development
    106,561-565.
Bornslaeger, E.A. et al. (1986) Effects of protein kinase
    C activators on germinal vesicle breakdown and polar
    body emission of mouse oocytes. Exp. Cell Res. 165,
    507-517.
Bornslaeger, E.A. & Schultz, R.M. (1985) Adenylate
    cyclase activity in zona-free mouse oocytes. Exp.Cell
    Res. 156,277-281.
Brook, J.D., Gosden, R.G. & Chandley, A.C. (1984) Mater-
    nal ageing and aneuploid embryos: evidence from the
    mouse that biological and not chronological age is the
    important influence. Hum. Genet. 66, 41-45.
Choi, T. et al. (1991) Activation of p34^{cdc2} protein ki-
    nase activity in meiotic and mitotic cell cycles in
    mouse oocytes and embryos. Development 113, 789-795.
Chirgwin, J.M. et al. (1979) Isolation of biologically
    active ribonuceic acid from sources enriched in ribo-
    nuclease. Biochemistry 18,5294-5299.
Eichenlaub-Ritter, U., Chandley, A.C. & Gosden, R.G.
    (1986) Alterations to the microtubular cytoskeleton
    and increased disorder of chromosome alignment in
    spontaneously ovulated mouse oocytes aged in vivo: an
    immunofluorescent study. Chromosoma 94, 337-345.
Eichenlaub-Ritter, U. Chandley, A.C. & Gosden, R.G (1988)
    The CBA mouse as a model system for age-related aneu-
    ploidy in man: studies of oocyte maturation, spindle
    formation and chromosome alignment during meiosis.
    Chromosoma 96,220-226.
Eichenlaub-Ritter, U. & Boll, I. (1989) Nocodazole sen-
    sitivity, age-related aneuploidy and alterations in
    the cell cycle of maturing mouse oocytes. Cytogenet.
    Cell Genet., 52, 170-176.
Endo, Y., Kopf, G. & Schultz, R.M. (1986) Stage-specific
    changes in protein phosphorylation accompanying meio-
    tic maturation of mouse oocytes and fertilization of
    mouse eggs. J. Exp. Zool. 239, 401-409.
Evans, R.M. (1988) The steroid and thyroid hormone recep-
    tor family. Science 240:889-895.
Gould, K.L. et al. (1991) Phosphorylation at Thr167 is
    required for *Schizosaccharomyces pombe* p34^{cdc2} func-
    tion. EMBO J. 10,3297-3310.

Gülle, M. & Eichenlaub-Ritter, U. (1990) Transient treatment of mouse oocytes with a tumor promoting phorbolester as model for maternal age-related aneuploidy in mammals. Eur. J. Cell Biol. Suppl. 30, P3.14 (abstr.).

Hassold, T.J. & Jacobs, P.A. (1984) Trisomy in man. Annu. Rev. Genet. 18, 69-97.

Hoyt, M.A., Totis, L., & Roberts, B.T. (1991) *S. cerevisiae* genes required for cell cycle arrest in response to loss of microtubule function. Cell 66,507-517.

Jonat, C. et al. (1990) Antitumor promotion and anti-inflammation: down modulation of AP-1 (Fos/Jun) activity by glucocorticoid hormone. Cell 62, 1189-1204.

Manejwala, F.M., Logan, C.Y. & Schultz, R.M. (1991) Regulation of hsp70 mRNA levels during oocyte maturation and zygotic gene activation in the mouse. Dev. Biol. 144, 301-308.

Nagata, A. et al. (1991) An additional homolog of the fission yeast cdc25+ gene occurs in humans and is highly expressed in some cancer cells. New Biol. 3, 959-968.

Nurse, P. (1990) Universal control mechanisms regulating onset of M-phase. Nature 344,503-508.

Sagata, N. et al. (1989) The c-*mos* proto-oncogene product is a cytostatic factor responsible for meiotic arrest in vertebrate eggs. Nature 342,512-518.

Schultz, R.M. (1988) Regulatory functions of protein phosphorylation in meiotic maturation of mouse oocytes *in vitro*, in Meiotic inhibition, Molecular control of meiosis (eds F.P. Haselötime & N.L. First), Alan Liss, Inc., New York, pp137-151.

Sherman, S.L. et al. (1991) Trisomy 21: association between reduced recombination and nondisjunction. Am. J. Hum. Genet. 49, 608-620.

Schweiger, F.J. & Zucker, H. (1988) Concentration of vitamin A, ß-carotin and vitamin in individual bovine follicles of different quality. J. Reprod. Fertil. 82, 575-579.

Sobek, I. & Eichenlaub-Ritter, U. (1991) Detection of c-mos and ß-actin mRNA in mouse oocytes by RT-PCR. Eur. J. Cell Biol. Suppl. 32, 5 (abstr.).

Urner, F. & Schorderet-Slatkine, S.(1984) Inhibition of denuded mouse oocyte meiotic maturation by tumor-promoting phorbol esters and its reversal by retinoids. Exp.Cell Res. 154, 600-605.

Zhao, K. et al. (1990) Requirement of the c-*mos* protein kinase for murine meiotic maturation. Oncogene 5, 1727-1730.

Zhou, R. et al. (1991) Ability of the c-*mos* product to associate with and phosphorylate tubulin. Science 251,671-675.

26 Minisatellite mutation and recombination

J.A.L. ARMOUR, D.G. MONCKTON, D.L. NEIL, M. CROSIER,
K. TAMAKI, A. MACLEOD and A.J. JEFFREYS

University of Leicester, UK

1 Introduction

1.1 Human tandemly repeated DNA

A prominent feature of the human genome, and many other
higher eukaryotic genomes, is the abundance of repeated
sequences. These multi-copy sequences may be dispersed around
the genome, like the Alu and L1 elements (Schmid and
Jelinek,1982; Singer and Skowronski,1985), or, by contrast,
found in tandemly repeated arrays. These tandem arrays fall
into a wide variety of size classes: the major alphoid
satellite sequences, for example (Waye and Willard,1986),
which exist in chromosome-specific subclasses, between them
account for about 5% of the human genome. The blocks of
alphoid satellite are composed of tandem arrays up to 5Mb in
length (Willard,1991), and appear to correspond to the
functional part of human centromeres (Willard,1990). At the
other end of the scale, "microsatellites", short arrays
(generally up to 60bp) of dinucleotide repeats, are extremely
abundant, and widely dispersed in the genome (Weber and
May,1989; Litt and Luty,1989; Weber,1990). Our work, and this
review, concern tandem arrays of intermediate size, the
"minisatellite" loci (Jeffreys et al.,1985a,b). These arrays
have a total length usually in the range 0.5-30kb, composed
of short repeated units. The repeat units at the loci we have
studied in detail range between 8 and 90bp in length.

1.2 Polymorphism at tandemly repeated loci

Many human minisatellite loci show multi-allelic polymorphism
(Wyman and White,1980; Higgs et al.,1981; Jeffreys et
al.,1985a,b; Wong et al.,1986,1987; Nakamura et al.,1987);
this polymorphism is due to allelic variation in the number
of tandemly repeated units, and hence results in variation in
the length of an array. This variation at minisatellite loci
can be assayed using a restriction enzyme which cuts outside
the tandemly repeated region, followed by Southern blot
hybridization using the minisatellite sequence as a probe at
high stringency (see, for example, Wong et al.,1986).
Polymorphism in the number of tandem repeats can also be
assayed at other types of repeated locus; PCR amplification

Chromosomes Today Volume 11. Edited by A.T. Sumner and A.C. Chandley. Published in
1993 by Chapman & Hall, London. ISBN 0 412 47670 3

of dinucleotide repeats, for example, has demonstrated the
frequent occurrence of usefully informative polymorphism
(Weber and May,1989; Litt and Luty,1989).

Among the minisatellite arrays, there is a range of
variability. Some loci appear to be monomorphic in the
populations studied (Armour et al.,1990), while others have
been isolated which show hypervariability. At the D1S7 locus,
for example, *Hin*fI alleles range between 1 and 20kb, and
99.5% of the population show allele length heterozygosity
(Wong et al.,1987; Jeffreys et al.,1988). This high level of
informativeness has made many of these loci very useful in a
wide variety of genetic analyses (see, for example, Reeders
et al.,1985; Wong et al.,1987; Malcolm et al.,1991). Our
concern in this review, however, is the origin of the high
levels of variability at human minisatellite loci. The
polymorphism at highly variable human minisatellites derives
ultimately from an extremely high mutation rate to new length
alleles in the germline; this germline mutation rate is so
high at some loci that it can be measured directly by
inspection of pedigrees (Jeffreys et al.,1988), and can be as
high as 15% per gamete (Vergnaud et al.,1991). While unequal
sister chromatid exchange, and unequal recombination (or gene
conversion) between alleles have long been implicated as a
mechanism of mutational change in tandemly repeated blocks of
DNA (Smith,1976; Dover,1982), other mechanisms, such as
replication slippage (Tautz et al.,1986; Levinson and
Gutman,1987) and deletion by intramolecular recombination are
also plausible mechanisms for mutation processes altering the
number of tandem repeats. This review will consider the
evidence concerning the mechanism(s) of length change
mutation at minisatellite loci, particularly in the light of
recent evidence implicating inter-allelic exchange in the
germline mutation process.

2 Correlations and conjectures

Although entirely intramolecular processes may contribute to
mutation at minisatellite loci, some lines of circumstantial
evidence point more explicitly towards the involvement of
recombinational mechanisms in minisatellite evolution.
Mapping by *in situ* hybridization shows that there is a
strong, though not exclusive, tendency for hypervariable
human minisatellites to cluster near the ends of chromosome
arms (Royle et al.,1988; Armour et al.,1989b). Where a
physical location is not known, linkage analysis frequently
places highly variable minisatellite loci at or near the ends
of linkage maps (Nakamura et al.,1988a; Lathrop et al.,1988;
Armour et al.,1990; Vergnaud et al.,1991). One consequence of
this clustering is that in these subtelomeric regions the
density of minisatellite loci can be very high indeed; the
high densities of hypervariable minisatellites in the
pseudoautosomal region (Cooke et al.,1985; Rouyer et

al.,1986; Page et al.,1987) and around the α-globin gene
cluster (Jarman and Wells,1989; Jarman and Higgs,1988) are
well documented, and in a number of instances a single clone
from a subtelomeric region has been shown to contain more
than one minisatellite array (Royle et al.,1988; Armour et
al.,1989b; Armour and Jeffreys,1991; Vergnaud et al.,1991).
Furthermore, there appears to be a correlation between
minisatellite loci and the presence of dispersed repeat
sequences, such as Alu elements, as deduced from the
unusually high frequency of these repeats in the DNA
immediately flanking minisatellites (Armour et al.,1989b;
Armour et al.,1992).

The concentration of minisatellites in the subtelomeric
regions correlates well with the elevated incidence of
chiasmata near chromosome ends (Hulten,1974; Laurie and
Hulten,1985; Chandley and Mitchell,1988). Some loci appear to
be equally unstable in the male and female germline (Jeffreys
et al.,1988; Armour et al.,1989a); since there are many more
mitotic divisions in the male germline, this equivalence
suggests a predominance of germline mutations at a restricted
stage in gametogenesis, possibly meiosis. However, it is
equally clear that at least one minisatellite locus is
preferentially unstable in the male germline (Vergnaud et
al.,1991). This sex-specific pattern of mutation may result
from the greater number of germline mitoses in the male, or
be connected with sex-specific rates of recombination in
subtelomeric regions, which have been observed in linkage
mapping (see, for example, Nakamura et al.,1988b; Nakamura et
al.,1989). Similar conclusions are suggested by the
chromosomal distribution of minisatellite loci. While all the
(non-acrocentric) autosomal telomeric regions appear
approximately equally rich in minisatellites, there is marked
contrast between the two ends of the X chromosome. Despite
the intensity of effort in mapping the X chromosome, only
very few polymorphic minisatellites have been isolated from
the sex-specific part of the X chromosome, a region which has
no partner in male meiosis (Donis-Keller et al.,1987; Fraser
et al,1989; Armour et al.,1990; Consalez et al.,1991). By
contrast, the pseudoautosomal region, a region of high
recombination in male meiosis, is very rich in minisatellite
loci (Cooke et al.,1985; Page et al.,1987).

These correlations suggest two general explanations (see
also Jarman and Wells,1989): first, that the subtelomeric
regions have high recombination rates (for reasons
unconnected with the presence of minisatellites), and it is
the high local recombination rates – or indeed other local
mechanisms – which drive the evolution of minisatellite loci
as highly unstable repeated sequences; second, that
minisatellites are "hot-spots" for meiotic recombination, and
that the evolution of hypervariable tandem arrays is at least
in part a consequence of the minisatellites' own
recombinational proficiency.

Our initial approach to investigating these possibilities has been by the structural analysis of mutational change in the germline, as a probe for the involvement of interallelic recombination. These studies therefore directly address the primary question of mutational mechanism; however, it is also possible that these studies may shed light on the recombinational "hot-spot" question, since the mutation rates to new length alleles at the loci studied are so high (0.4-1.2% per gamete). Thus if only a relatively small proportion (say 10%) of length change mutations were due to true interallelic exchanges resulting in exchange of flanking markers, then the inferred rate of recombination would indeed be elevated over the extremely small physical distances involved.

3 Experimental approaches

3.1 Mechanisms of mutation: basic parameters

Some of the basic parameters of minisatellite evolution may be inferred from the general properties of minisatellite mutation as observed in pedigree analysis (Jeffreys et al.,1988). Studies of mutation events at five highly variable loci showed, for example, that mutation rates appeared to be approximately equal in the male and female germline; however, as noted above, at least one locus appears to have a pronounced bias towards mutations in the male germline (Vergnaud et al.,1991). Analysis of allele size changes showed that small changes, involving gain or loss of a few repeats, were responsible for most mutations (Jeffreys et al.,1988); mutations causing an increase in repeat unit copy number were as common as those reducing the size of the array. Such a mutation process, apparently symmetrical with respect to allele length, could not be due exclusively or predominantly to intramolecular recombination, which can only decrease allele size. Mechanisms which would be predicted to produce mutant alleles of increased or decreased length with equal frequency include replication slippage and unequal sister chromatid exchange (both intraallelic), and unequal recombination and gene conversion (both interallelic).

Early studies on the mechanisms of minisatellite mutation looked for exchange of flanking markers as predicted by a mechanism of unequal meiotic recombination. Thus analysis of a single mutation event at the minisatellite YNZ22 (D17S5) showed that the new mutant allele had parental (non-recombinant) phase for immediately flanking polymorphisms, suggesting that the event was intra-allelic (Wolff et al.,1988). Similarly, analysis of a number of mutations at MS1 (D1S7) showed that the generation of new mutant alleles was not significantly associated with recombination events between (quite distant) flanking polymorphic markers in CEPH pedigrees (Wolff et al.,1989). Both these early studies, therefore, suggested that unequal recombination between alleles was not the predominant mechanism for mutation at the two loci examined; neither, however, excluded the possibility of unequal recombination as one of several contributing mechanisms. Moreover, neither study addressed the pattern of repeat unit turnover, and therefore the data could not rule out the possibility of unequal gene conversion, a process of inter-allelic exchange which does not result in the exchange of distal flanking markers.

Further, indirect support for a model excluding interallelic recombination as a major mechanism comes from studies of linkage disequilibrium. Any frequently recombining locus would be predicted to act as a boundary between local domains of linkage disequilibrium; across such a boundary recombinant associations of haplotypes would frequently arise. Indeed, this general principle underlies the evidence for a recombinational "hot-spot" - not due to a minisatellite - in the β-globin cluster (Antonarakis et al.,1982). Studies of haplotypes surrounding the minisatellite near the insulin gene have failed to provide convincing support for a locally elevated recombination rate (Cox et al.,1988).

3.2 Internal mapping of minisatellite alleles

More detailed evidence on the mechanisms of allele length mutation relies on a second level of variation at minisatellite loci (Jeffreys et al.,1990,1991; see also Pemberton and Amos,1990; Dover,1992). In addition to variation in the number of tandem repeats in an allele, alleles at most minisatellite loci also show subtle variation in the sequence of the repeated units. Thus an allele at such a locus would be composed of an interspersed mixture of two or more different repeat unit sequences. The positions of the minisatellite variant repeats (MVRs) along an array define a map ("MVR map") which represents the internal structure of an allele. The changes in the MVR map which accompany mutations can be used to deduce possible mutational mechanisms.

Initial work on MVR mapping used restriction mapping to define variant repeats. For example, the minisatellite MS32 (D1S8) has repeat units which always contain a site for *Hin*fI, but only some repeat units contain the recognition sequence for *Hae*III (Jeffreys et al.,1990). Thus the positions of variant repeats, containing or lacking sites for *Hae*III could be mapped by restriction mapping of PCR-amplified alleles. These studies showed that variation and mutation at D1S8 was highly polar; while one end of the locus showed relatively little variation between alleles, the other end was extremely variable. Furthermore, mapping of the break-points of *de novo* deletion mutants recovered *in vitro* showed a bias towards involvement of the "ultravariable" end of the array.

This approach, while very productive, was limited both by the technical difficulty of the mapping procedure, as well as by size constraints: since mapping an allele required that the allele be first amplified in its entirety, the process was limited to the analysis of alleles small enough (<6kb) to amplify relatively efficiently by PCR. Thus many alleles at D1S8 were simply too large to analyse. The mapping procedure has since been modified in two ways (Jeffreys et al.,1991). Firstly, mapping is done not over the entire length of an allele, but in from one end, that previous defined as "ultravariable". Secondly, the distinction between repeat unit types is not assayed with restriction enzymes, but directly by using PCR primers matching one repeat sequence or the other. This modified mapping technique is applicable to mapping individual alleles, and, at loci such as D1S8 which show no variation in repeat unit length, also applicable directly to genomic DNA to give a profile resulting from the superimposition of contributions from both alleles (Jeffreys et al.,1991). This diploid code may have important contributions to make to the future of forensic DNA typing, as the result can be encoded in digital form, thereby simplifying database construction and sample comparison. Since simple Boolean rules may be applied to their segregation in pedigrees, they have also provided a rapid method for the examination of variation and germline mutation at D1S8. In the following sections we summarize the results of detailed analysis by MVR-PCR at D1S8, and two other highly polymorphic minisatellites.

3.3 Variation and mutation at D1S8

The hypervariable locus MS32 is probably the best-characterized human minisatellite to date. The advent of a rapid PCR based approach to MVR mapping (MVR-PCR, Jeffreys et al.,1991) has enabled us to generate large numbers of single allele codes. These latest studies have extended those discussed above and continue to reflect the huge amount of information hidden in the internal structure of minisatellite alleles. Over 500 Caucasian and Japanese alleles have now been mapped and the vast majority (522/538) of these are different. This typing system has a theoretical capacity to distinguish over 2^{50} ($>1\times10^{15}$) different alleles mapped over the first 50 repeats, whilst current figures allow us to estimate the *minimum* number of different alleles present in Caucasian populations as in excess of 3,500. However, using the known mutation rate and world population size we predict in excess of 10^8 different alleles to exist world-wide (Jeffreys et al.,1991). Nevertheless, as reported previously (Jeffreys et al.,1990, Jeffreys et al.,1991, Tamaki et al., 1992) many of these alleles do share zones of haplotypic identity and may be sorted into groups of aligned alleles assumed to share a recent common ancestor. Variation between aligned alleles shows extreme polarity being largely confined to the 5' end of alleles, consistent with a mutational hotspot active in this region (Figure 1).

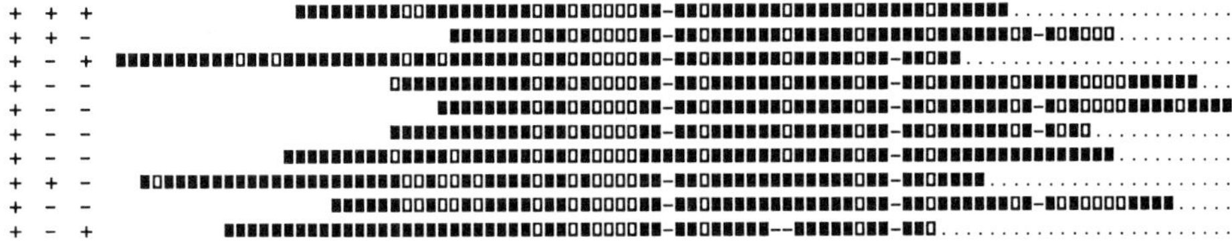

Fig.1. Alignment of structurally similar alleles at D1S8. MVR maps extending over the 5' ends of aligned alleles are shown; alleles at D1S8 average about 200 repeat units in total length. Black and white boxes represent the two types of repeat unit. Gaps (-) have been introduced to improve alignments, and flanking polymorphisms (+/-) are indicated.

MVR-PCR pedigree analysis, as well as providing valuable single allele data, has enabled us to examine germline mutation events in terms of internal structure changes. MVR-PCR analysis of parents and offspring in the 40 large CEPH families has revealed seven germline mutation events. Four of these had previously been identified by Southern blot analysis (Jeffreys et al.,1988), but three new events involving single repeat unit additions (+29bp) were also detected (Jeffreys et al.,1991). Interestingly all seven events so far characterised result in an increase in repeat unit copy number, possibly representing a bias toward small length increases. Defining the "recipient" allele as the parental allele contributing the relatively invariant end of the locus to the resulting new mutant allele, patterns of unequal exchange (either presumptive sister chromatid exchanges or recombinations) between "donor" and "recipient" can be defined. Six (out of seven) mutations have "donor" breakpoints within the first 24 repeats; the remaining event is a large subterminal duplication, with a donor breakpoint between repeats 48 and 52. All seven "recipient" breakpoints are within the first 23 repeats, indicating a direct correlation between observed allelic variability and *de novo* mutation events. These seven events are also entirely conservative with no loss of information from the recipient allele. Two of the events (e.g. Figure 2), for the first time, suggest the involvement of interallelic exchange in the generation of the new allele, revitalising speculation that minisatellite loci may be involved in homologous recombination (Jeffreys et al.,1991).

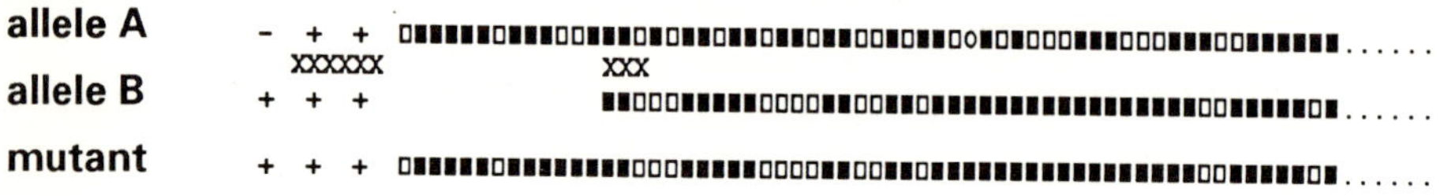

allele A   - + + ▫▪▪▪▪▪▫▪▪▪▫▫▪▪▪▫▫▫▪▪▫▪▪▫▪▪▫▪▪▫▪▪▫▪▪▫▫▫▫▫▪▪▫▪▫▪▫▫▫▪▪▪▫▫▫▪▪▫▫▫▪▪▪▪▪......
            xxxxxx   xxx

allele B   + + +   ▪▪▫▫▫▪▪▪▪▪▫▫▫▫▪▪▫▫▪▪▫▪▪▫▪▪▪▪▪▪▪▪▪▪▪▪▪▪▪▪▪▪▫▫▪▪▪▪▫▪......

mutant   + + + ▫▪▪▪▪▪▫▪▪▪▪▪▪▪▪▪▫▫▫▪▪▫▫▪▪▫▫▫▫▪▪▫▫▪▫▪▪▪▪▪▪▪▪▪▪▪▪▪▪▪▪▪▪▫▫▪▪▪▪▫▪......

Fig.2. Analysis of a spontaneous germline mutation at D1S8 by
internal mapping. Parental alleles (A and B) are shown
aligned with the new mutant allele and the possible positions
of presumed conversion breakpoints (XXX).

Detailed sequence analysis of the DNA flanking the
minisatellite has enabled us to identify three commonly
polymorphic base substitutions within 400bp of the terminal
repeat (Monckton et al., manuscript in preparation). These
flanking polymorphisms provide another approach to single
allele mapping in heterozygotes via the use of allele
specific flanking primers in "knockout" MVR-PCR (Monckton et
al., manuscript in preparation), facilitating the rapid
acquisition of further single allele codes from unrelated
individuals. These flanking markers should also assist our
investigations into the role of homologous recombination in
maintaining MS32 variability. Initial studies analysing
haplotypes at these polymorphic sites has revealed that
linkage disequilibrium between the flanking markers does
exist, but is not absolute as might be expected for such
closely spaced loci. Haplotyping of the flanking
polymorphisms relative to the minisatellite array shows that
in general groups of aligned alleles share flanking
haplotypes, but occasional switching within allele groups is
observed, indicative of inter-allelic recombination events
(Figure 1); this type of analysis, however, cannot
distinguish true recombinations from shorter "patches" of
gene conversion. Details of this work will be published
elsewhere, but preliminary data suggest an elevated
recombination frequency around 200 fold greater than the
genomic average operating over the 5' terminus of the
minisatellite array and extending into the flanking DNA.
Haplotyping of flanking polymorphisms in one of the
presumptive interallelic mutants does not show exchange of
immediate flanking markers (Figure 2; unfortunately flanking
markers are uninformative for the second presumptive
interallelic event). These results suggest that unequal
homologous conversion may have played a role in the
generation of at least one observed *de novo* mutation event

at MS32 and that recombination in general has played a
significant role in MS32 evolution.

3.4 Variation and mutation at D16S309
The minisatellite MS205 is a highly variable minisatellite on
chromosome 16 (Royle et al.,1992). This locus has the unusual
combination of high variability and restricted (<6kb) allele
size; this allows the prospect of mapping most alleles in
their entirety in a highly polymorphic system. Sequence
analysis has shown that repeat unit sequence variation
exists, and this has been exploited in the MVR mapping of
alleles at D16S309. This work will be published in detail
elsewhere (manuscript in preparation); however, a selection
of illustrative results is shown.

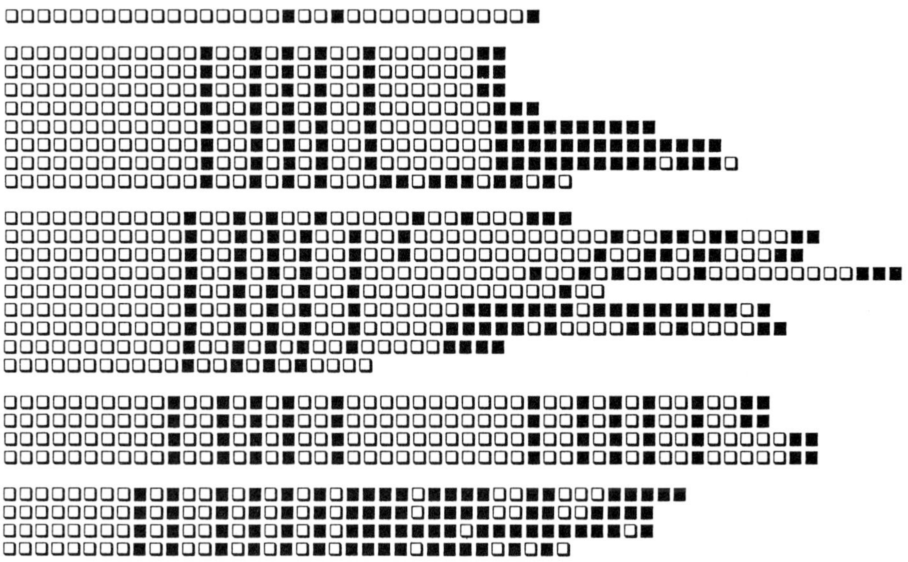

Fig.3. Internal maps of 25 alleles from unrelated Caucasians
at D16S309. Alleles are shown in their entirety, and grouped
by common structures at the relatively invariant (left) end.

Analysis of alleles from unrelated Caucasians shows that
at this locus too there is a polarity of variation (Figure
3). Nearly all alleles mapped fall into four groups, which
can be defined by a common structure at the relatively
invariant end of the array. By contrast, the other end of the

array shows great diversity between alleles, with nearly all alleles having different structures. As predicted from this pattern of variation between alleles, the mutation events analysed so far all involve changes at this extreme "ultravariable" end, and one shows clear evidence of unequal exchange between alleles (data not shown).

3.5 Variation and mutation at D7S21

The D7S21 locus (MS31A) is a minisatellite with an allele size range of 2-13kb. It exhibits very high (99%) allele length heterozygosity that reflects extreme variability in tandem repeat copy number (Wong et al.,1987; Armour et al.,1989b). Sequence analysis of MS31A alleles reveals that, like most minisatellites, there are polymorphic positions within the repeat units generating minisatellite variant repeat units (MVRs). However, MS31A is unusual in that all repeat units so far characterized at this locus have the same length (20bp) (Wong et al.,1987). These attributes make MS31 an ideal locus for internal mapping of minisatellite repeat unit variation by applying the same MVR-PCR technique as used for D1S8 (MS32:Jeffreys et al.,1991).

Analysis of single alleles from unrelated Caucasians reveals very high levels of allelic variability at this locus. However a few of the alleles sampled so far do appear to be related. As with related alleles at MS32 and MS205, these alleles are almost identical along most of their length with the major region of difference between them confined to one end. The few mutation events studied so far also involve this end of MS31 alleles. Both of these observations provide further evidence for polarity in the mutation events giving rise to new length minisatellite alleles at some hypervariable loci.

3.6 General conclusions and prospects

The three minisatellite loci which we have examined in detail, as outlined above, all have in common the feature of polarity of variation in allele structure. Where new mutations have been analysed, these too, as predicted, are restricted to one end of the array. This extreme polarity may reflect a locally acting element in *cis*, such as an initiation site for recombination or replication, or even reflect longer range polarity such as orientation within the chromosome as a whole. In most cases of germline mutation, MVR mapping can suggest probable mechanisms of mutation; of the cases studied, most remain entirely intraallelic events. A significant minority of mutations, however, require the involvement of both alleles to produce the mutant allele. Nevertheless, demonstration that these interallelic exchanges are true recombination events requires the exchange of distal flanking markers; in those interallelic mutation events where flanking polymorphisms are informative, exchange of distal flanking markers has not yet been observed. In some mutations, sequence analysis across breakpoints in donor, recipient and new mutant alleles may allow more detailed inferences about the mechanisms involved, particularly with respect to "gap-repair" models of homologous recombination. Thus the generation of new alleles by inter-allelic exchange is not always a process of unequal recombination, but unequal gene conversion may be an equal, or even a predominant, mechanism for this class of mutation.

4 Acknowledgements

J.A.L.A. is a Wellcome Trust Postdoctoral Fellow. The work from our group summarized in this review was supported by grants to A.J.J. from the MRC and the Wolfson Foundation.

5 References

Antonarakis, S.E., et al. (1982) Nonrandom association of polymorphic restriction sites in the β-globin gene cluster. **Proc.Nat.Acad.Sci.U.S.A.**, 79,137-141.

Armour, J.A.L., et al. (1989a) Analysis of somatic mutations at human minisatellite loci in tumors and cell lines. **Genomics**, 4, 328-334.

Armour, J.A.L., et al. (1989b) Sequences flanking the repeat arrays of human minisatellites: association with tandem and dispersed repeat elements. **Nucleic Acids Res.**, 17,4925-4935.

Armour, J.A.L., et al. (1990) Systematic cloning of human minisatellites from ordered array Charomid libraries. **Genomics**, 8,501-512.

Armour, J.A.L. and Jeffreys, A.J. (1991) STS for minisatellite MS607 (D22S163). **Nucleic Acids Res.**, 19,3158.

Armour, J.A.L., Crosier, M. and Jeffreys, A.J. (1992) Human minisatellite alleles detectable only after PCR amplification. **Genomics**, 12, 116-124.

Chandley, A.C. and Mitchell, A.R. (1988) Hypervariable minisatellite regions are sites for crossing-over at meiosis in man. **Cytogenet. Cell. Genet.**, 48,152-155.

Consalez, G.G., et al. (1991) Assignment of Emery-Dreyfuss muscular dystrophy to the distal region of Xq28 - the results of a collaborative study. **Am.J.Hum.Genet.**, 48, 468-480.

Cooke, H.J., Brown, W.R.A. and Rappold, G.A. (1985) Hypervariable telomeric sequences from the human sex chromosomes are pseudoautosomal. **Nature**, 317, 687-692.

Cox, N.J., Bell, G.I. and Xiang, K. (1988) Linkage disequilibrium in the human insulin/insulin-like growth factor II region of human chromosome 11. **Am.J.Hum.Genet.**, 43, 495-501.

Donis-Keller, H. et al. (1987) A genetic linkage map of the human genome. **Cell**, 51, 319-337.

Dover, G. (1982) Molecular drive: a cohesive mode of

species evolution. **Nature**, 299, 111–117.

Dover, G.A. (1992) A digital DNA dipstick for probing human diversity. **Trends Genet.**, 8, 45–47.

Fraser, N.J., Boyd, Y. and Craig, I. (1989) Isolation and characterization of a human variable copy number tandem repeat at Xcen-p11.22. **Genomics**, 5,144–148.

Higgs, D.R., <u>et al</u>. (1981) Highly variable regions of DNA flank the human globin genes. **Nucleic Acids Res.**, 9, 4213–4224.

Hulten, M. (1974) Chiasma distribution at diakinesis in the normal human male. **Hereditas**, 76, 55–78.

Jarman, A.P. and Wells, R.A. (1989) Hypervariable minisatellites: recombinators or innocent bystanders? **Trends Genet.**, 5, 367–371.

Jarman, A.P. and Higgs, D.R. (1988) A new hypervariable marker for the human alpha-globin gene-cluster. **Am.J.Hum.Genet.**, 43, 249–256.

Jeffreys, A.J., Wilson, V. and Thein, S.L. (1985a) Hypervariable "minisatellite" regions in human DNA. **Nature** 314,67–73.

Jeffreys, A.J., Wilson, V. and Thein, S.L. (1985b) Individual-specific "fingerprints" of human DNA. **Nature**, 316, 76–79.

Jeffreys, A.J., <u>et al</u>. (1988) Spontaneous mutation rates to new length alleles at tandem-repetitive hypervariable loci in human DNA. **Nature**, 332,278–281.

Jeffreys, A.J., Neumann, R. and Wilson, V. (1990) Repeat unit sequence variation in minisatellites: a novel source of DNA polymorphism for studying variation and mutation by single molecule analysis. **Cell**, 60,473–485.

Jeffreys, A.J., <u>et al</u>. (1991) Minisatellite repeat coding as a digital approach to DNA typing. **Nature**, 354, 204–209

Lathrop, G.M., <u>et al</u>. (1988) A mapped set of genetic markers for human chromosome 9. **Genomics**, 3,361–366.

Laurie, D.A. and Hulten, M.A. (1985) Further studies on chiasma distribution and interference in the human male. **Ann.Hum.Genet.**, 49, 203–214.

Levinson, G. and Gutman, G.A. (1987) Slipped-strand mispairing: a major mechanism for DNA sequence evolution. **Mol.Biol.Evol.**, 4, 203–221.

Litt, M. and Luty, J.A. (1989) A hypervariable microsatellite revealed by in vitro amplification of a dinucleotide repeat within the cardiac muscle actin gene. **Am.J.Hum.Genet.**, 44,397–401.

Malcolm, S., <u>et al</u>. (1991) Uniparental paternal isodisomy in Angelman's syndrome. **Lancet**, 337,694–697.

Nakamura, Y., <u>et al</u>. (1987) Variable number of tandem repeat (VNTR) markers for human gene mapping. **Science**, 235, 1616–1622.

Nakamura, Y., <u>et al</u>. (1988a) A primary map of ten DNA markers and two serological markers for human chromosome 19. **Genomics**, 3,67–71.

Nakamura, Y., <u>et al</u>. (1988b) A mapped set of markers for human chromosome 17. **Genomics**, 2,302-309.

Nakamura, Y., <u>et al</u>. (1989) Frequent recombination is observed in the distal end of the long arm of chromosome 14. **Genomics**, 4, 76-81.

Page, D.C., <u>et al</u>. (1987) Linkage, physical mapping and DNA sequence analysis of pseudoautosomal loci on the human X and Y chromosomes. **Genomics**, 1, 243-256.

Pemberton, J. and Amos, B. (1990) DNA fingerprinting: a new dimension. **Trends Genet.**, 6, 101-103.

Reeders, S.T., <u>et al</u>. (1985) A highly polymorphic DNA marker linked to adult polycystic kidney disease on chromosome 16. **Nature**, 317, 542-544.

Rouyer, F., <u>et al</u>. (1986) A gradient of sex linkage in the pseudoautosomal region of the human sex chromosomes. **Nature**, 319, 291-295.

Royle, N.J., <u>et al</u>. (1988) Clustering of hypervariable minisatellites in the proterminal regions of human autosomes. **Genomics**, 3,352-360.

Royle, N.J., <u>et al</u>. (1992). A hypervariable locus D16S309 located at the distal end of 16p. **Nucleic Acids Res.**, 20,1164.

Schmid, C.W. and Jelinek, W.R. (1982) The Alu family of dispersed repetitive sequences. **Science**, 216,1065-1070.

Singer, M.F. and Skowronski, J. (1985) Making sense out of LINES: long interspersed repeat sequences in mammalian genomes. **Trends Biochem.Sci.**, 10,119-122.

Smith, G.P. (1976) Evolution of repeated DNA sequences by unequal crossover. **Science**, 191, 528-535.

Tamaki, K., <u>et al</u>. (1992) Minisatellite variant repeat (MVR) mapping: analysis of "null" repeat units at D1S8. **Hum. Mol. Genet.** (in the press).

Tautz, D., Trick, M. and Dover, G.A. (1986) Cryptic simplicity in DNA is a major source of genetic variation. **Nature**, 322, 652-656.

Vergnaud, G., <u>et al</u>. (1991) The use of synthetic tandem repeats to isolate new VNTR loci: cloning of a human hypermutable sequence. **Genomics**, 11, 135-144.

Waye, J.S. and Willard, H.F. (1986) Nucleotide sequence heterogeneity of alpha satellite repetitive DNA: a survey of alphoid sequences from different human chromosomes. **Nucleic Acids Res.**, 15,7549-7568.

Weber, J.L. (1990) Informativeness of human (dC-dA).(dG-dT) polymorphisms. **Genomics**, 7, 524-530.

Weber, J.L. and May, P.E. (1989) Abundant class of human DNA polymorphisms which can be typed using the polymerase chain reaction. **Am.J.Hum.Genet.**, 44,388-396.

Willard, H.F. (1990) Centromeres of mammalian chromosomes. **Trends Genet.**, 6,410-416.

Willard, H.F. (1991) Evolution of alpha satellite. **Curr. Opin. Genet. Dev.**, 1,509-514.

Wolff, R.K., Nakamura, Y. and White, R. (1988) Molecular characterization of a spontaneously generated new allele at a VNTR locus: no exchange of flanking DNA sequence. **Genomics**, 3, 347–351.

Wolff, R.K., et al. (1989) Unequal crossingover between homologous chromosomes is not the major mechanism involved in the generation of new alleles at VNTR loci. **Genomics**, 5, 382–394.

Wong, Z., et al. (1986) Cloning a selected fragment from a human DNA "fingerprint": isolation of an extremely polymorphic minisatellite. **Nucleic Acids Res.**, 14, 4605–4616.

Wong, Z., et al. (1987) Characterization of a panel of highly variable minisatellites cloned from human DNA. **Ann.Hum.Genet.**, 51, 269–288.

Wyman, A.R. and White, R. (1980) A highly polymorphic locus in human DNA. **Proc.Nat.Acad.Sci.U.S.A.**, 77, 6754–6758.

27 The *M26* homologous recombination hotspot: sequences, factors and chromosomal context

W.P. WAHLS and G.R. SMITH
Fred Hutchinson Cancer Research Center, Washington, USA

1 Introduction

At some point in the life cycle of virtually every organism there is an exchange of genetic information between segments of DNA with sequence homology. This process of homologous recombination serves three functions. First, homologous recombination exchanges genetic information between homologous chromosomes to generate new combinations of alleles. This permits evolution to occur much more rapidly than if newly arising mutations maintained their original physical linkages. Second, recombination is required for the proper pairing and segregation of chromosomes during meiosis (reviewed by Hawley, 1988). Third, homologous recombination provides a mechanism for the repair of some types of DNA damage (Resnick, 1976; Szostak *et al.* 1983).

Although the frequency of recombination between two markers is generally proportional to their physical distance along the DNA, certain regions of chromosomes undergo recombination much more frequently than others. Such regions, called recombination hotspots, were originally identified by genetic studies in several fungi and have been subsequently found in a variety of organisms from bacteriophages to humans (reviewed by Smith, 1988). Identified recombination hotspots act locally, within a few kbp to a few tens of kbp, and stimulate recombination up to 30-fold in their vicinity. This type of action is reminiscent of transcriptional enhancers and suggests that recombination hotspots are involved in a central, rate-limiting step of recombination. Presumably, as for transcriptional enhancers, there are proteins that bind to recombination hotspots to mediate their biological activity. By studying recombination hotspots, we expect to elucidate discrete steps in the pathway of homologous recombination.

The *M26* recombination hotspot, located in the *ade6* gene of the fission yeast *Schizosaccharomyces pombe*, is in some ways the best understood eukaryotic recombination hotspot. We review here the characteristics of the *M26* recombination hotspot. Although a unique heptameric DNA sequence is required for hotspot activity, and we have identified and purified to near homogeneity a protein that interacts with the heptamer, a specific chromosomal context, or second sequence element, is required for hotspot activity. This

Chromosomes Today   Volume 11. Edited by A.T. Sumner and A.C. Chandley. Published in 1993 by Chapman & Hall, London. ISBN 0 412 47670 3

context-dependent hotspot activity is compared to other examples of chromosomal position effects.

2 The *ade6-M26* recombination hotspot

2.1 Genetic characterization

Among the more than 300 identified alleles of the *ade6* gene on chromosome III in *S. pombe*, the *M26* allele is unique. Gutz (1971) isolated the *M26* allele as a nonsense suppressible adenine auxotrophic mutation that increased the frequency of gene conversion at the *ade6* locus from approximately 0.5% to approximately 5%. Although the requirement for adenine was suppressible, the recombination hotspot activity was not, suggesting that hotspot activity was not related to expression of the *ade6* gene product. All other *ade6* alleles, including the *M375* allele that maps close to *M26*, exhibit a lower (less than 0.5%) frequency of gene conversion (Gutz, 1971). Both the *M26* and the *M375* alleles are due to simple G to T transversions that create stop condons located 135 bp and 132 bp downstream of the initiation codon, respectively (Fig. 1) (Szankasi *et al.*, 1988; Ponticelli *et al.* 1988). Thus, the *M375* allele serves as an excellent control for studies of recombination promoted by *M26*.

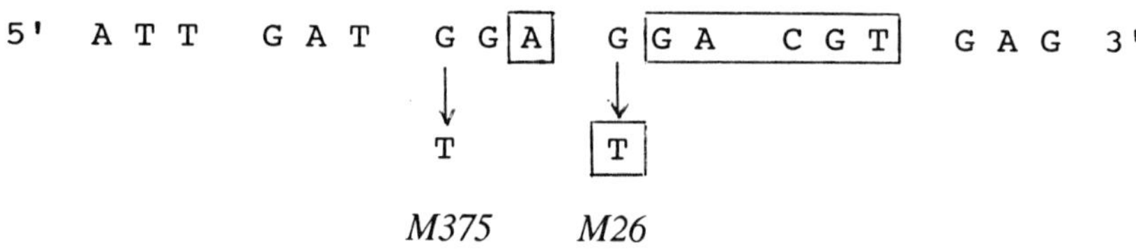

Fig. 1. DNA sequence in the vicinity of the *M26* recombination hotspot (Szankasi *et al.*, 1988; Ponticelli *et al.*, 1988). The single-base-pair substitutions of the *ade6-M375* (normal recombination) and *ade6-M26* (recombination hotspot) alleles are shown below the wild-type *ade6* sequence. Both mutations generate identical chain termination codons. Shown in boxes is the 7 bp sequence required for *M26* recombination hotspot activity *in vivo* (Schuchert *et al.*, 1991). Mts1 protein binds to the heptameric sequence (data not shown).

When *M26* is crossed with other *ade6* alleles the frequency of recombinant prototrophs is increased up to 15-fold relative to crosses containing the *M375* allele (Gutz, 1971). This recombination hotspot activity is meiosis-specific; *M26* does not detectably stimulate mitotic recombination (Ponticelli *et al.*, 1988; Schuchert and Kohli, 1988). *M26* stimulates both gene conversion and reciprocal exchange events to an equal extent (Schuchert and Kohli, 1988), and during gene conversion the *M26* allele is preferentially converted to wild-type about 90% of the time (Gutz, 1971). Thus, the chromosome bearing the *M26* allele preferentially acts as the recipient of information during gene conversion events. This disparity of conversion is not found with any of the other *ade6* alleles analyzed. Gene conversion events at *M26* often extend several hundred base pairs 5' and 3' of the *M26* site, and result in co-conversion of other *ade6* alleles (Gutz, 1971). These data suggest that recombination is initiated in the vicinity of *M26* and can extend in both directions. The hotspot activity is observed whether *M26* is located in one, the other, or both of the recombining chromosomes (Ponticelli *et al.*, 1988). Thus, although mismatch correction may play a role in hotspot activity, hotspot activity does not result from mismatch correction at the *M26* site. The hotspot activity is exerted only locally; *M26* promotes recombination at *ade6* but not at *arg1*, located approximately 80 cM to the right of *ade6* on chromosome III (Ponticelli *et al.*, 1988). Furthermore, it has no effect upon recombinant frequencies in several intervals located on chromosomes I and II (Ponticelli *et al.*, 1988). Taken together, these results indicate that *M26* is involved in a rate-limiting step of meiotic recombination at the *ade6* locus, perhaps by acting as a site for the initiation of recombination.

Based upon his genetic data, Gutz proposed a model for how *M26* might promote recombination (Gutz, 1971). In this scenario, the single base-pair substitution at the *M26* locus created a site that is recognized by an endonuclease. Single-strand cleavage of the DNA at or near *M26* and exonucleolytic degradation of this strand resects the DNA to one or both sides of *M26* and, depending upon the extent of the degredation, perhaps beyond flanking alleles. These resulting ends interact with the homologous chromosome which serves as a template for repair synthesis. This interaction of the chromosomes leads to a Holliday junction which could be resolved as in other models for homologous recombination.

2.2 *Cis* sequences

In one approach to determine what DNA sequence elements are required for hotspot activity at *M26*, Schuchert and colleagues (1991) analyzed the ability of variants of the *M26* site to promote recombination. They constructed many *S. pombe* strains that differed only by single base-pair substitutions in the vicinity of *M26*. The relative hotspot activities for each of these mutations is summarized in Fig. 2. A specific 7 bp sequence, 5'-ATGACGT-3' (*M26* mutation underlined), is required for hotspot activity *in vivo*, while alterations outside of this heptamer had no effect upon hotspot activity. These results support the hypothesis that the *M26* mutation creates a protein recognition sequence.

If nucleotides were randomly associated in the *S. pombe* genome, one would expect 31 copies of the *M26* heptamer to be present in the 254 kbp of *S. pombe* DNA that has been sequenced (Genbank version 69). However, our homology search detected the heptamer only three times, while ten randomized heptamer sequences were found an average of 27 times. Thus, the heptameric sequence is about 10-fold under-represented in the *S. pombe* genome. The reason for this disparity is not clear, but it is possible that the distribution of the heptamer is related to its function. For example, if the *M26* recombination hotspot is a central component of recombination throughout the genome, 220 copies (approximately 1/10 the number expected) might be sufficient to ensure appropriate rates of meiotic recombination.

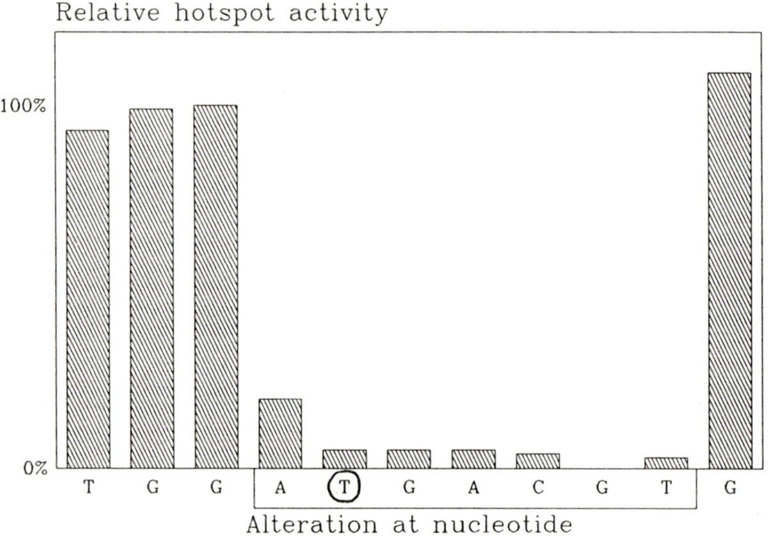

Fig. 2. Mutational analysis of the *M26* recombination hotspot. Single base-pair substitutions in the vicinity of *M26* were assayed for hotspot activity *in vivo*. Shown is the average hotspot activity for the three possible nucleotide substitutions at each position. The circle indicates the *M26* mutation and the box shows the 7 base pair sequence required for *M26* recombination hotspot activity. The data are from Schuchert *et al.* (1991). Mts1 protein binding *in vitro* correlates precisely with these relative hotspot activities *in vivo* (data not shown).

One interesting and speculative possibility is that all homologous recombination initiates at specific DNA sequences such as *M26*. The *S. pombe* genome contains 18 Mbp of DNA on three chromosomes with a total meiotic map size of 2000 cM; approximately 40 exchanges occur during the average meiosis. From the observed frequency of *M26* heptamers we can predict that there are likely to be about 220 *M26* heptamers in the genome. Although the *ade6-M26* heptamer functions as a hotspot in 5% of tetrads, at some loci *M26* might stimulate recombination more often and some *M26* loci might be inactive (as discussed below). Thus, 220 copies of *M26* could be sufficient to confer appropriate rates of exchange during meiosis. It will be interesting to determine whether the *M26* heptamers at loci other than *ade6* exhibit recombination hotspot activity.

2.3 *Trans* factors

To investigate the possibility that the *M26* mutation created a protein binding site, we have used a gel mobility shift assay to search for such a protein in extracts of cells undergoing synchronous meiosis. A protein that binds probes bearing the *ade6-M26* site, but not to probes bearing either wild-type *ade6* DNA or the *ade6-M375* site, was detected (W.P. Wahls and G.R. Smith, manuscript in preparation). We have called this protein Mts1 (M twenty-six binding protein 1).

Using conventional biochemical approaches we have purified Mts1 to near homogeneity. SDS-PAGE analysis reveals that Mts1 is a 71 kDa polypeptide. Methylation interference analyses reveal that Mts1 makes close protein-DNA contacts within the seven base pair sequence required for hotspot activity (Fig. 2). Methylation of any base pairs outside of the heptameric sequence does not interfere with protein binding. These data suggest that the Mts1 protein is involved in hotspot activity.

To confirm that the Mts1 protein is required for *M26* recombination hotspot activity, we tested the ability of Mts1 to bind to DNA molecules that contained single base-pair substitutions in the vicinity of the *M26* site. The amount of protein-DNA binding *in vitro* correlated precisely with the degree of hotspot activity observed *in vivo* (Fig. 2) for each single base-pair substitution. We conclude that binding of Mts1 protein to the heptanucleotide *M26* DNA sequence is required for the homologous recombination hotspot activity.

The *M26* recombination hotspot functions only during meiosis (Ponticelli *et al.*, 1988; Schuchert and Kohli, 1988). However, the *M26* binding protein Mts1 is present in extracts from both mitotic cells and in extracts of cells from all timepoints of meiosis. Therefore, although mutational analyses indicate that binding of Mts1 at *M26* is required for hotspot activity, some other factor(s) must confer the meiotic specificity upon the *M26* recombination hotspot.

2.4 Chromosomal context

Table 1 presents typical results of genetic crosses containing the *M26* recombination hotspot (Ponticelli *et al*, 1988). *M26* stimulates meiotic recombination between chromosomes approximately 9-fold and, when present as the chromosomal allele, stimulates recombination 14-fold between a chromosomal *ade6* gene and an *ade6* gene on a multi-copy plasmid. In an attempt to define the DNA sequence elements responsible for hotspot activity, Ponticelli and Smith measured the hotspot activity of the *M26* allele transplaced to other locations in the genome (Ponticelli and Smith, 1992). Intriguingly, as shown in Table 2, the *M26* hotspot stimulates recombination between a plasmid-born *ade6* gene and a chromosomal *ade6* gene *only* when the *M26* allele was located in the chromosomal copy of *ade6*. Therefore, the

M26 recombination hotspot requires the proper "chromosomal context" for activity.

Table 1. Hotspot activity of *M26* located at *ade6* during mitosis and meiosis

Cross	Allele	Ade$^+$ recombinants per 10^6 mitotic cells or meiotic spores	
		Mitotic	Meiotic
Chromosome x chromosome			
M26 or *M375* +	*M26*	73 (0.9X)	5300 (9X)
+ *M210*	*M375*	84	590
Chromosome x plasmid			
M26 or *M375* +	*M26*	350 (1X)	14000 (14X)
+ *L469*	*M375*	350	1000

Recombination was measured with either the *M26* or the *M373* allele at the endogenous *ade6* locus. The frequency of Ade$^+$ recombinants was determined in standard genetic crosses. Crosses including the multi-copy plasmid were conducted by self-mating of homothallic strains. The degree of *M26* recombination hotspot activity (relative to *M375* activity) is indicated in parentheses. The data are from Ponticelli *et al.* (1988).

To further analyze the context-dependence of the hotspot activity, two other moving experiments were conducted. In the first experiments the *M26* allele was transplaced to the *ura4* locus, located about 250 cM from *ade6* on the other arm of chromosome III. Placing 354 bp of *ade6*, containing either *M26* or *M375*, into the 5' end of the *ura4* coding region rendered the cells Ura⁻. Recombination was measured between these *ura4::ade6* disruption alleles and the *ura4-595* allele, located 620 bp from the insertions. Ura$^+$ recombinants were generated at the same frequency whether the inserted DNA contained the *M26* or the *M375* allele (Table 3, cross 2). Thus, a 354 bp fragment bearing *M26* does not exhibit hotspot activity when transplaced to the *ura4* locus.

Table 2. Hotspot activity of *M26* when moved to a multicopy plasmid

Cross	Allele	Ade$^+$ recombinants per 10^6 spores	*M26* hotspot activity
Chromosome (*M26*) x plasmid			
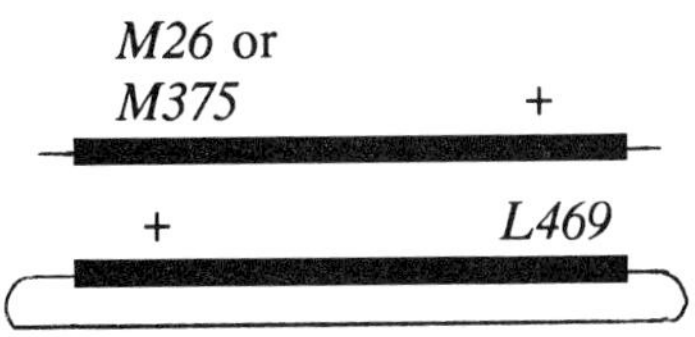 *M26* or *M375* + / + *L469*	*M26* *M375*	14000 1000	14X
Chromosome x plasmid (*M26*)			
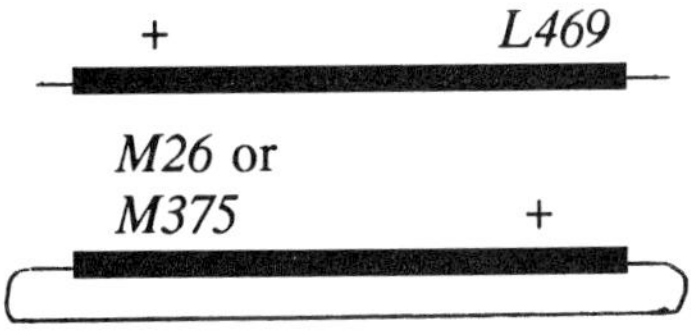 + *L469* / *M26* or *M375* +	*M26* *M375*	310 360	0.9X

The frequency of Ade$^+$ recombinants was determined in homothallic strains. The multicopy plasmids used contained 3 kbp of *ade6* DNA, including the entire open reading frame, and the indicated *ade6* alleles. The data are those of Ponticelli and Smith (1992).

In the second experiment 3.2 kbp of DNA bearing the entire *ade6* gene, plus approximately 500 bp of flanking DNA on either side of the coding region, was inserted into the *ura4* locus. A large deletion of the endogenous *ade6* gene was introduced to eliminate the possibility of recombination between the endogenous *ade6* and the transplaced *ade6* genes. This arrangement permitted measuring the effect of *M26* upon recombination between complete *ade6* genes transplaced to the *ura4* locus. Furthermore, heteroallelic insertions were

Table 3. Activity of *M26* transplaced to the *ura4* locus

Cross	Allele	Ade$^+$ or Ura$^+$ per 10^6 spores	*M26* hotspot activity
Endogenous at *ade6* locus			
M26 or	*M26*	10600	16X
M375	*M375*	680	
354 bp inserted at *ura4* locus			
M26 or	*M26*	610	1X
M375	*M375*	580	
3 kbp inserted at *ura4* locus			
M26 or	*M26*	1900	0.5X
M375	*M375*	3500	

Standard genetic crosses were conducted to determine the effect of *M26* upon recombination at various loci. In the third set of crosses the normal *ade6* locus was deleted to prevent ectopic recombination between the two loci. The data are from Ponticelli and Smith (1992).

constructed to provide complete homology across the recombining chromosomes. In those experiments (summarized in Table 3, cross 3) no hotspot activity was observed [the two-fold higher activity of *M375* in these crosses may be due to different rates of mismatch correction for the *M26* and *M375* alleles (Gutz, 1971)].

These experiments can be summarized as follows: *M26* has no homologous recombination hotspot activity when either 354 bp or 3.2 kbp fragments are inserted into the *ura4* locus. Because the *M26* mutation was located 1010 bp away from the nearest end of the transplaced *ade6* gene, we can presume that at least one nucleotide located at least 1011 bp away from *M26* is required for hotspot activity.

There are two possible explanations for the context-dependent hotspot activity exhibited by *M26*. First, although the *M26* mutation does create a protein binding site required for hotspot activity (see above), activity may also require a second, *cis*-acting sequence element. The Chi recombination hotspot in *Escherichia coli* provides precedence for such a two element system (reviewed by Smith, 1988). RecBCD enzyme, required for the major recombination pathway of *E. coli*, enters a double-strand DNA break in a chromosome and travels unidirectionally along the chromosome until it encounters an 8 bp Chi site. If RecBCD enzyme encounters Chi from the correct direction (with regard to the nucleotide sequence of Chi), it nicks the DNA to generate a recombinogenic DNA strand, "activating" the Chi recombination hotspot. If RecBCD enzyme approaches Chi from the other direction, Chi does not function as a hotspot. A distant, *cis*-linked DNA sequence, *cos*, acts as an entry site for RecBCD enzyme in phage λ crosses. Therefore, the correct relative orientations of *cos* and Chi (two *cis*-linked sequences) are required for hotspot activity of Chi. Because *M26* behaves genetically like Chi, it may function in a similar fashion: perhaps a recombination enzyme recognizes and binds a distant site, translocates along the DNA, and exerts its effect upon encountering *M26*. Alternatively, a second sequence might be involved in looping or otherwise positioning chromosomes for recombination.

The second possibility is that the *M26* recombination hotspot requires a specific structural context to function appropriately. Chromosome structural elements, such as local superhelical density, assembly of histones, *et cetera*, can be influenced by the primary DNA sequence. In this possibility the bulk of the DNA in the vicinity of *M26*, not simply a discrete sequence, may dictate whether or not the hotspot functions. Several examples support such a possibility. They include finding of DNAase I hypersensitive sites at a mouse recombination hotspot (Shenkar *et al.*, 1991) and the observation that eukaryotic transcription can be markedly influenced by the location of a test gene in the genome (for *e.g.*, see Laurie-Ahlberg and Stam, 1987). Although chromatin structure can clearly influence gene expression and recombination, it may be difficult to prove that a certain structural configuration is *required* for *M26* hotspot activity.

Distinguishing between the two possible explanations for position-dependant activity of the *M26* hotspot may be possible by further moving experiments. For example, moving a second "activating" sequence into the appropriate position might restore hotspot activity. Alternatively, moving the inactive 3.2 kbp fragment to various other chromosomal locations might reveal active and inactive locations, depending upon the surrounding chromatin state. Several experimental approaches are currently being used to investigate these possibilities (J. Virgin, personal communication).

A similar example of context-dependent recombination hotspot activity has been observed in the *ARG4* gene of *Saccharomyces cerevisiae*. Both deletion analyses (Nicolas *et al.* 1989; Schultes and Szostak, 1991) and moving experiments (A. Nicolas, personal communication) suggested that the *ARG4* recombination hotspot may require chromosomal context for activity; no discrete nucleotide sequence with hotspot activity could be identified when DNA sequences surrounding the hotspot were moved or altered. However, upon transplacement of a 12.5 kbp piece of DNA containing the *ARG4* locus to a replicating linear plasmid, the *ARG4* recombination hotspot functioned (Ross *et al.* 1992). Those results suggest that the signals which dictate recombination behavior, including the recombination hotspot, are intrinsic to the 12.5 kbp and can function autonomously. It should therefore be possible to determine the *cis*-sequences or chromosomal contexts within the 12.5 kbp that are responsible for conferring full hotspot activity.

3 Conclusions

3.1 Chromosomal context

At first approximation, the *M26* recombination hotspot appears to be well-defined. Since a specific seven bp sequence (5' ATGACGT 3') is required for hotspot activity, and we have purified a protein (Mts1) that binds to the heptameric sequence to mediate hotspot activity, we could imagine relatively simple models for how *M26* functions biologically. However, moving experiments indicate that *M26* requires either a second, *cis*-linked sequence located at least 1011 bp away or a particular chromosomal context for biological activity. Similar position effects have been described for gene expression and the activity of other recombination hotspots, suggesting that chromosome context may be important to many biological processes. This urges caution in the interpretation of experiments, such as deletion studies or moving experiments, that may alter chromatin structure or chromosomal context. Single-base-pair substitutions, which are less likely to alter local chromosomal structure, may be more appropriate for the identification and analysis of sequence elements involved in chromosome dynamics.

3.2 Role for recombination hotspots

One common property of chromosomes is the disparity between their physical and genetic maps: Although the amount of recombination that occurs is generally a function of physical distance, certain intervals recombine much more or much less frequently than might be expected. The presence of specific DNA sequences (such as *M26*) that stimulate homologous recombination provides a basis for this discrepancy. The amount of recombination that occurs in a particular physical interval may depend more upon the relative proximity of recombination hotspots than upon the actual physical size of the interval.

This supports the intriguing and speculative possibility that much, perhaps all recombination is initiated at a finite number of sites in the eukaryotic genome. This proposal, previously made for Chi-stimulated RecBCD enzyme-dependent recombination in *E. coli* (Smith, 1991), also makes good sense for the regulation of recombination in eukaryotes. Homologous recombination is required for the proper pairing and segregation of chromosomes during meiosis (reviewed by Hawley, 1988). Specific initiation sites would serve two important functions: First, if the initiation of recombination were simply a statistical function of distance along the chromosome (as is often assumed), then with average rates of exchange a significant proportion of homolog pairs would fail to recombine and would suffer nondisjunction. Controlled initiation at specific sites could eliminate such nondisjunction by ensuring that recombination occurs at least once on each chromosome pair. Second, the assembly of recombination factors at particular locations could simplify regulation of recombination. Assembly of multi-subunit recombination complexes, confirming that each homologous pair has the machinery required to produce crossovers, and sending the appropriate temporal start signals to enzyme complexes already bound to DNA are examples of hypothetical processes that might be easier to accomplish if limited to a finite number of sites along the chromosome. If all recombination initiates at specific sites, then the highest recombination frequencies would be observed near hotspots (initiation sites) and then would decrease with increasing distance from the hotspot until another initiation site is approached. As more recombination hotspots are identified and characterized, this should become a testable hypothesis.

The proposal that all homologous recombination initiates at specific sequences such as *M26* is likely to remain speculative for quite some time. What is clear at this point, however, is that recombination hotspots and their DNA binding proteins associate to somehow stimulate homologous recombination in adjacent DNA intervals. Because these sequences and their binding proteins are involved in a rate-limiting step of recombination they provide a unique opportunity to study the recombination process. Analysis of recombination hotspots and their associated proteins will elucidate discrete steps of homologous recombination pathways.

4 Acknowledgements

We are grateful to our colleagues Susan K. Amundsen and Philippe Szankasi for their comments. Supported by a postdoctoral fellowship (DRG-1110) from the Damon Runyon-Walter Winchell Cancer Research Fund to W.P.W. and a research grant (GM31693) from the United States Public Health Service to G.R.S.

5 References

Gutz, H. (1971) Site specific induction of gene conversion in *Schizosaccharomyces pombe. Genetics*, **69**, 317-337.

Hawley, R.S. (1988) Exchange and chromosomal segregation in eucaryotes, in *Genetic Recombination*, (eds. R. Kucherlapati and G.R. Smith), American Society for Microbiology, Washington, DC.

Laurie-Ahlberg, C.C. and Stam, L.F. (1987) Use of P-element-mediated transformation to identify the molecular basis of naturally occurring variants affecting *Adh* expression in *Drosophila melanogaster. Genetics*, **115**, 129-140.

Nicolas, A. *et al.* (1989) An initiation site for meiotic gene conversion in the yeast *Saccharomyces cerevisiae. Nature*, **338**, 35-39.

Ponticelli, A.S. *et al.* (1988) Genetic and physical analysis of the *M26* recombination hotspot of *Schizosaccharomyces pombe. Genetics*, **119**, 491-497.

Ponticelli, A.S. and Smith, G.R. (1992) Context dependence of a eukaryotic recombination hotspot. *Proc. Natl. Acad. Sci. USA*, **89**, 227-231.

Resnick, M.A. (1976) The repair of double-strand breaks in DNA: a model involving recombination. *J. Theor. Biol.*, **59**, 97-106.

Ross, L.O. *et al.* (1992) Meiotic recombination on artificial chromosomes in yeast. *Genetics*, **131**, 541-550.

Schuchert, P. and Kohli, J. (1988) The *ade6-M26* mutation of *Schizosaccharomyces pombe* increases the frequency of crossing over. *Genetics*, **119**, 507-515.

Schuchert, P. *et al.* (1991) A specific DNA sequence is required for high frequency of recombination in the *ade6* gene of fission yeast. *EMBO J.*, **10**, 2157-2163.

Schultes, N.P. and Szostak, J.W. (1991) A poly(dA.dT) tract is a component of the recombination initiation site at the ARG4 locus in *Saccharomyces cerevisiae. Mol. Cell. Biol.*, **11**, 322-328.

Shenkar, R. *et al.* (1991) DNase I-hypersensitive sites and transcription factor-binding motifs within the mouse E beta meiotic recombination hot spot. *Mol. Cell. Biol.*, **11**, 1813-1819.

Smith, G.R. (1988) Homologous recombination sites and their recognition, in *The Recombination of Genetic Material* (ed. K.B. Low), Academic Press, New York, pp. 115-154.

Smith GR (1991) Conjugational recombination in *E. coli*: Myths and mechanisms. *Cell*, **64**, 19-27

Szankasi, P. *et al.* (1988) DNA sequence analysis of the *ade6* gene of *Schizosaccharomyces pombe*. Wild-type and mutant alleles including the recombination hot spot allele *ade6-M26*. *J. Mol. Biol.*, **204**, 917-925.

Szostak, J.W. *et al.* (1983) The double-strand-break repair model for recombination. *Cell*, **33**, 25-35.

Imprinting and methylation

28 Parental imprinting and epigenetic programming of the mouse genome: long lasting consequences for development and phenotype

W. REIK, H. SASAKI, A. FERGUSON-SMITH, R. FEIL, L. BOWDEN, J. PENBERTH, A. SURANI
Institute of Animal Physiology and Genetics Research, Cambridge, UK
I. GURTMANN and J. KLOSE
University of Berlin, Germany

The importance of epigenetic genome modifications for mammalian development is well established. In general, it appears to be useful for any multicellular organism to convey a common phenotype to a population of descendants of a single progenitor cell, and thus to establish a collective phenotype of clonally related cells. One of the most powerful mechanisms to achieve this is by heritable epigenetic modifications of chromosomes that are introduced into the progenitor genome, and are clonally stable thereafter, potentially for many cell generations.

In the mouse, epigenetic modifications are introduced into the chromosomes at different stages of development (and presumably for different purposes and using different mechanisms). Parental imprinting is initiated in gametes, but imprints presumably evolve through different molecular phases in the embryo (Chaillet 1992, Allen and Mooslehner 1992). A variety of gene sequences also undergo dramatic changes of epigenetic modification such as DNA methylation in the preimplantation mouse embryo, most of them becoming strikingly demethylated by the blastocyst stage (Sanford et al. 1987, Monk et al. 1987, Howlett and Reik 1991, Kafri et al. 1992). A number of transgenes in the mouse acquire specific methylation patterns after fertilisation, some of them possibly at a relatively early stage (Sapienza et al. 1989, McGowan et al. 1989, Allen et al. 1990, Reik et al. 1990, Chaillet et al. 1991, Engler et al. 1991). This epigenetic programming can be genetically controlled and presumably involves genotype-specific modifier genes, different alleles of which are present in inbred strains of mice. Some of this early programming may also be influenced by interaction between the egg cytoplasm and the parental genomes, thus leading to parental effects (Surani et al. 1990, Reik et al. 1990). Following implantation and just prior to gastrulation, the level of DNA methylation increases genome-wide and many sequences become methylated *de novo* (Monk et al. 1987, Kafri et al. 1992). That this process of modification is essential for normal development has recently been established in mice that carry

Chromosomes Today   Volume 11. Edited by A.T. Sumner and A.C. Chandley. Published in 1993 by Chapman & Hall, London. ISBN 0 412 47670 3

mutant alleles of the enzyme methyltransferase with substantially reduced activity. Homozygous embryos were found to die at around mid-gestation, with no overt morphological abnormalities but with extensive cell death occurring throughout the embryo (Li et al. 1992). Whether epigenetic modifications acquired during gastrulation and organogenesis are important for maintaining the differentiated state of tissues remains to be established.

Two well studied examples of epigenetic gene control in the mouse are the inactivation of genes on the X-chromosome in females, and the parental imprinting and genotype-specific modification of transgenes. In both systems, modification can be parent specific (as for the X-chromosome in extra-embryonic tissues), or alternatively can arise after fertilisation with no parental influence (random X- inactivation in somatic tissues). With both the X-chromosome and transgenes, relatively extensive differences in DNA methylation and in nuclease sensitivity of chromatin have been observed between active and inactive forms (Monk 1986, Grant and Chapman 1988, Pourcel et al. 1990, Chaillet 1992, Allen and Mooslehner 1992). However, there is no knowledge at present of the primary epigenetic signals that initiate modification in the germline or after fertilisation, although in two examples of imprinted transgenes it was observed that the methylation pattern characteristic of maternal transmission was already present in ovulated oocytes before fertilisation (Chaillet et al. 1991, H. Sasaki unpubl.). This suggests that there may be circumstances under which DNA methylation is used as a primary imprinting signal.

Here we examine the molecular properties of epigenetic modification of an imprinted autosomal gene, the Insulin-like growth factor 2 gene (*Igf2*), and compare these properties to those of X-chromosome genes and of imprinted transgenes. We also consider the different phases that epigenetic programming goes through, in particular soon after fertilisation, and describe instances where this could lead to parental effects. We describe an experimental system in which such effects can be identified.

2 Molecular properties of imprinting of the *Igf2* gene

At least 3 autosomal genes in the mouse have now been shown to undergo parent-specific imprinting (DeChiara et al. 1991, Ferguson-Smith et al. 1991, Bartolomei et al. 1991, Barlow et al. 1991, Rappolee et al. 1992). The *Igf2* gene is imprinted to be repressed on the maternal chromosome (termed maternal imprinting) in various foetal tissues, with the exception of the leptomeninges and choroid plexus where both alleles are expressed. Deficiency of IGF-II polypeptide causes foetal growth retardation. The *Igf2* receptor gene and the *H19* gene, by contrast, are paternally imprinted. So far, no differences in epigenetic modification between the expressed and the repressed copies have been reported for these genes.

Phenotypes controlled by imprinted genes have been investigated in experimental mouse embryos that carry different combinations of maternally and paternally derived chromosomes (such as androgenones and parthenogenones, Barton et al. 1991). Subsets of imprinted genes on particular chromosome regions have been examined through the use of monoparental disomies, produced by intercrossing translocation heterozygotes, in which a particular chromosome (or a part thereof) is duplicated from one parent and deficient from the other parent (Cattanach and Beechey 1990). The *Igf2* gene, together with the closely linked *H19* gene, is located on distal chromosome 7 (Bartolomei and Tilghman 1992). Embryos maternally disomic for the distal part of chromosome 7 thus show a greatly reduced level of *Igf2* transcripts (Ferguson-Smith et al. 1991), and an elevated (2-fold) level of *H19* transcripts (A. Ferguson-Smith unpubl.). Biochemical properties of the repressed allele of *Igf2* can therefore be directly analysed in these maternal disomies (before they die at later foetal stages), and by comparison to normal embryos, the characteristics of the paternal allele can be deduced.

An extensive analysis of DNA methylation and chromatin structure, comparing disomic to normal embryos, was therefore carried out (Sasaki et al. 1992). Methylation was analysed in the transcribed region as well as in the two major foetal promoters P2 and P3. Surprisingly, careful comparison of a substantial number of methylation-sensitive restriction sites revealed no differences between disomic and normal DNA (Sasaki et al. 1992). In particular, the extensive CpG island associated with promoter 2 was unmethylated on both the expressed and repressed chromosome. Several DNAseI hypersensitive sites near the promoters were also present on both chromosomes and sensitivity of chromatin to MspI digestion was the same (Sasaki et al. 1992). The repressed *Igf2* allele, therefore, exists in an open chromatin conformation without extensive island methylation. Indeed, a low level of transcription (2-4% of that of the paternal allele) was detectable from the repressed allele (Sasaki et al. 1992). An extensive search in more 5' regions of the gene has shown that there is an addition CpG island, 2.5 Kb upstream of the first promoter, which is also unmethylated on both chromosomes (W. Reik unpubl.). However, just upstream of this island, in a region relatively rich in CpG's a difference was discovered (Fig. 1). Surprisingly, some of the MspI sites in this region were more methylated on the paternal chromosome from which *Igf2* is expressed (Sasaki et al. 1992). Further analysis has shown that in this region methylation drops down from 5' to 3' over several MspI sites until reaching the completely unmethylated state in the island. This 'shoulder' looks different on the maternal and paternal chromosomes, with the paternal chromosome showing a steeper and more precipitous drop. Sites on the other side (3') of the island have not been analysed in detail, but it is possible that they behave in a similar way.

These observations raise several questions. First, it appears that epigenetic modification of imprinted transgenes and X-linked genes is far more extensive than that of the imprinted *Igf2* gene. It usually involves several CpG's in the case of transgenes, and extensive island methylation in the case of X inactivation; moreover, where this has been tested, imprinted transgenes are more DNAseI resistant as are the inactive X-linked genes (Pourcel et al. 1990, Grant and Chapman 1988). It is of course possible that we have overlooked some very specific methylation differences present in the promoters of *Igf2*, which are not assayable by restriction enzyme digest. We are now addressing this point by using a recently developed *in vitro* mutagenesis technique that tests methylation of every CpG residue (Frommer et al. 1992). It is not clear yet whether repression of *Igf2* is at the level of transcription initiation, or whether other mechanisms of regulation are involved. The low level of transcription from the repressed allele may be higher than that from X inactivated genes (at least in somatic lineages, Brown et al. 1990); it may be interesting, however, to compare residual expression with that of X-linked genes in the extraembryonic tissues (which may not undergo extensive island methylation).

Is the methylation difference detected in the 5' region of the *Igf2* gene (Fig. 1) of any functional relevance for the imprinting of the gene? This question, of course, can not be answered at present. Extensive analyses have not been done for non-imprinted genes to see whether they might have parent specific methylation patterns. Evidence available to date only indicates that on a very general basis this is not the case: methylation differences that exist between sperm and oocyte genomes appear to be lost in the early embryo (Howlett and Reik 1991); an experiment designed to pick up sequences with methylation differences between the parental chromosomes has failed to do so (Chaillet 1992); and some other gene sequences analysed on chromosome 7 have not shown any parent specific methylation patterns (our unpubl. observations), although admittedly these experiments have not been done in any detail comparable to those with *Igf2*. On the other hand parent specific methylation patterns have been observed for *DN34* in the human, a candidate cDNA sequence for Prader-Willi/Angelman syndrome (which show clear signs of imprinting), and these differences are quite extensive with hypermethylated sites on both maternal and paternal chromosomes (in different regions of the gene, Nicholls et al. 1992).

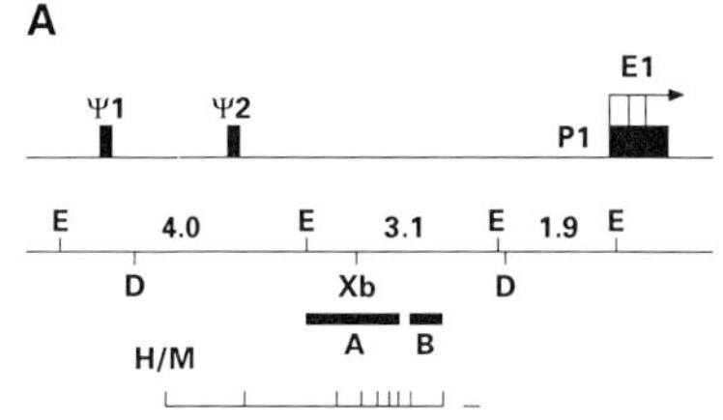

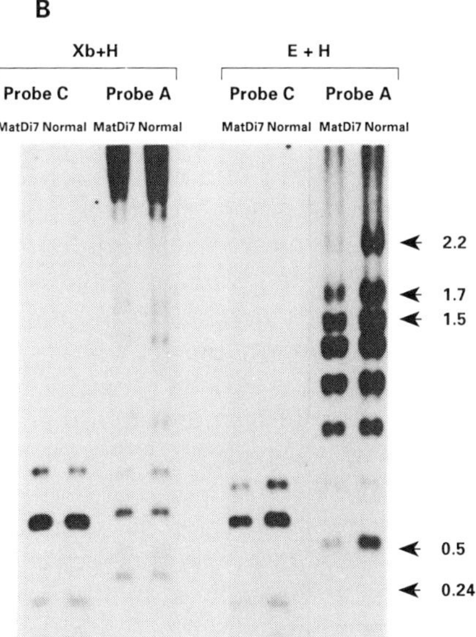

Fig. 1 Parent-specific methylation pattern upstream of the *Igf2* gene. (A) shows the 5′ flanking region of the gene; exon1 (E1), promoter1 (P1) and pseudoexons 1 and 2 (ψ1,2) are indicated together with restriction sites for EcoRI (E), XbaI (Xb), HpaII and MspI (H/M). The position of hybridization probe A is also shown. (B) DNAs from embryos maternally disomic for chromosome 7 (MatDi7) and from normal embryos was digested with XbaI + HpaII, and with EcoRI + HpaII, respectively. Blots were hybridized with probe A, and with probe C which is further downstream in the *Igf2* gene, to control for complete digestion. Size markers are in Kb. From Sasaki et al. 1992.

Whether or not there are parent specific differences anywhere near the promoter is not known yet. Likewise, parent specific methylation differences have been observed in the promoter region of the *H19* gene with the repressed copy being more methylated (A. Ferguson-Smith unpubl.). It is therefore possible that some imprinted genes have more extensive (and more easily detectable) epigenetic modifications than others, which may reflect a different mechanism of repression. Of particular interest is the possibility that maternally and paternally imprinted genes may utilise different mechanisms as they are available in the respective germlines. The possible differential stability of imprints in parthenogenetic and androgenetic embryonic stem cells would then not come as a total surprise (Mann 1992).

In conclusion, it looks that it is possible to detect parent specific epigenetic modifications of some kind in most, if not all, imprinted genes, although it has yet to be demonstrated to what extent such modifications are absent in non-imprinted genes. The next big, and difficult, step will be to demonstrate that these modifications are of functional significance. It will also be of some interest to study their evolution during development, beginning with the parental germlines. For the *Igf2* 5′ region, the higher (paternal specific) methylation pattern is already present in mature spermatozoa, but the pattern present in oocytes is not known yet (W. Reik unpubl.).

3 Epigenetic programming after fertilisation: Modifier genes and nucleocytoplasmic interactions

As pointed out before, there are several experiments that show that epigenetic modifications go through major changes in the early embryo before reaching the final, and somatically stable pattern. This programming may be influenced by genotype specific modifier genes, as for example a variety of transgenes show different methylation patterns on different inbred backgrounds. There are at least two ways in which this programming can result in parental effects, which are of considerable interest for development, quantitative genetics, and for the understanding of parental effects in genetic disease. First, parental influences can result from a genotype specific interaction of egg cytoplasm with either of the parental genomes, with the consequence of epigenetic modification of that genome. Genotype specific interactions between egg cytoplasm and the paternal genome have been demonstrated in the DDK syndrome in the mouse (Babinet et al. 1990), and with the differential developmental potential of paternal genomes exposed to different egg cytoplasms (Latham and Solter 1991). Second, parental effects can arise from modifier genes that are themselves imprinted. For example, maternal transmission of the BALB/c allele of a modifier gene has been found to increase methylation of an unlinked transgene locus, whereas paternal transmission of the BALB/c allele or maternal transmission of the DBA/2 allele of the same modifier does not lead to increased methylation (Allen et al. 1990). All these mechanisms will lead to differences in phenotype when the offspring from reciprocal crosses are compared between the inbred strains that carry different alleles of modifiers. For example, the BALB/c x DBA (BALB/c female x DBA male) offspring will have high methylation at the transgene locus described above, whereas the reciprocal F1 offspring, DBA x BALB/c, will show low methylation.

To detect these parental effects in a more systematic way, we have therefore begun to analyse F1 reciprocal hybrids from a cross between C57BL/6 (B6) and DBA2 (D2). To examine gene expression and phenotype on a broad basis we are using high resolution two-dimensional (2D) protein electrophoresis (Klose and Reik 1992). Many exceptions to Mendelian rule of inheritance have already been detected in our reciprocal cross. The criterion for a parental effect is that gene expression is different in the two types of crosses. The polypeptide spots that are differently expressed fall mainly into two classes (Fig. 2), those that show a matroclinous pattern (resembling the phenotype of the mother) and those that show a patroclinous pattern (resembling the father). The occurrence of these parental effects is surprisingly high: 11% of all genetic variants in the parents show an influence of parental inheritance on their expression in the offspring, with, perhaps not surprisingly, a preponderance of matroclinous inheritance (Klose and Reik 1992, Vogel and Klose 1992).

Our system identifies of course genetic traits that are susceptible to different sources of parental effects, ranging from classical maternal (intra-uterine) effects to nucleocytoplasmic interactions and imprinting. A combination of embryo transfer and nuclear transplantation techniques is now necessary to examine the relative influence of these factors.

In a preliminary series of experiments, we have begun to address the role of nucleocytoplasmic interactions in genome programming, and their effects on development and gene expression. Nucleocytoplasmic hybrids were made by pronuclear transplantation between B6 and D2 zygotes in which the genotype of the egg cytoplasm differed from that of the maternal genome (a situation that of course never occurs in natural hybrids). In offspring developing from such experimental zygotes we have observed (by 2D analysis) that a group of liver polypeptides (which was strongly expressed in control animals) was largely repressed. It appears therefore that inducing aberrant nucleocytoplasmic interactions in these experiments has a profound and longlasting effect on later gene expression. Hence, nucleocytoplasmic interactions at an early stage of development may have a significant influence on epigenetic programming of the genome.

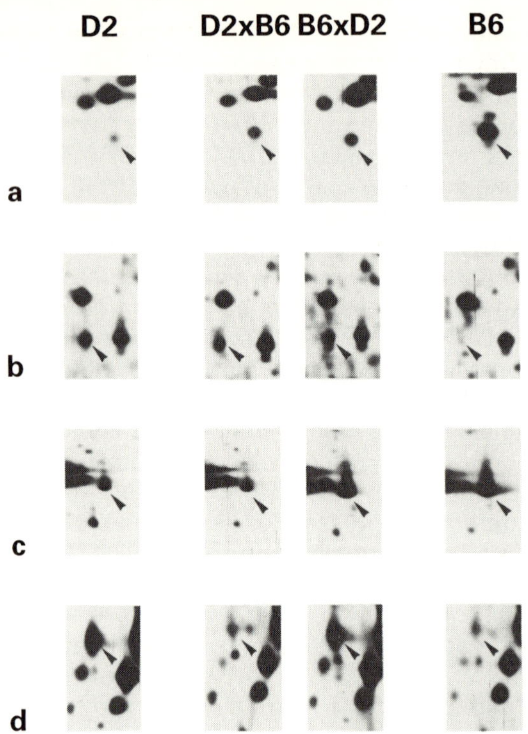

Fig. 2 Inheritance of polypeptide variants in a reciprocal cross. Liver proteins from the two inbred mouse strains DBA/2 (D2) and C57BL/6 (B6) and their reciprocal crosses (D2♀xB6♂ and B6♀xD2♂) were separated by two-dimensional electrophoresis, and the patterns obtained were compared. Selected sections from these patterns are shown. Most of the quantitative variants found can be attributed to one of four categories of inheritance: (a) codominant; (b) dominant/recessive; (c) matroclinous; (d) patroclinous. From Klose and Reik 1992.

4 References

Allen, N.D. and Mooslehner, K.A. (1992) Imprinting, transgene methylation and genotype-specific modification. **Semin. Dev. Biol.**, 3, 87-98.

Allen, N.D., Norris, M.L. and Surani, M.A.H. (1990) Epigenetic control of transgene expression and imprinting by genotype-specific modifiers. **Cell**, 61, 853-861.

Babinet, C. et al. (1990) The DDK inbred strain as a model for the study of interactions between parental genomes and egg cytoplasm in mouse preimplantation development. **Development Suppl.**, 81-87.

Barlow, D.P. et al. (1991) The mouse insulin-like growth factor type-2 receptor is imprinted and closely linked to the *Tme* locus. **Nature, 349**, 84-87.

Bartolomei, M.S. and Tilghman, S.M. (1992) Parental imprinting of mouse chromosome 7. **Semin. Dev. Biol., 3**, 107-117.

Bartolomei, M.S., Zemel, S. and Tilghman, S.M. (1991) Parental imprinting of the mouse H19 gene. **Nature, 351**, 153-155.

Barton, S.C. et al. (1991) Influence of paternally imprinted genes on development. **Development, 113**, 679-688.

Brown, C.J. et al. (1990) X chromosome inactivation of the human TIMP gene. **Nucl. Acids Res., 18**, 4191-4195.

Cattanach, B.M. and Beechey, C.V. (1990) Autosomal and X-chromosome imprinting. **Development Suppl.**, 63-72.

Chaillet, J.R. (1992) DNA methylation and genomic imprinting in the mouse. **Semin. Dev. Biol., 3**, 99-105.

Chaillet, J.R. et al. (1991) Parental specific methylation of an imprinted transgene is established during gametogenesis and progressively changes during embryogenesis. **Cell, 66**, 77-84.

DeChiara, T.M., Robertson, E.J. and Efstratiadis, A. (1991) Parental imprinting of the mouse insulin-like growth factor II gene. **Cell, 64**, 849-859.

Engler, P. et al. (1991) A strain-specific modifier on mouse chromosome 4 controls the methylation of independent transgene loci. **Cell, 65**, 939-948.

Ferguson-Smith, A.C. et al. (1991) Embryological and molecular investigations of parental imprinting on mouse chromosome 7. **Nature, 351**, 667-670.

Frommer, M. et al. (1992) A genomic sequencing protocol that yields a positive display of 5-methylcytosine residues in individual DNA strands. **Proc. Natl. Acad. Sci. USA, 89**, 1827-1831.

Grant, S.G. and Chapman, V.M. (1988) Mechanisms of X-chromosome regulation. **Annu. Rev. Genet., 22**, 199-233.

Howlett, S.K. and Reik, W. (1991) Methylation levels of maternal and paternal genomes during preimplantation development. **Development, 113**, 119-127.

Kafri, T. et al. (1992) Developmental pattern of gene-specific DNA methylation in the mouse embryo and germ line. **Genes Dev., 6**, 705-714.

Klose, J. and Reik, W. (1992) Expression of maternal and paternal phenotypes at the protein level. **Semin. Dev. Biol., 3**, 119-126.

Latham, K.E. and Solter, D. (1991) Effect of egg composition on the developmental capacity of androgenetic mouse embryos. **Development**, 113, 561-568.

Li, E., Bestor, T.H. and Jaenisch, R. (1992) Targeted mutation of the DNA methyltransferase gene results in embryonic lethality. **Cell**, 69, 915-926.

Mann, J.R. (1992) Properties of androgenetic and parthenogenetic mouse embryonic stem cell lines; are genetic imprints conserved? **Semin. Dev. Biol.**, 3, 77-85.

McGowan, R. et al. (1989) Cellular mosaicism in the methylation and expression of hemizygous loci in the mouse. **Genes Dev.**, 3, 1669-1676.

Monk, M., Boubelik, M. and Lehnert, S. (1987) Temporal and regional changes in DNA methylation in the embryonic, extraembryonic and germ cell lineages during mouse embryo development. **Development**, 99, 371-382.

Monk, M. (1986) Methylation and the X chromosome. **BioEssays**, 4, 204-208.

Nicholls, R.D., Rinchik, E.M. and Driscoll, D.J. (1992) Genomic imprinting in mammalian development: Prader-Willi and Angelman syndromes as disease models. **Semin. Dev. Biol.**, 3, 139-152.

Pourcel, C., Toillais, P. and Farza, H. (1990) Transcription of the S gene in transgenic mice is associated with hypomethylation at specific sites and with DNAse I sensitivity. **J. Virol.**, 64, 931-935.

Rappolee, D.A. et al. (1992) Insulin-like growth factor II acts through an endogenous growth pathway regulated by imprinting in early mouse embryos. **Genes Dev.**, 6, 939-952.

Reik, W., Howlett, S.K. and Surani, M.A. (1990) Imprinting by DNA methylation: from transgenes to endogenous gene sequences. **Development Suppl.**, 99-106.

Sanford, J.P. et al. (1987) Differences in DNA methylation during oogenesis and spermatogenesis and their persistence during early embryogenesis in the mouse. **Genes Dev.**, 1, 1039-1046.

Sapienza, C. et al. (1989). Epigenetic and genetic factors affect transgene methylation imprinting. **Development**, 107, 165-168.

Sasaki, H. et al. (1992) Parental imprinting: Potentially active chromatin of the repressed maternal allele of the mouse insulin-like growth factor II (*Igf2*) gene. **Genes Dev.**, (in press).

Surani, M.A. et al. (1990) Genome imprinting and development in the mouse. **Development Suppl.**, 89-98.

Vogel, T. and Klose, J. (1992). Two-dimensional electrophoretic protein patterns of reciprocal hybrids of the mouse strains DBA and C57BL. **Biochem. Genet.**, (in press).

29 Methylated DNA-binding proteins and chromatin structure

R. MEEHAN
University of Edinburgh, UK
J. LEWIS
University of Edinburgh, UK

P. JEPPESEN
MRC Human Genetics Unit, Edinburgh, UK
A. BIRD
University of Edinburgh, UK

1 Introduction.

DNA in the eukaryotic nucleus is highly compacted by its association with histones and many other proteins to form chromatin. Mapping studies indicate that genes are clustered in chromosomal regions that are structurally distinct from inactive regions (Bickmore and Sumner, 1989). This is reflected at the biochemical level by the differential accessibility in chromatin of selected regions of DNA to nucleases. For example transcriptionally active genes show enhanced sensitivity to micrococcal nuclease (MNase) compared to inactive regions (Bellard et al, 1978). One of the two X-chromosomes in female mammals remains condensed and is distinguished as a densely staining heterochromatic body (the Barr body) throughout interphase (Grant and Chapman, 1988). The vast majority of genes on the inactive X-chromosome (Xi) are non-transcribed in comparison to their counterparts on the active X-chromosome (Xa) (Grant and Chapman, 1988). In addition the Xi has reduced sensitivity to DNase-I as compared to its active counterpart (Kerem et al 1983). This type of compact heterochromatin is also found in *Drosophila melanogaster* and has been associated with position effect variegation (Tartoff and Bremer, 1990), whereby the position of a gene relative to heterochromatin can affect its activity. These phenomena imply that the stable assembly of DNA into different types of higher order chromatin structures can be a major determinant of genome activity.

The basic repeating unit of chromatin is the nucleosome (van Holde, 1989) and in mammals consists of approximately 146 bp of DNA wrapped on an octamer of lysine rich histones (consisting of two copies each of histones H2A, H2B, H3 and H4). Histone H1 can bind to the DNA in the linker region and contribute to the higher order folding of chromatin (Thoma et al, 1979). Nucleosomes have been well characterized biochemically and their structure solved by X-ray crystallography (see van Holde, 1989). Despite this wealth of information it is not clear how different higher order chromatin structures are formed. The positioning of a nucleosome over a promoter can inhibit transcription initiation but not necessarily chain elongation of the nascent RNA (Felsenfeld, 1992; Morse, 1992). This suggests that there is competition between trans-acting factors in the nucleus and histones for sites on DNA which will have a

Chromosomes Today   Volume 11. Edited by A.T. Sumner and A.C. Chandley. Published in 1993 by Chapman & Hall, London. ISBN 0 412 47670 3

bearing on the formation of cytologically observable chromatin structures.

Transcriptionally inactive regions such as centromeric heterochromatin are stained more densely with AT-specific fluorochromes then the lighter staining euchromatic regions (Bickmore and Sumner, 1989). These different banding patterns are stably inherited in somatic cells but in some cases may be optional as in the case of X-inactivation (Grant and Chapman, 1988). Banding patterns are probably established during replication. For example DNA replication in mammals is temporally bimodal with active (housekeeping and expressing tissue specific) genes replicating early in the S phase of the cell cycle (Goldman et al, 1984). The incorp- oration of 5-BrdU early in S phase can give rise to chromosome bands that contain most genes (Bickmore and Sumner, 1989). A different banding technique, R-banding, can also gives rise to euchromatic bands that can be associated with transcription units. At the molecular level it is known that transcription factors can facilitate initiation of replication in some eukaryotic systems (Jones et al, 1987; DePamphilis, 1988; O'Neill et al, 1988; Wilcock and Lane,1991). It would be of interest to know whether there is a relationship between the binding of nuclear factors early in the replication of DNA and the formation of euchromatic bands. There is an obvious problem of resolution here in trying to relate the replication (or transcription etc) of relatively small pieces of DNA, 200-300 kb (Brown et al, 1987), with the formation of chromosome bands (or structures) that are orders of magnitude bigger. A band probably represents an average of many different chromatin domains.

A feature of mammalian DNA that may allow us to relate gene expression and higher order structure (as observed down the microscope) is methylation (Lewis and Bird, 1991). Animal genomes can be divided into two categories on the basis of the presence or absense of the modified base 5-methylcytosine (m^5C). This modification occurs post-replicatively, is restricted to the dinucleotide CpG and approximately 70% of all CpGs are methylated in mammals (Cedar, 1988). Methylated DNA is associated with transcriptional repression and the formation of nuclease insensitive chromatin (Keshet et al, 1986). For example the promoters of genes on the Xi are hyper-methylated relative to their Xa counterparts (Hansen et al, 1988; Beggs and Migeon, 1989). The remaining non-methylated CpGs are clustered into small GC-rich regions (approximately 1.4 kb long) termed 'CpG' islands which have been correlated with transcriptionally active DNA and the promoters of genes (Tazi and Bird, 1990; Bird et al, 1985; Bird,1987). The CpG islands (with the exception of those on the Xi) of somatic cells are always non- methylated (Bird, 1987). In bulk terms CpG islands comprise about 2% of the genome which means that the vast majority of chromatin is associated with methylated DNA. In this paper we will discuss the function of methylated DNA and how it may relate to the formation of higher order chromatin structures in animals.

2 Function of methylation.

There is overwhelming evidence that methylation near promoters is incompatible with gene expression (Stein et al, 1982; Busslinger et al, 1983; Murray and Grosveld, 1987; Boyes and Bird, 1991). For example retroviral proviruses are stably repressed by de novo methylation of CpG in embryonic cells (Jahner and Jaenisch, 1984). In the case of X-inactivation we know that the stability of repression of genes on the Xi is due in part to methylation (Grant and Chapman, 1988; Riggs and Pfeifer, 1992). In both of these cases there is some indication that the genes become inactivated prior to or at the onset of methylation (Gautsch and Wilson, 1983; Lock et al, 1987; Singer-Sam et al, 1990; Bartlett et al, 1991). DNA methylation appears to reinforce prior events that have lead to gene (and chromosome) suppression. In marsupials DNA methylation is not required for X-inactivation (Kaslow and Migeon, 1987) and in opossum cell lines X-linked genes are frequently derepressed (Migeon et al, 1989). Thus DNA methylation may not be essential for the initiation and/or propagation of repressed chromatin but its presence stablizes the repressed state. There may well be other cases where the repressive effects of methylation have been utilized for the benefit of controlling gene expression during development. At present we know of few examples but the recent demonstration that mice with reduced levels of DNA methyltransferase (MTase) cannot develop to term suggests that methylation may have a significant developmental role (Li et al, 1992), perhaps by preventing the inappropiate expression of normally tissue specific genes.

3 Unmethylated CpG and chromatin structure.

The non-methylated CpG island fraction has a distinct open chromatin structure that is different from that found associated with methylated-CpGs in bulk chromatin (Tazi and Bird, 1990). Firstly CpG islands contain a region of nucleosome free DNA and are hypersensitive to nucleases. This is in line with the nuclease sensitivity of specific sequences such as the X-linked HPRT, G6PD and PGK-1 genes (Wolf and Migeon, 1985; Pfeifer et al, 1990a; Pfeifer and Riggs, 1991). The nucleosome free region of these promoters on the Xa apparently bind transcription factors (Pfeifer et al, 1990a; Pfeifer and Riggs, 1991). Their inactive methylated counterparts on the Xi are nuclease insensitive and are probably wrapped up in nucleosomes (Pfeifer et al, 1990a,b). As stated earlier this difference in nuclease sensitvity between genes on the Xa and Xi is reflected at the whole chromosome level. The Xi chromosome is largely DNase-I insensitive except for the pseudoautosomal region (Kerem et al, 1983) whereas the whole Xa chromosome is sensitive to DNase-I.

Treatment of rodent cell lines with the potent inhibitor of DNA methylation, 5-azacytidine (5-aza-C), can lead to an increase in DNase-I sensitivity of the Xi correlated with an advance in replication timing from late to early S phase (Jablonka et al, 1985). There is also a significant decrease in the frequency of detectable sex heterochromatin after 5-aza-C treatment (Mukherjee et al, 1986), whilst the normally represed genes on the Xi are activated (Pfeifer et al, 1990b). In general methylation-free CpG islands are associated with active transcription of genes which probably map to negative G-bands (or R-bands) (Bickmore and Sumner,1989). The artificial removal of m^{5}C from previously methylated genes on the Xi can lead to derepression and associated changes in higher chromatin structures.

5-aza-C also induces undercondensation of heterochromatin associated with G-band positive regions on autosomes (Schmid et al, 1984; Haaf and Schmid, 1989). The amount of observed undercondensation is dependent on the dose and timing of 5-aza-C treatment. Prolonged treatment leads to so many undercondensations that the chromosomes appear 'pulverized' (Viegas-Pequignot and Dutrillaux, 1976; Schmid et al, 1984). However undermethylation *per se* may not lead to a change in chromatin structure as determined by nuclease sensitivity. The naturally undermethylated satellite DNA sequence in the embryonal carcinoma (EC) cell line, F9, is still nuclease insensitive (Selig et al, 1988). Its adoption of an inactive chromatin structure may be related to the transcriptional inactivity of satellite DNA. In this context it is also worth noting that the hypomethylated major satellite DNA in F9 cells is shifted in its replication timing from late to early S in comparison to its methylated counterpart (Selig et al, 1988). This implies that it is the absence of methylation that leads to an advance in the timing of replication of the mouse major satellite. This effect may be mimiced in somatic cells by the the binding of transcription factors to particular DNA sequences which is associated with the demethylation of those sequences. It has been shown that tissue specific genes replicate early in expressing tissues and later in S-phase in non-expressing tissues (Goldman et al, 1984). These observations suggest that the absense of transcription factors in somatic cells for a particular promoter sequence leads to the default (or ground state) of inactive chromatin and hence late replication.

The protein components of the CpG island fraction are also distinct from that found in bulk chromatin (Tazi and Bird, 1990). The linker histone, H1, is significantly under-represented and the histones H3 and H4 were found to be hyperacetylated in island chromatin when compared with non-island chromatin. Both of these observations are compatible with the CpG island fraction having an open and transcriptionally active chromatin structure (Allfrey, 1980; Hebbes et al, 1988). Interestingly antibodies to acetylated forms of H4 detect a non-uniform distribution on human metaphase chromosomes (Jeppesen et al, 1991). Some sites on the chromosome arms are more intensely labelled with the antibody whilst heterochromatic domains were found to be under-acetylated. Control antibodies against total histone fractions H4 and H1 produce relatively homogeneous fluorescence.

Further analysis of the distribution of acetylated H4 in mammalian metaphase chromosomes shows that, in general, sites of hyper-acetylation map to R-bands, i.e. the actively transcribed fraction of the genome. In contrast, the transcriptionally inactive Xi is underacetylated compared to the Xa and the autosomes (P.J., unpublished observations). Thus acetylation of H4, and possibly other histones, marks generally active chromatin, whereas CpG methylation is associated with gene repression. Whether the apparently complementary distribution of histone acetylation and DNA methylation in mammals are reflections of intrinsically different conformations of active compared to inactive chromatin remains to be seen. Similar observations have been made on the well characterized polytene chromosomes of *Drosophila melanogaster* (Turner et al, 1992). Beta-heterochromatin is depleted in these acetylated isoforms whilst the hyperactive male X chromosome is enriched in H4 acetylated at lysine 16 in comparison to the male autosomes or any chromosome in female cells.

The presence of acetylated histones at CpG island chromatin may be causally linked to the absense of histone H1. Acetylation of histones may effect their protein/protein interactions, perhaps with H1. Highly acetylated histone core particles also exhibit structural changes consistent with the unwinding of DNA (Norton et al, 1989). Since H1 participates in the higher order folding of chromatin (Thoma et al, 1979) it might imply that CpG island chromatin is not arrayed in a higher order structure. In summary the chromatin structure of the non-methylated CpG island fraction is consistent with it being a repository of active gene promoters which may will reflect the higher order chromatin structures associated with it.

4 Methylated CpGs in bulk chromatin.

As stated previously the majority of CpGs are found in bulk chromatin, these are methylated and occur at a frequency of about 1 CpG per 100 bp (Bird, 1986). The first suggestion that methylated CpGs (Me-CpG) are in a specific chromatin structure came from antibody studies by Miller and co-workers (Miller et al, 1974). Using anti-m^{5}C antibodies they could localize 5-methylcytosine to mouse centric heterochromatin. Biochemical studies also indicated that m^{5}C in native chromatin is refractory to digestion by MNase when compared to the digestion kinetics of thymidine (Solage and Cedar, 1978). This difference was not observed with naked DNA. When methylated DNA constructs are transfected into mouse L cells they adopt a nuclease insensitive structure (Keshet et al, 1986). Non-methylated constructs were relatively hypersensitive to nucleases which implies that it is possible for methylation to specify an inactive chromatin structure. It has been demonstrated biochemically that m^5C in chromatin has a preferential association with histone H1 containing nucleosomes (Ball et al, 1983). The accessibility of MeCpGs to nucleases that can cleave at MeCpG is also very much reduced in chromatin (Hansen et al, 1988; Antequera et al, 1989). In contrast bulk chromatin is extensively digested by non-CpG recognizing restriction enzymes (Antequera et al, 1989; Tazi and Bird; 1990), which implies that the observed differential sensitivity is specific to MeCpGs. The simplest explanation for these biochemical results is that there are factors in the nucleus that can bind m^5C thereby preventing cleavage of methylated sites.

5 How methylation exerts its effect.

Our current working model is that <u>m</u>ethylated-<u>C</u>pG binding <u>p</u>roteins (MeCPs) in the nucleus bind to methylated DNA leading to an altered chromatin structure. This in turn would deny access by the transcription machinery to a methylated gene and lead to the formation of inactive chromatin (Lewis and Bird, 1991; Bird, 1992). Binding of MeCPs in this model could also account for the other effects associated with DNA methylation such as late replication and inhibition of recombination (Selig et al, 1988; Hsieh and Leiber, 1992). It can be appreciated that in this model it is the interplay between different types of nuclear factors (activators and/or repressors) and DNA that determines the resulting type of chromatin. Binding of MeCPs would presumably encourage nucleosome formation on methylated DNA.

6 Methylated DNA binding proteins (MeCPs).

We have detected two proteins (MeCPs 1 and 2) which have a preference for methylated DNA in mammalian and avian nuclei (Meehan et al, 1989; 1992; Lewis et al,1992). MeCP1 binds *in vitro* to DNA containing at least 12 methylated CpGs symmetrically distributed on both strands (Meehan et al, 1989), while MeCP2 can bind to a single methylated CpG pair (Lewis et al, 1992). Only m^5C in the context of CpG is a good DNA binding substrate for these proteins. Hemi-methylated DNA is a poor substrate. No other dinucleotide pair, including TpA, has a significant affinity for these proteins. The relaxed sequence specificity and wide-spread tissue distribution of these proteins makes them likely candidates as the mediators of the effects of methylation. MeCP1 is of relatively low abundance (about 5000 molecules per nucleus), and is loosely bound, whereas MeCP2 is comparatively abundant (about 200,000 molecules per nucleus) and tightly bound to chromatin. Both proteins are deficient in embryonal carcinoma (EC) and stem (ES) cells (Meehan et al, 1992). MeCP2 has been purified, its cDNA cloned and sequenced (Lewis et al,1992). Its deduced amino acid sequence indicates that it is a highly basic protein with some similarity to histone H1, which is consistent with the biochemical properties of MeCP2.

7 MeCPs and gene repression.

Studies of MeCP1 have implicated it (rather than MeCP2) in methylation-associated gene inactivation (Boyes and Bird,1991;1992; Levine et al, 1991). The expression of the X-linked mouse PGK1 gene promoter can be inhibited by methylation in transcription extracts and in cell lines which contain MeCP1 (Boyes and Bird,1991). Expression can be restored by competition with methylated DNA substrates which are specific for MeCP1. In addition methylated genes are not efficiently repressed in ES and EC cell lines or extracts which lack MeCPs (Boyes and Bird, 1991; Levine et al, 1991). As an aside it has also been reported that histone H1 binding to methylated DNA *in vitro* can inhibit the normally methylation-insensitive restriction enzyme MspI (Higurashi and Cole, 1991). Histone H1 is present in ES and EC cell lines and so is unlikely to be a methylation specfic transcriptional repressor.

MeCP2 cannot selectively inhibit transcription from CpG rich methylated DNA templates *in vitro* (Meehan et al, 1992). Thus the biological significance of MeCP2 is presently uncertain. It is still possible that MeCP2 is involved in transcriptional repression. Nuclease treatment of nuclei indicates that MeCP2 is bound to chromatin, but the transcription experiments were carried out in histone-free extracts in which chromatin cannot form (Meehan et al, 1992). If the natural ligand for MeCP2 is chromatin, as is the case for histone H1 (Wolffe, 1990), this could explain our failure to observe specific effects.

8 MeCPs and heterochromatin.

Unfortunately purified preparations of MeCP1 are not available for antibody production. However its low abundance and requirement for methyl-CpG rich substrates might suggest that it is not responsible for the genome-wide protection of methyl-CpGs against nucleases. This role could be undertaken by MeCP2. Brain nuclei, which have the highest levels of MeCP2, show particularly striking protection of methyl-CpGs against nucleases (Antequera et al, 1989). Conversely PC13 cells, which have very reduced levels of MeCP1 and MeCP2, show markedly reduced levels of protection. Antibodies against MeCP2 localise preferentially to centromeric heterochromatin regions in interphase nuclei and metaphase spreads from mouse L cells (Lewis et al, 1992). Although mouse major satellite DNA is AT-rich, more than half of the methyl-CpGs in the mouse genome are located in this repeated sequence (Manuelides, 1981; Horz and Altenberger, 1981). Satellite DNA is concentrated at centromeric heterochromatin. The asymmetrical distribution of MeCP2 parallels the pattern that was found for m^{5}C antibodies which preferentially stain regions of pericentromeric heterochromatin (Miller et al, 1974). Thus MeCP2 colocalizes with chromosomal regions that are known to be rich in methyl-CpG and heterochromatin. It could also be demonstrated that the methylated form of the 234bp satellite monomer DNA was a good substrate for MeCP2 (Lewis et al, 1992). However it should be stressed that MeCP2 is not exclusively located to centromeres. Euchromatic chromosome arms are also stained with MeCP2 antibodies but with a much reduced intensity compared to centromeric heterochromatin in mouse L cells (Lewis et al, 1992). This can be visualized more easily in rat metaphase spreads which have a less prominent satellite DNA fraction (Lewis et al, 1992). In this case prominent staining in the euchromatic chromosome arms can be seen. These studies did not detect a pattern of fluorescence that could be correlated with any experimentally induced banding pattern such as G-banding. Methylation is important for mouse centromeric chromatin formation as removal of methyl groups by 5-aza-C treatment of mouse L cells leads to stretched or decondensed chromatin at mouse centromeres (Joseph et al, 1989)

The MeCP2 protein does contain sequence motifs that are present in proteins that interact with the minor groove of B-form DNA. Notably (R)GRP(K) which is found in the high mobility group proteins, HMG-I and -Y (Johnson et al, 1989), which may be responsible for their specificity for AT-rich domains (Reeves and Nisen,1990). However these motifs are probably not responsible for methylated DNA binding as antibodies to HMG-I do not preferentially localize to MeCpGs on mouse metaphases (Disney et al, 1989). Interestingly HMG-I could be localized to G-bands which are thought to be depleted in transcribed DNA sequences.

9 Conclusions and future prospects.

From the preceding discussion it is evident that methylation of DNA in mammals is associated with inactive chromatin both at the sequence and cytological level. Accumulating evidence suggests that this effect is mediated by the binding of MeCPs which act as repressors of chromatin activity by participating in the formation of higher order chromatin structures. The similarities and differences between MeCP1 and 2 suggest a working model for their roles in methylation-mediated repression. MeCP1 may compete with transcription (or replication) factors in the nucleoplasm for binding to methylated DNA, the outcome depending on the density of methylation and the affinity of the factors (Boyes and Bird,1992; Bird,1992). It is possible that binding to MeCP1 then guides the DNA into a heterochromatic structure involving stable association with MeCP2. How this might happen is unknown, but it probably occurs at DNA replication, since the resistance of methylated DNA to nucleases is known to increase dramatically at this time (Hsieh and Lieber, 1992).

A future area of study would appear to be a detailed cytological and biochemical study of ES and EC cells. They provide a developmental system with which to study the onset of X-inactivation and changing patterns of DNA methylation (Lock et al, 1987; Bartlett et al, 1991). A 2/3 reduction of m^{5}C in the DNA of ES cells had no effect on their ability to survive in tissue culture whereas embryos with reduced levels of m^{5}C fail to come to term (Li et al, 1992). This suggests that the repressive effects of methylation are not required in these cell lines. This may be due to that fact that they correspond to early embryo cells which are undergoing wholesale reprogramming of their methylation and transcription profiles (Kafri et al, 1992). An interesting question is whether there are associated changes at the cytological level that can be visualized with antibodies to defined nuclear factors, such as MeCPs, acetylated histones etc.

10 Acknowledgements

We thank Sally Cross, Xinsheng Nan and Peri Tate for critical reading of the manuscript through many drafts, and Christine Struthers for help with the manuscript. This work was supported by the Imperial Cancer Research Fund and the Wellcome trust. R.M. and J.L. are members of the ICRF epigenetics laboratory.

11 References

Antequera, F., MacLeod, D. and Bird, A.P. (1989). Specific protection of methylated CpGs in mammalian nuclei. Cell **58**, 509-517.

Allfrey, V.G. (1980). Molecular aspects of eukaryotic transcription-nucleosomal proteins and their postsynthetic modification in the control of DNA conformation and template function. In Cell Biology : A comprehensive treatise, Vol 3. L. Goldstein and D.M. Prescott, eds. (New York : Academic Press).

Ball, D.J., Gross, D.S. and Garrard, W.T. (1983). 5-Methylcytosine is localized in nucleosomes that contain histone H1. Proc. Natl. Acad. Sci. USA **80**, 5490-5494.

Barlett, H., et al (1991) DNA methylation of two X chromosome genes in female somatic and embryonal carcinoma cells. Som. Cell Mol. Genet. **17**, 35-47.

Beggs, A.H. and Migeon, B.R. (1989). Chromatin loop structure of the human X chromosome : relevance to X inactivation and CpG clusters. Mol. Cell. Biol. **9**, 2322-2331.

Bellard, M., Gannon, F. and Chambon, P. (1978). Nucleosome structure III: the structure and transcriptional activity of chromatin containing the ovalbumin and globin genes in chicken oviduct nuclei. Cold Spring Harbor Symp. Quant. Biol. **42**, 779-791.

Bickmore, W.A. and Sumner, A.T. (1989). Mammalian chromosome banding - an expression of genome organization. TIG **5**, 144-148.

Bird, A. (1986). CpG rich islands and the function of DNA methylation. Nature **321**, 209-213.

Bird, A.P. (1987). Cp[G islands as gene markers in the vertebrate nucleus. TIG **3**, 342-347.

Bird, A.P. (1992). The essentials of DNA methylation. Cell **70**, 5-8.

Bird, A.P., et al (1985). A fraction of the mouse genome that is derived from islands of non-methylated, CpG-rich DNA. Cell **40**, 91-99.

Boyes, J. and Bird, A. (1991). DNA methylation inhibits transcription indirectly via a methyl-CpG binding protein. Cell **64**, 1123-1134.

Boyes, J. and Bird, A. (1992). Repression of genes by DNA methylation depends on CpG density and promoter strength: evidence for involvement of a methyl-CpG binding protein. EMBO **11**, 327-333.

Brown, E.H., et al (1987). Rate of replication at the murine immunoglobulin heavy-chain locus: Evidence that the region is part of a single replicon. Mol. Cell. Biol. **7**, 450-457.

Busslinger, M., Hurst, J. and Flavell, R.A. (1983). DNA methylation and the regulation of globin gene expression. Cell **34**, 197-206.

Cedar, H. (1988). DNA methylation and gene activity. Cell **53**, 3-4.

DePamphilis, M.L. (1988). Transcriptional elements as components of eukaryotic origins of DNA replication. Cell **52**, 635-638.

Disney, J.E., et al (1989). High mobility group proteins HMG-I localizes to G/Q- and C-bands of human and mouse chromosomes. J Cell Biol. **109**, 1975-1982.

Felsenfeld, G. (1992) Chromatin as an essential part of the transcriptional mechanism. Nature **355**, 219-214.

Gautsch, J.W. and Wilson, M.C. (1983). Delayed *de novo* methylation in tetracarcinoma cells suggests additional tissue-specific mechanisms for controling gene expression. Nature Lond. **301**, 32-37.

Goldman, M.A., et al(1984). Replication timing of genes and middle repetitive sequences. Science **224**, 686-692.

Grant, S.G. and Chapman, V.M. (1988). Mechanisms of X-chromosome regulation. Annu. Rev. Genet. **22**, 199-233.

Haaf, T. and Schmid, M. (1989). 5-Azadeoxycytidine induced under-condensation in the giant chromosome of *Microtus agrestis*. Chromosoma **98**, 93-98.

Hansen, R.S., Ellis, N.A. and Gartler, S.M. (1988). Demethylation of specific sites in the 5' region of the inactive X-linked human phosphoglycerate kinase gene correlates with appearance of nuclease sensitivity and gene expression. Mol. Cell Biol. **8**, 4692-4699.

Hebbes, T.R., Thorne, A.W. and Crane-Robinson, C. (1988). A direct link between core histone acetylation and transcriptionally active chromatin. EMBO J. **7**, 1395-1402.

Higurashi, M. and Cole, R.D. (1991). The combination of DNA methylation and H1 histone binding inhibits the action of a restriction nuclease on plasmid DNA. J. Biol. Chem. **266**, 8619-

Horz, W. and Altenberger, W. (1981). Nucleotide sequence of mouse satellite DNA. Nucleic Acids Res. **9**, 683-676.

Hsieh, C-L. and Lieber, M.R. (1992). CpG methylated mini-chromosomes become inaccessible for V(D)J recombination after undergoing replication. EMBO J. **11**, 315-325.

Jablonka, E., et al (1985). DNA hypomethylation causes an increase in DNase I sensitivity and an advance in the time of replication of the entire X chromosome. Chromosoma **93**, 152-156.

Jahner, D. and Jaenisch, R. (1984) DNA methyaltion in early mammalian development. In DNA methyaltion: Biochemistry and Biological significance. (Razin, A., Cedar, H. and Riggs, A.D., eds.) pp 189-219. (Springer-Verlag, New York)

Jeppesen, P.N., et al (1991). Anti-bodies to defined histone epitopes reveal variations in chromatin conformation and underacetylation of centric heterochromatin in human metaphase chromsomes. Chromosoma **101**, 322-332.

Joseph, A., Mitchell, A.R. and Miller, O.J. (1989). The organization of mouse satellite DNA at centromeres. Exp. Cell Res. **183**, 494-500.

Johnson, K., Lehn, D. and Reeves, R. (1989). Alternative processing of mRNAs encoding mammalian chromosomal high mobility group proteins, HMG I and HMG Y. Mol. Cell. Biol. **9**, 2114-2123.

Jones, K.A., et al (1987). A cellular DNA-binding protein that activates eukaryotic transcription and DNA replication. Cell **48**, 79-89.

Kafri, T., et al (1992). Developmental pattern of gene-specific DNA methylation in the mouse embryo and germ line. Genes Dev. **6**, 705-714.

Kaslow, D.C. and Migeon, B.R. (1987). DNA methylation stabilizes X chromosome inactivation in eutherians but not in marsupials: Evidence for multistep maintenance of mammalian X dosage compensation. Proc. Natl. Acad. Sci. USA **84**, 6210-6214.

Kerem, B-S., et al (1983) *In situ* nick-translation distinguishes between active and inactive X-chromosomes. Nature **304**, 88-89.

Keshet, I., Lieman-Hurwitz, J. and Cedar, H. (1986). DNA methylation affects the formation of active chromatin. Cell **44**, 535-543.

Levine, A., Cantoni, G-L. and Razin, A. (1991). Inhibition of promoter activity by methylation: possible involvement of protein mediators. Proc. Natl. Acad. Sci. USA, **88**, 6515-6518.

Lewis, J.D. and Bird, A.P. (1991). DNA methylation and chromatin structure. FEBS Lett. **285**, 155-159.

Lewis, J.D., et al (1992). Purification, sequence and cellular localization of a novel chromosomal protein that binds to methylated DNA. Cell **69**, 905-914.

Li, E., Bestor, T.H. and Jaenisch, R. (1992). Targeted mutation of the DNA methyltransferase gene results in embryonic lethality. Cell 915-926.

Lock, L.F., Takagi, N. and Martin, G.R. (1987). Methylation of the HPRT gene on the inactive X chromosome occurs after chromosome inactivation. Cell **48**, 39-46.

Manuelides, L. (1981). Consensus sequence of mouse satellite DNA indicates it is derived from tandem 116 base pair repeats. FEBS Lett. **129**, 25-28.

Meehan, R.R., et al (1989). Identification of a mammalian protein that binds specifically to DNA containing methylated CpGs. Cell 499-507.

Meehan, R.R., Lewis, J.D. and Bird, A.P. (1992). Characterization of MeCP2, a vertebrate DNA binding protein that binds methylated DNA. Nucleic Acids Res. **20**, 5085-5092.

Migeon, B.R., DeBeur, S.J. and Axelman, J. (1989). Frequent derepression of G6PD and HPRT on the marsupial inactive X chromosome associated with cell proliferation *in vitro*. Exp. Cell Res. 597-609.

Miller, L.L., et al (1974). 5-methylcytosine localized in mammalian constitutive heterochromatin. Nature **251**, 636-637.

Morse, R.H. (1992) Transcribed chroamtin. TIB **17**, 23-26.

Mukerjee, A.B., Luckett, D.C. and Hereru, R.J. (1986). 5-Azacytidine induced decrease in the frequency of Barr body in human fibroblasts. Genet. Res. **47**, 199-203.

Murray, E.J. and Grosveld, F. (1987). Site specific demethylation in the promoter of human gamma globin gene does not alleviate methylation mediated suppression. EMBO J. **6**, 2329-2335.

Norton, V.G., et al (1989). Histone acetylation reduces nucleosome core particle linking number change. Cell **57**, 449-457.

O'Neill, E.A., et al (1988). Transcription factor OTF-1 is functionally identical to the DNA replication factor NF-III. Science **241**, 1210-1213.

Pfeifer, G.P., et al (1990a). In vivo footprint and methylation analysis by PCR-aided genomic sequencing: comparison of active and inactive X-chromosomal DNA at the CpG island and promoter of PGK-1. Genes Dev. **4**, 1277-1287.

Pfeifer, G.P., et al (1990b). Polymerase chain reaction aided genomic sequencing of an X-chromosome linked CpG island: Methylation patterns suggest clonal inheritance, CpG site autonomy and an explanation of activity state stability. Proc. Natl. Acad. Sci. USA **87**, 8252-8256.

Pfeifer, G.P. and Riggs, A.D. (1991). Chromatin differences between active and inactive X chromosomes revealed by genomic footprinting of permeabilized cells using DNaseI and ligation mediated PCR. Genes Dev. **5**, 1102-1113.

Reeves, R. and Nissen, M.S. (1990). The AT-rich binding domain of mammalian high mobility group I chromosomal proteins. J. Biol. Chem. **109**, 1975-1982.

Riggs, A.D. and Pfeifer, G.P. (1992). X-chromosome inactivation and cell memory. TIG **8**, 169-174.

Schmid, M., Haaf, T. and Grummet, D. (1984). 5-Azacytidine induced undercondensation of human chromosomes. Hum. Genet. **67**, 252-263.

Selig, S., et al (1988). Regulation of mouse satellite DNA replication time. EMBO J. **7,** 419-426.

Singer-Sam, J., et al (1990). Use of HpaII-polymerase chain reaction assay to study DNA methylation in the Pgk-1 CpG island of mouse embryos at the time of X-chromosome inactivation. Mol. Cell Biol. 4987-4989.

Solage, H. and Cedar, H. (1978). Organization of 5-methylcytosine in chromosomal DNA., Biochemistry **17,** 2934-2938.

Stein, R., Razin, A. and Cedar, H (1982) *In vitro* methylation of the hamster APRT gene inhibits its expression in mouse L-cells. Proc. Natl. Acad. Sci. USA **79,** 3413-3422.

Tartoff, K.D. and Bremer, M. (1990). Mechanisms for the construction and developmental control of heterochromatin formation and imprinted chromosome domains. Development **suppl.** 35-45.

Tazi, J. and Bird, A. (1990). Alternative chromatin structure at CpG islands. Cell **60,** 909-920.

Thoma, F., Koller, T. and Klug, A. (1979). Involvement of histone H1 in the organization of the nucleosome and of the salt dependent structures of chromatin. J. Cell Biol. **83,** 403-427.

Turner, B.H., Birley, A.J. and Lavender, J. (1992). Histone H4 isoforms acetylated at specific lysine residues define individual chromosomes and chromosome domains in Drosophila polytene nuclei. Cell **69,** 375-384.

van Holde, K.E. (1989). Chromatin. Springer series in molecular biology. A. Rich, series ed. (Springer-Verlag, New York).

Viegas-Pequignot, E. and Dutrillaux, B. (1976). Segmentation of human chromosomes induced by 5-ACR (5-azacytidin). Hum. Genet. **34,** 247-254.

Wilcock, D. and Lane, D.P. (1991) Localization of p53, retinoblastoma and host replication proteins at sites of viral replication in Herpes-infected cells. Nature **349,** 429-431.

Wolf, S.F. and Migeon, B.R. (1985). Clusters of CpG dinucleotides implicated by nuclease hypersensitivity as control elements of housekeeping genes. Nature **314,** 467-469.

Wolffe, A.P. (1990). New approaches to chromatin function. New Biologist **2,** 211-218.

30 Rye B chromosome transmission depends on the B, the carrier of the B and the mother of the carrier

M.J. PUERTAS, M.M. JIMÉNEZ and F. ROMERA
Universidad Complutense, Madrid, Spain

1 Introduction

B chromosomes are not required for the normal processes of growth and development, they lack homology with any members of the basic set, and show irregular and non-Mendelian modes of inheritance. There are two main reasons for our interest in Bs. Firstly they are a widespread phenomenon in nature and they are known to occur in many species. Secondly, they raise questions about genome organisation and evolution. They provide us with avenues for exploring the significance of "selfish" genetic elements which are promoted by processes of drive, and which transgress our normal experience of evolutionary forces which operate by the processes currently known for Mendelian populations.

The mode of transmission of Bs is different in different species. In many cases they can maintain their own polymorphism by mechanisms of drive, usually balanced by their harmful effect on fitness (Jones, 1991). We are interested in knowing the mode of inheritance of Bs because this is the first step to understanding their role in the genome of a species and its evolution. In the present paper we review the main results of our work on this subject with reference to rye.

In rye locally adapted varieties are frequently found to be polymorphic for Bs. A single rye plant has a constant number of Bs in all its cells, but for the population as a whole the numbers range from O to 4 Bs with a high preponderancy of O and 2B plants. In laboratory samples up to 8B plants have been reported although in all cases the OB and 2B classes are the most important, the rest being subproducts of the intrinsic instability of the Bs. Therefore, we mainly study the transmission and the effects of Bs in 2B plants since this is the B class mainly responsible for maintaining the polymorphism.

Rye B chromosome inheritance is non-Mendelian. At meiosis the picture is complicated because Bs can form univalents, bivalents and multivalents, producing microspores with 0, 1, 2 or more Bs (Kishikawa, 1965 gives a rather complete review). Besides this instability, which creates a permanent variation in B number, there is also a structural instability of the B which produces variants of the standard type, although only the standard B forms the long term polymorphism in nature. We have found about 2% of structural variants per B per generation arising *de novo* in progenies of OBx2B crosses (Jiménez et al.,

Chromosomes Today   Volume 11. Edited by A.T. Sumner and A.C. Chandley. Published in 1993 by Chapman & Hall, London. ISBN 0 412 47670 3

in press), while they rarely appear when the Bs are transmitted by the female.

The main event of the B chromosome story unfolds during the postmeiotic mitoses to form both the male and female gametophytes: the pollen grain and the embryo sac. At anaphase of the first postmeiotic mitosis the B undergoes nondisjunction and migrates to the pole from which the gametes will form (Hasegawa 1934, Müntzing 1946, Hakanson 1948). In this way the B is "driven" to the next generation.

2. The mechanism of drive, and the effect of Bs on fitness, depends on the Bs themselves

The rate of nondisjunction is constant, occurring with the same high frequency in all rye populations studied (Romera et al., 1989). It also happens when rye Bs are transferred to other species such as *Triticum aestivum* (Müntzing, 1970) and *Secale vavilovii* (Puertas et al., 1979, 1986, 1987). The constancy in the mechanism of drive reveals firstly that the process is controlled by the B itself, probably by a major gene since it is not influenced by the genetic or environmental background, and secondly it shows the importance of the process to the B chromosome system.

The mechanism of drive would lead to an endless increase of the number of Bs, were it not balanced by the harmful effect of Bs on the fitness of carriers. In order to evaluate whether this effect depends on the Bs themselves, or on an evolutionary modification of the background genotype, we made two types of studies. Firstly we introduced the Bs from *S. cereale* into the related species *S. vavilovii* (Puertas et al., 1985), and various components of fitness were estimated for plants having O to 4 Bs. In all cases the variables of fertility were significantly different with respect to B number both for *S. cereale* and *S. vavilovii.*, but not for the components of viability. In table 1 the results obtained for grains per plant and flowers per plant are given as an example. The regression and correlation analyses demonstrate that the decrease of fertility with the increase of the number of Bs is similar in both species, and also that this variable behaves similarly in both species. It can therefore be deduced that Bs affect fitness irrespective of the genome of the carriers tested. A genetic co-adaptation between Bs and the *cereale* genome does not seem to be the main cause for the maintenance of B chromosome polymorphisms.

Table 1. Variables of fertility (grains per plant, G/P) and viability (flowers per plant, F/P) in *S. cereale* and *S. vavilovii* with different numbers of Bs.

Bs in the plant	*S. cereale*		*S. vavilovii*	
	G/P	F/P	G/P	F/P
0B	42.51	115.52	58.30	85.88
1B	51.88	151.18	-	-
2B	27.73	124.83	42.00	78.43
3B	15.89	142.64	23.35	91.91
4B	8.85	141.85	6.22	111.04

In a second experiment (Romera et al.,1989) we studied the same components of fitness in 0B, 2B and 4B rye plants from four natural populations which have different frequencies of Bs in nature. The results were similar *i.e.,* the mean number of grains per plant decreases as the number of Bs increases, while the variables of viability do not depend on the B-number. Rye Bs therefore fall into the category of "selfish"chromosomes due to their main effect of decreasing fertility together with their tendency to increase in number by "drive". They were proposed to be selfish firstly by Jones (1985), and all our results confirm his hypothesis.

3. The B transmission depends on the carrier genotype

In different crosses involving 0B and 2B plants the mean number of Bs in the progenies varied from 0.57 to 3.21 per plant, in 2Bx0B crosses from 0 to 2.84, and in 2Bx2B ones from 0.75 to 3.48. Therefore some of the progenies tended to lose Bs while others tended to gain them.

We selected the seed of progenies of 2Bx0B crosses with about 0.5 Bs per plant (low transmission rate class, or "L class") and those of progenies with about 1.5 (high transmission rate class or "H class") and repeated the 2Bx0B crosses again (Romera et al., 1991). The results are summarised in figure 1, which compares the distribution of the variable "mean number of Bs per plant in the progeny" in the parental population and in the descendants of both the L and H classes. The selection gain in both lines was high, and the heritability close to 1, demonstrating a strong genetic component of the observed variability in the transmission of Bs on the female side in this population.

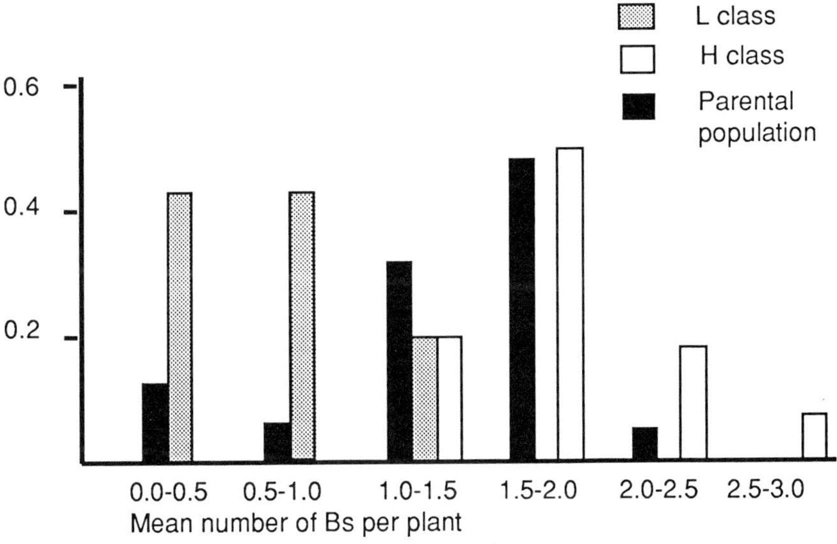

Figure 1. Distribution of the mean number of Bs per plant in the progenies of 2Bx0B crosses, in the parental population and in the descendants of both the low (L) and high (H) transmission rate class of B chromosomes.

Again, plants from the L and the H lines were used for studying the effect of these genotypes on the male side in 0Bx2B crosses (Jiménez et al., in press). Every single 2B male was crossed with four different 0B females and the results are summarised in figure 2. The progenies of both types of males differed significantly, and the analyses demonstrated that the males from progenies with low or high mean number of Bs had progenies with low or high mean number of Bs, respectively. In other words, the variation for the mean number of Bs transmitted by the male has also a genetic component.

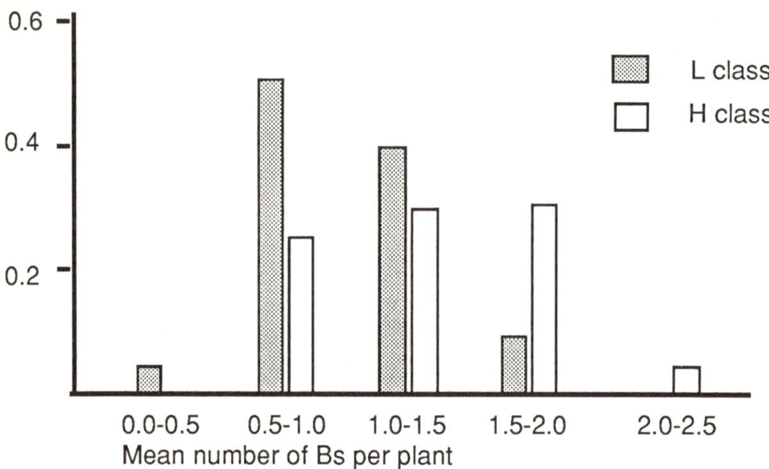

Figure 2. Distribution of the mean number of Bs per plant in the progenies of 0Bx2B crosses of the low (L) and high (H) transmission rate class.

4. B transmission depends on the mother of the carrier: an imprinting effect

Genomic imprinting, as stable, heritable and differential modification of the paternal and maternal genomes has been reported to occur in mammals and in angiosperms. Haig and Westoby (1989) made the appealing hypothesis that imprinting has evolved in these organisms where the offspring genome is active at the stage when the mother provisions her offspring.

There are similarities in the imprinting effects in both types of organisms, but there are some main differences due to their different biology. Mammals are diplontic organisms while angiosperms are haplodiplontic with a long and conspicuous sporophytic phase. The sporophyte in many species carries hermaphroditic flowers where the male and female gametophytes are formed; therefore, there are no male or female individuals. However, in experimental crosses a plant can be used as a male using only its pollen grains, or as a female by removing its anthers. In rye, where many spikes per plant are usually available, the same plant can be used as female and/or as male.

In angiosperms the seed is formed by a double fertilisation: the embryo is diploid and the endosperm is triploid with two maternal and one paternal genomes. Therefore, the imprinting effect can occur both in the embryo and the subsequent tissues of the sporophyte, and also in the endosperm with important consequences on seed development. The effect of the genomic imprinting of the endosperm was reviewed by Haig and Westoby (1991).

In order to study imprinting in relation to B chromosomes we classify the plants according to their maternal parent in four types: 0/0 are 0B plants whose maternal parent had 0B, 2/0 are 0B plants whose maternal parent had 2B, 0/2 are 2B plants whose maternal parent had 0B, and 2/2 are 2B plants whose maternal parent had 2B. With these plants we made the following experiments:

4.1 Maternal imprinting effect: a 2B or 0B maternal parent produces differential fertilisation

We firstly crossed plants in all the 12 possible combinations involving 0Bx2B, 2Bx0B and 2Bx2B plants, considering their maternal parents (0/0 x 0/2, 0/0 x 2/2, etc.) (Puertas et al., 1990). The results are summarised in table 2.

Table 2. Mean number of Bs per plant in 0Bx2B, 2Bx0B and 2Bx2B crosses, according to the maternal parent of the plants involved.

		Females			
		0/0	2/0	0/2	2/2
	2/2	1.62±0.03	2.62±0.25	3.16±0.09	2.89±0.12
	0/2	1.29±0.11	1.41±0.10	2.29±0.30	2.82±0.12
Males	2/0			1.60±0.21	1.26±0.21
	0/0			1.39±0.14	1.29±0.20

The four 0Bx2B crosses (shaded) are different. It is evident that the B status of the maternal parent on both the male and the female side has a role in accounting for the differences: 2/2 males produce more Bs in the offspring than 0/2 males on any 0B female, and more Bs were produced in 2/0 females than in 0/0 ones. The joint effect of 0/2 males and 0/0 females produces a remarkable decrease of descendants with Bs, while the joint effect of 2/2 males and 2/0 females produces descendants with many Bs. So that, (i) when pollen grains are formed in 2B plants whose maternal parent had 2B they transmit Bs with high efficiency; (ii) when stigmas belong to 0B plants whose maternal parent had 2B they accept Bs with high efficiency; (iii) when these two effects occur together, the number of Bs transmitted to offspring is very high; (iv) the opposite is also true, when pollen is formed in 2B plants whose maternal parent had 0B, Bs are transmitted with low efficiency. When the styles belong to 0B plants whose maternal parent had 0Bs they accept 0B pollen more readily, and therefore, the minimum frequency of 0B is obtained in 0/0 x 0/2 crosses.

There are no significant differences among the 2Bx0B crosses (down right side of table 2), but a large variation within the four types of crosses is observed, which was later demonstrated to be due to genotypic differences (L or H classes) of the plants involved (Romera et al., 1991).

The progeny obtained in 2Bx2B crosses is of course influenced by the interactions that we have just discussed. Theoretically, the frequencies of the gametes carrying Bs transmitted by a 2B plant either as male or female can be estimated from the progenies obtained in 0Bx2B and 2Bx0B crosses respectively, and in this way the expected progeny of a 2Bx2B cross can be calculated. Let m_0, m_1 and m_2 be the gametes with 0, 1 and 2 Bs transmitted by a 2B plant in a 0Bx2B cross, and f_0, f_1 and f_2 the gametes with 0, 1, or 2 Bs transmitted by a 2B plant in a 2Bx0B cross. The expected distribution of Bs in a 2Bx2B cross is $m_0 x f_0$ of 0B, $(m_0 x f_1 + m_1 x f_0)$ of 1B, and so on.

The distribution of the mean number of Bs per plant observed in the 2Bx2B crosses (upper right part of table 2) does not differ significantly from either the total expected, or from that expected considering only the male gametes transmitted by a 2B male on a 0/0 female, while it does differ significantly when we consider the pollen transmitted by a 2B male on a 2/0 female. This again shows the interaction of 2/0 females with 2B pollen and the effect of the male maternal parent on the transmission of B chromosomes.

In addition to the total observed in 2Bx2B crosses we have considered the partial results of either 2Bx0/2 or 2Bx2/2. The observed distributions of 2Bx0/2 crosses fits well with any expectation in which a 0/2 male is considered, irrespective of the female; however, the distribution is very different when a 2/2 male is considered. This again indicates the strong interaction produced due to the 2B maternal parent.

4.2 The same male produces different progeny on 0/0 or 2/0 females

In the experiment described in 4.1, we found a large variation within the same type of crosses. In order to discriminate the possible sources of variation we did another experiment in which twenty 2B plants were used as males, crossing each 2B male with four different 0B females, the males always being 2/2 and with ten belonging to the low and ten to the high transmission rate class. The 0B females were either 0/0 or 2/0, and belonged to the L or H classes (Jiménez et al., in press). The effect of the L or H genotypes on the transmission of Bs was described in the third part of this paper, and we will mention now the differences due to the 2/0 or 0/0 status of the female. Table 3 shows the results of one of these crosses as an example, where it is evident that the number of Bs transmitted by a single male to its progeny is different on different 0B females, the 2/0 females being those which receive the higher percentage of gametes with Bs.

Figure 3 shows the distribution of the frequencies of plants with 0-more than 3 Bs in the progenies of ten 2B males of the H class obtained in twenty 2/0 and twenty 0/0 females, where it is shown that 2/0 females produce progeny with more Bs. Both distributions are significantly different.

Table 3. Progenies of one single 2B plant used as male, crossed with four different 0B female plants.

Female	Number of Bs in the progeny					Mean
	0	1	2	3	4	
2/0 H	1		18	1	5	2.36
2/0 L	4		21			1.68
0/0 H	16		7			0.61
0/0 L	7	1	16			1.37
Total	28	1	62	1	5	1.53

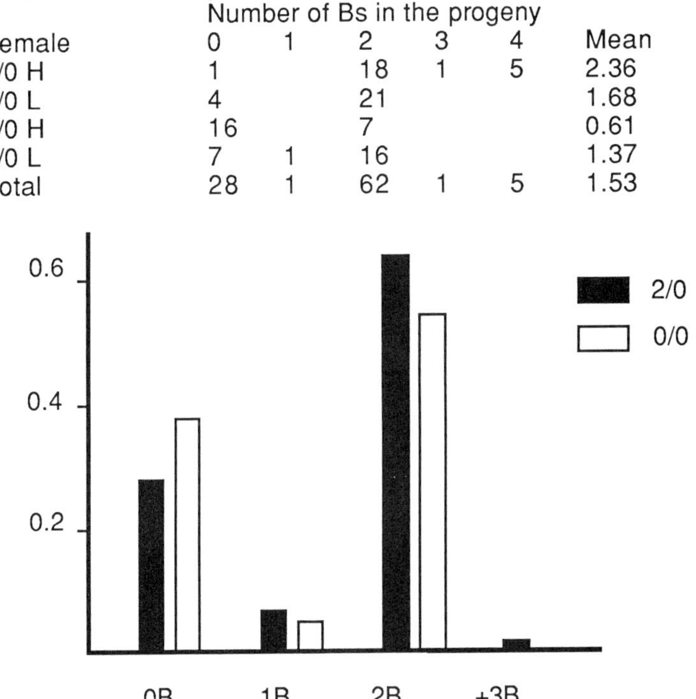

Figure 3. Frequency distribution of Bs in progenies of 0/0x2B and 2/0 x 2B crosses, using as males plants of the H class.

4.3 The same 2B plant behaves differently as male or as female depending on its 2/0 or 0/0 partner.

We made another type of cross in which each individual plant was used as male and as female in 2Bx0B, 0Bx2B and 2Bx2B crosses. All 2B plants were 2/2 and the 0B were either 0/0 or 2/0. We also considered the H or L genotypes of the plants involved. The results are summarised in figure 4. In crosses involving 0/0 plants (2Bx0/0 or 0/0x2B) the plants of the H genotypes produced descendants with many Bs, while those of the L genotypes produced few Bs as expected. There is also a significant positive correlation between the mean number of Bs transmitted by the same plant as male or female, indicating that the alleles controlling the transmission of Bs are the same and act both on the male and on the female side.

The results of the crosses in which the 0B plants were 2/0 show strong interactions . For example, the plants of the L line produce as many Bs in the offspring as those of the H line; so that, there is no significant correlation between the Bs in the parents and in the offspring. There is a negative correlation (although nonsignificant) between the mean number of Bs transmitted by the same plant as male and as female, and in 2Bx2B crosses the deviations between observed and expected progenies were highly significant in some cases.

All these results show that the B status of the maternal parent of plants carrying Bs is, together with the genotypes that control the B transmission rate, essential in the control of the transmission of Bs.

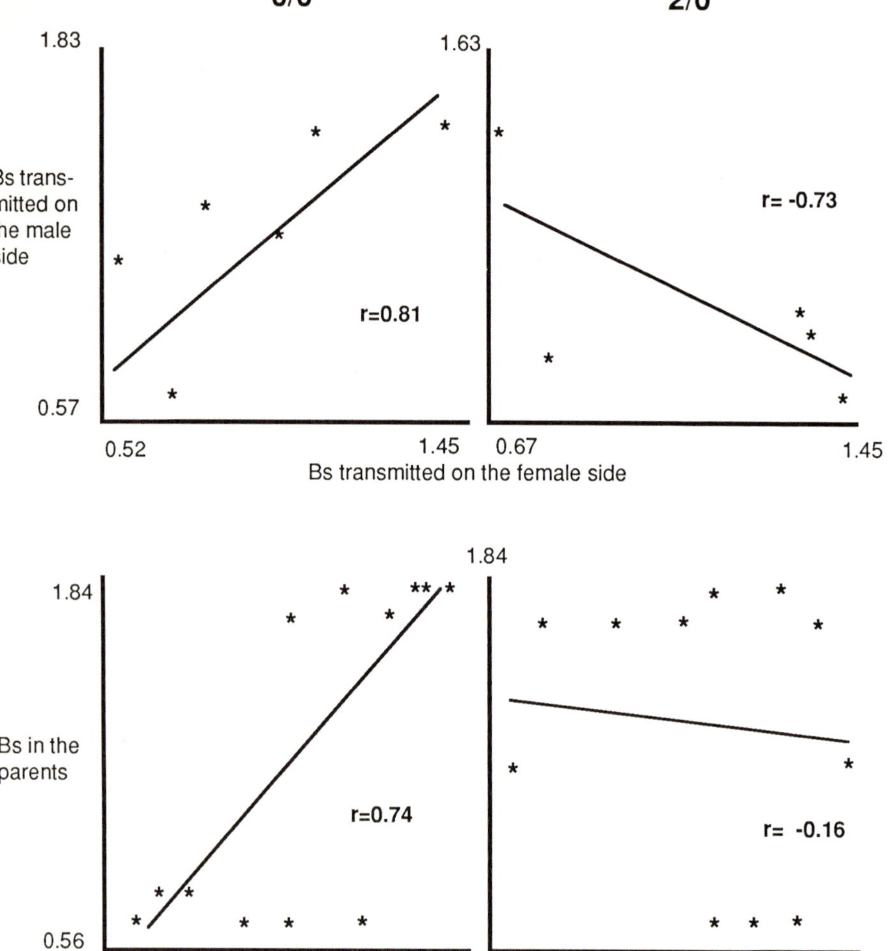

Figure 4. Correlations between the variables "Bs in the parents" and "Bs in the offspring" in crosses 2Bx0B and 0Bx2B, and between the variables "Bs transmitted as females in 2Bx0B crosses" and "Bs transmitted as males in 0Bx2B crosses" when the 0B plant involved is 0/0 (on the left), and when the 0B plant involved is 2/0 (on the right).

5. References

Haig, D. Westoby, M. (1989) Parent specific gene expression and the triploid endosperm. **Am Nat**. 134, 147-155

Haig, D. Westoby, M. (1991) Genomic imprinting in endosperm:its effects on seed development in crosses between species, and between different ploidies of the same species, and its implications for the evolution of apomixis. **Phil. Trans. R. Soc. Lond. B** 333, 1-13

Hakanson, A. (1948) Behaviour of accessory rye chromosomes in the embryo sac. **Hereditas** 34, 35-59

Hasegawa, N. (1934) A cytological study on 8-chromosome rye. **Cytologia,** 6, 68-77

Jiménez, M.M. Romera, F. and Puertas, M. J. (in press) Genetic control of the rate of transmission of B chromosomes. II Genotypic and maternal effects in 0Bx2B crosses.

Jones, R.N. (1985) Are B chromosomes selfish? in **The Evolution of Genome size** (ed. Cavalier Smith) John Wiley and Sons Ltd. pp 397-425

Jones, R.N. (1991) B chromosome drive. **Am. Nat.,** 137, 430-442.

Kishikawa, H (1965) Cytogenetic studies of B chromosomes in rye, *Secale cereale* in Japan. **Agric. Bull Saga Univ.,** 21, 1-81.

Müntzing, A. (1946) Cytological studies of extra fragment chromosomes in rye. III. The mechanism of non-disjunction at the pollen mitosis. **Hereditas** 32, 97-119

Müntzing, A (1970) Chromosomal variation in the Lindstrom strain of wheat carrying accessory chromosomes of rye. **Hereditas** 66, 279-286

Puertas, M.J. Diez, M. and Carmona, R. (1979) Rye B chromosome behaviour at first and second pollen mitosis and its relationship with anther maturity. **Theor and Appl. Genet**. 54, 65-68

Puertas, M.J. Romera, F. and de la Peña, A. (1985) Comparison of B chromosome effects on *Secale cereale* and *Secale vavilovii.* **Heredity** 55, 229-234

Puertas, M.J. Baeza, F. and de la Peña, A. (1986) The transmission of B chromosomes in populations of *Secale cereale* and *Secale vavilovii.* I. Offspring obtained from 0B and 2B plants. **Heredity** 57, 389-394

Puertas, M.J. Ramirez, A. and Baeza, F. (1987) The transmission of B chromosomes in populations of *Secale cereale* and *Secale vavilovii.* II. Dynamics of populations. **Heredity** 58, 81-85

Puertas, M.J. Jiménez, M.M. Romera, F. Vega, J.M. and Diez, M. (1990) Maternal imprinting effect on B chromosome transmission of rye B chromosomes in rye. **Heredity** 64. 197-204

Romera, F. Jiménez, M.M. and Puertas M.J. (1991) Genetic control of the rate of transmission of B chromosomes. I. Effects in 2Bx0B crosses. **Heredity** 66, 61-65

Romera, F. Vega, J.M. Diez, M. and Puertas, M.J. (1989) B chromosome polymorphism in Korean rye populations. **Heredity** 62, 117-121.

Acknowledgements. This work was supported by the grant PB88-0122 of the DGICYT of Spain.

Index